N

cDNA Library Protocols

METHODS IN MOLECULAR BIOLOGY™

John M. Walker, SERIES EDITOR

cDNA Library Protocols

Edited by

Ian G. Cowell
and Caroline A. Austin

University of Newcastle, UK

Humana Press ✳ **Totowa, New Jersey**

© 1997 Humana Press Inc.
999 Riverview Drive, Suite 208
Totowa, New Jersey 07512

This publication is printed on acid-free paper. ∞
ANSI Z39.48-1984 (American Standards Institute) Permanence of Paper for Printed Library Materials.

For additional copies, pricing for bulk purchases, and/or information about other Humana titles, contact Humana at the above address or at any of the following numbers: Tel.: 201-256-1699; Fax: 201-256-8341; E-mail: humana@interramp.com

Photocopy Authorization Policy:

Printed in the United States of America. 10 9 8 7 6 5 4 3 2 1

Library of Congress Cataloging in Publication Data

Main entry under title:

Methods in molecular biology™.

cDNA library protocols / edited by Ian G. Cowell and Caroline A. Austin.
        p.   cm. -- (Methods in molecular biology™ ; 69)
        Includes index.
        ISBN 0-89603-383-X (alk. paper)
        1. Gene libraries--Laboratory manuals. 2. DNA--Laboratory manuals. 3. Antisense DNA--Laboratory manuals. 4. Molecular cloning--Laboratory manuals.   I. Cowell, Ian G.   II. Austin, Caroline A.   III. Series.
        [DNLM: 1. Gene Library.   2. DNA, Complementary.   3. Genetic Techniques.   W1 ME9616J v.69 1997 / QH 442.4 C386 1997]
        QH442.4.C36   1997
        574.87'3282--dc20
        DNLM/DLC
        for Library of Congress   96-36127
                                                                               CIP

Preface

The first libraries of complementary DNA (cDNA) clones were constructed in the mid-to-late 1970s using RNA-dependent DNA polymerase (reverse transcriptase) to convert polyA$^+$ mRNA into double-stranded cDNA suitable for insertion into prokaryotic vectors. Since then cDNA technology has become a fundamental tool for the molecular biologist and at the same time some very significant advances have occurred in the methods for constructing and screening cDNA libraries.

It is not the aim of *cDNA Library Protocols* to give a comprehensive review of all cDNA library-based methodologies; instead we present a series of up-to-date protocols that together should give a good grounding of procedures associated with the construction and use of cDNA libraries. In deciding what to include, we endeavored to combine up-to-date versions of some of the most widely used protocols with some very useful newer techniques. *cDNA Library Protocols* should therefore be especially useful to the investigator who is new to the use of cDNA libraries, but should also be of value to the more experienced worker.

Chapters 1–5 concentrate on cDNA library construction and manipulation, Chapters 6 and 7 describe means of cloning difficult-to-obtain ends of cDNAs, Chapters 8–18 give various approaches to the screening of cDNA libraries, and the remaining chapters present methods of analysis of cDNA clones including details of how to analyze cDNA sequence data and how to make use of the wealth of cDNA data emerging from the human genome project. In most cloning projects the library screening method to be employed will be the most important factor in the cloning strategy. We have therefore included protocols for a wide range of approaches to screening cDNA libraries, including contemporary protocols for screening by hybridization of labeled DNA probes to filter-bound cDNA clones (Chapters 9, 11, and 12), protocols for screening expression libraries with different types of probe (Chapters 13–15), and some more specialized screening techniques (Chapters 16–18).

Except where specific examples are useful, the protocols are presented with minimum reference to particular experimental systems. The format of the chapters will be familiar to those who have used other volumes in the

Methods in Molecular Biology series. Additional information that is not usually present in original publications, such as details of potential problems, troubleshooting tips, and alternative methods, are included in the Notes sections found at the end of each chapter.

Once again, we hope that this volume will benefit the beginner and the cDNA aficionado alike. Finally, we would like to thank all of the contributing authors for providing the material for this volume as well as John Walker and the staff at Humana Press for their assistance in putting this volume together.

Ian G. Cowell
Caroline A. Austin

Contents

Contributors

HANS-CHRISTIAN AASHEIM • *Department of Immunology, The Norwegian Radium Hospital, Oslo, Norway*

MICHELLE ALTING-MEES • *Stratagene Cloning Systems, La Jolla, CA*

CAROLINE A. AUSTIN • *Department of Biochemistry and Genetics, The Medical School, University of Newcastle, Newcastle Upon Tyne, UK*

MARTIN J. BISHOP • *HGMP, Cambridge, UK*

PAUL R. CARON • *Vertex Pharmaceuticals, Cambridge MA*

IAN G. COWELL • *Department of Biochemistry and Genetics, The Medical School, University of Newcastle, Newcastle upon Tyne, UK*

ALISON DUNN • *Department of Biochemistry and Genetics, The Medical School, University of Newcastle, Newcastle upon Tyne, UK*

MICHAEL A. FROHMAN • *Department of Pharmacology, School of Medicine, SUNY at Stony Brook, NY*

ANDREI V. GUDKOV • *Department of Genetics, College of Medicine, University of Illinois, Chicago, IL*

TASUKU HONJO • *Department of Medical Chemistry, Faculty of Medicine, Kyoto University, Kyoto, Japan*

BENT HONORÉ • *Danish Center for Human Genome Research and Department of Medical Biochemistry, Aarhus University, Aarhus, Denmark*

SHENG-HE HUANG • *Division of Infectious Diseases and Hematology/Oncology, Departments of Pediatrics and Biochemistry, University of South California, Childrens Hospital, Los Angeles, CA*

HELEN C. HURST • *Imperial Cancer Research Fund Oncology Unit, Hammersmith Hospital, London, UK*

AMBROSE JONG • *Division of Infectious Diseases and Hematology/Oncology, Departments of Pediatrics and Biochemistry, University of South California, Childrens Hospital, Los Angeles, CA*

ANTHONY R. KERLAVAGE • *The Institute for Genomic Research, Rockville, MD*

EWEN F. KIRKNESS • *The Institute for Genomic Research, Rockville, MD*

KRIS N. LAMBERT • *Department of Nematology, University of California, Davis, CA*

FRANK LARSEN • *Department of Immunology, The Medical School, University Hospital Utrecht, The Netherlands*

TON LOGTENBERG • *Department of Immunology, The Medical School, University Hospital, Utrecht, The Netherlands*

PEDER MADSEN • *Danish Center for Human Genome Research and Department of Medical Biochemistry, Aarhus University, Aarhus, Denmark*

CHRISTOPHER R. MUNDY • *HGMP, Cambridge, UK*

TORU NAKANO • *Department of Medical Chemistry, Faculty of Medicine, Kyoto University, Kyoto, Japan*

YASUHIKO NAKATA • *DNA Bank, Gene Bank, Tsukuba Life Science Center, Ibaraki, Japan*

GREGORY M. PRESTON • *Departments of Medicine and Biological Chemistry, Johns Hopkins University School of Medicine, Baltimore, MD*

JOHN C. REED • *The La Jolla Cancer Research Foundation, Cancer Center, La Jolla, CA*

PATRICIA RODRIGUEZ-TOMÉ • *The EBI Data Library, Cambridge, UK*

IGOR B. RONINSON • *Department of Genetics, College of Medicine, University of Illinois, Chicago, IL*

D. ROSS SIBSON • *J. K. Douglas Laboratories, Clatterbridge Cancer Research Trust, Clatterbridge Hospital, Wirral, UK*

JAY M. SHORT • *Recombinant BioCatalysis Inc., La Jolla, CA*

MARJORY A. SNEAD • *Recombinant BioCatalysis Inc., La Jolla, CA*

MICHAEL P. STARKEY • *HGMP, Cambridge, UK*

SHINICHI TAKAYAMA • *Cancer Center, The La Jolla Cancer Research Foundation, La Jolla, CA*

XIAOREN TANG • *Department of Medical Genetics, China Medical University, Shenyang, China*

KEI TASHIRO • *Department of Medical Chemistry, Faculty of Medicine, Kyoto University, Kyoto, Japan*

YAGNESH UMRANIA • *HGMP, Cambridge, UK*

VALERIE M. WILLIAMSON • *Department of Nematology, University of California, Davis, CA*

KAZUSHIGE YOKOYAMA • *DNA Bank, Gene Bank, Tsukuba Life Science Center, Ibaraki, Japan*

YUE ZHANG • *Department of Pharmacology, School of Medicine, SUNY at Stony Brook, NY*

1

cDNA Library Construction from Small Amounts of RNA Using Paramagnetic Beads and PCR

Kris N. Lambert and Valerie M. Williamson

1. Introduction

In the study of biological systems, often the amount of starting material available for molecular analysis is limiting. In some situations, only a few cells in a tissue are expressing genes of interest or the tissue is in limited supply. A cDNA library from the targeted cells is preferable to a library constructed from the entire tissue containing these cells, because in the latter case, genes expressed in the targeted cells may be too rare to be detected. cDNA clones of genes expressed in small amounts of material are often hard to obtain because the construction of conventional cDNA libraries requires microgram amounts of polyA$^+$ RNA *(1)*.

The polymerase chain reaction (PCR) is commonly used to amplify tiny amounts of DNA *(2,3)*. This technique also has been adapted to facilitate cloning of 3'- and 5'-ends of specific cDNAs from low amounts of RNA *(4,5)*. The strategies used to clone specific cDNAs were extended to allow the construction of cDNA libraries from small quantities of polyA$^+$ RNA *(6–9)*. For PCR amplification, cDNAs must possess a known DNA sequence (~20 bp) at each end. These end sequences can be generated by homopolymer tailing with terminal transferase *(10)*, by ligating adapters/linkers to the cDNAs *(11)*, or by using primers that anneal by means of random hexamers at their 3'-ends *(12)*.

Most cDNA library construction methods require multiple purification or precipitation steps to remove primers and change buffers. These steps result in significant loss of material and compromise the quality of the final library. It is especially important when working with small amounts of RNA and cDNA to minimize such steps. The cDNA amplification methods presented here elimi-

From: *Methods in Molecular Biology, Vol. 69: cDNA Library Protocols*
Edited by: I. G. Cowell and C. A. Austin Humana Press Inc., Totowa, NJ

nate all precipitation and chromatography steps so that all cDNA synthesis and modification reactions can be conducted in a single tube. The polyA$^+$ RNA is purified using oligo dT paramagnetic beads, and the synthesis of the first-strand cDNA is primed with the oligo dT that is covalently attached to the beads *(13,14)*. This results in first-strand cDNA covalently attached to the beads, minimizing cDNA loss in subsequent enzymatic manipulations *(14)*. The attraction of the beads to magnets allows rapid solution change by placing the sample tube in a magnetic stand to hold the beads against the side of the tube as the solution is pipeted off.

Two methods for amplification of the first-strand cDNA are presented. In one method, an adapter is ligated to the 5'-end of the cDNA to generate the priming site. The second procedure uses terminal transferase to generate polyA tails for primer annealing sites *(15)*. Using either of these methods, microgram amounts of amplified cDNA can be generated in 1–2 d from 5–200 ng of polyA$^+$ RNA. Advantages of the adapter method are the ease of distinguishing the 5'- and 3'-ends of the cDNA and the capability to alter the procedure for production of directional libraries. However, the terminal transferase method is simpler to carry out, and, in our hands resulted in a comparable, if not better library.

PCR-amplified DNAs are sometimes difficult to clone, because restriction enzymes may cut unreliably in terminal linkers *(16)* and because the 3'-ends of the PCR product can be A-tailed by the terminal transferase activity of *Taq* polymerase *(17)*. A library construction procedure is presented that overcomes these problems. This cDNA library construction approach should be suitable for generating plasmid or phage vector cDNA libraries in systems where RNA is limiting.

2. Materials

2.1. Generation of First-Strand cDNA Covalently Linked to Paramagnetic Beads

1. Diethyl pyrocarbonate (DEPC) H$_2$O: Stir distilled water with 0.1% DEPC for 12 h or longer, and then autoclave.
2. Dynabeads® mRNA purification kit (product no. 610.05; Dynal, Great Neck, NY). The kit contains oligo (dT)$_{25}$ Dynabeads, 2X binding buffer, wash buffer, and the magnetic microcentrifuge tube stand. Store at 4°C.
3. Superscript RNaseH$^-$ reverse transcriptase (200 U/µL; Gibco BRL, Grand Island, NY) and 5X first-strand buffer (Gibco BRL).
4. Reverse transcriptase mix: 4 µL 5X first-strand buffer, 10 µL DEPC-treated H$_2$O, 2 µL 0.1M dithiothreitol (DTT), 1 µL 10 mM dNTP mix (10 mM each dATP, dGTP, dCTP, and dTTP), 1 µL RNasin (10 U/µL; Gibco BRL); prepare just before use.

Table 1
Primers and Adapter Used in cDNA Library Construction

Name	Sequence
AL adapter[a]	5'-TTGCATTGAC<u>GTCGAC</u>TATCCAGG-3'[b]
	3'-GCTGATAGGTCC-5'
L-primer[a]	5'-TTGCATTGACGTCGACTATCCAGG-3'
T-primer[c]	5'-TTGCATTGACGTCGACTATCCAGGT-TTTTTTTTTTTTTT-3'

[a]The 24-mer strand of the adapter is identical to the L-primer. Our L-primer was phosphorylated at the 5'-end, but this is probably not necessary.
[b]Underline indicates *Sal*I endonuclease cleavage site.
[c]By designing a different T-primer that does not share homology with L-primer, one could make a directional library with the olignucleotide adapter protocol.

2.2. cDNA Amplification Using Oligonucleotide Adapters

1. 5X Second-strand cDNA synthesis buffer: 94 mM Tris-HCl, pH 6.9, 453 mM KCl, 23 mM MgCl$_2$, 750 mM β-NAD, 50 mM (NH$_4$)$_2$SO$_4$; prepare fresh.
2. Second-strand cDNA reaction mix: 91.6 μL distilled H$_2$O, 32 μL 5X second-strand cDNA synthesis buffer, 3 μL of 10 mM dNTP mix, 6 μL 0.1M DTT, 1.4 μL *Escherichia coli* RNase H (1 U/μL; Boehringer Mannheim, Indianapolis, IN); 4 μL of *E. coli* DNA polymerase I (10 U/μL; New England Biolabs, Beverly, MA), 2 μL (20 U) of *E. coli* DNA ligase (10 U/μL; Gibco BRL); prepare just before use.
3. T4 DNA polymerase reaction mix I: 42.4 μL H$_2$O, 5 μL 10X T4 DNA polymerase reaction buffer (Epicentre Technologies, Madison, WI), 2.5 μL 10 mM dNTP mix, 0.1 μL T4 DNA polymerase (10 U/μL; Epicentre Technologies); prepare just before use.
4. Polynucleotide kinase reaction: mix 42.4 μL of distilled water, 5 μL of 10X T4 DNA polymerase buffer, 2.5 μL of 10 mM adenosine triphosphate (ATP), and 0.1 μL polynucleotide kinase (10 U/μL; New England Biolabs); prepare just before use.
5. AL-adapter, 34 pmol/μL; T-primer, 0.5 pmol/μL; L-primer 50 pmol/μL. Adapter and primer sequences are shown in Table 1.
6. 10X Blunt-end ligation buffer: 660 mM Tris-HCl, pH 7.6, 50 mM MgCl$_2$, 50 mM DTT, 1 mg/mL bovine serum albumin (BSA).
7. Adapter ligation mix: 6.5 μL H$_2$O, 1 μL AL adapter, 7.5 μL 40% polyethylene glycol (average mol wt 8000; Sigma, St. Louis, MO), 2.5 μL 10 mM ATP, 2 μL 10X blunt-end ligation buffer, 0.5 μL T4 DNA ligase (400 U/μL, New England Biolabs); prepare just before use.
8. 10X PCR reaction buffer: 200 mM Tris-HCl, pH 8.3, 25 mM MgCl$_2$, 250 mM KCl, 0.5% Tween-20, 1 mg/mL autoclaved gelatin.

9. PCR reaction mix: 0.5 µL *Taq* polymerase (5 U/µL; Promega, Madison, WI), 0.4 µL 1*M* tetramethylammonium chloride (TMAC) (Note 1), 5 µL 10X PCR reaction buffer, 1 µL 10 m*M* dNTP mix, 1 µL (50 pmol) of L-primer.
10. Perkin-Elmer Cetus DNA thermocycler 480.

2.3. cDNA Amplification Using Terminal Transferase

1. T4 DNA polymerase reaction mix II: 41.5 µL H$_2$O, 5 µL of 10X T4 DNA polymerase reaction buffer, 2.5 µL 10 m*M* dNTP mix, 1 µL (10 U) of T4 DNA polymerase.
2. RNaseH buffer: 20 m*M* Tris-HCl, pH 8.0, 50 m*M* KCl, 10 m*M* MgCl$_2$, 1 m*M* DTT; prepare fresh.
3. RNaseH reaction mix: 20 µL of RNaseH buffer, 0.5 µL (0.5 U) of RNaseH; prepare just before use.
4. 500 m*M* EDTA, pH 7.5, sterile stock.
5. Terminal transferase mix: 14 µL H$_2$O, 2 µL of 10X One-Phor-All *PLUS* buffer (Pharmacia, Uppsala, Sweden), 3 µL of 1.5 m*M* dATP, 1 µL terminal deoxynucleotidyl transferase (22 U/µL, Pharmacia); prepare just before use.

2.4. Cloning Amplified cDNA

1. pBluescript II SK-plasmid, 1 mg/mL (Stratagene, La Jolla, CA), and competent *E. coli* (Strain SURE, Stratagene).
2. Chromaspin-100 spin column (Clontech, Palo Alto, CA).
3. GeneClean® (BIO 101, La Jolla, CA).
4. Ligation mix: 5 µL distilled water, 2.5 µL 10 m*M* ATP, 2 µL 10X blunt-end ligation buffer, 0.5 µL (200 U) T4 DNA ligase; prepare just before use.

3. Methods

3.1. Generation of First-Strand cDNA Covalently Linked to Superparamagnetic Beads

The starting material is dried total nucleic acid prepared by an appropriate method (Note 2). PolyA$^+$ RNA is isolated using oligo dT Dynabeads, but is not eluted from the beads. The first-strand cDNA is synthesized using the oligo dT covalently attached to the bead as a primer *(14)*.

1. PolyA$^+$ RNA isolation: Resuspend dried total nucleic acid in 25 µL of DEPC-treated water, heat to 65°C for 2 min, and then cool on ice. Add 20 µL (100 µg) of Dynabeads to a 0.5-mL microcentrifuge tube and place the tube in the magnetic stand. The beads will bind to the side of the tube adjacent to the magnet. Remove the supernatant with a micropipet without disturbing the beads. Remove the tube from the stand and resuspend the beads in 25 µL of 2X binding buffer. Remove the buffer, and add a fresh 25 µL of 2X binding buffer to the beads. Add 25 µL of the resuspended nucleic acids to the beads and allow the polyA$^+$ RNA to hybridize to the oligo dT beads for 15 min at 22°C. Remove the binding buffer

and unhybridized nucleic acids from the beads. Wash the beads twice with 50 μL of wash buffer, and then remove wash buffer.

2. First-strand cDNA synthesis: Wash the beads in 50 μL of 2X first-strand buffer to remove residual wash buffer. Remove buffer, and add 19 μL of reverse transcriptase mix, and heat to 37°C for 2 min. Add 1 μL reverse transcriptase and mix. Continue to incubate the reaction at 37°C for 15 min to allow extension from the 3'-end of the oligo dT primer, and then increase the temperature to 42°C for 45 min to help disrupt secondary structure in the RNA. Mix the tube every 15 min to keep the beads suspended. The first-strand cDNAs are now covalently linked to the beads. The RNA is still attached noncovalently to the cDNA and beads.

3.2. cDNA Amplification Using Oligonucleotide Adapters

In this method (outlined in Fig. 1), the second-strand cDNA is synthesized on the beads, and an adapter (Note 3) is ligated to the free end of the cDNA *(18,19)*. The intact second-strand cDNA is removed and amplified using PCR.

1. Second-strand synthesis: Add 140 μL of second-strand cDNA reaction mix to the first-strand cDNAs, which are attached to the beads and in 20 μL reverse transcriptase mix. Incubate the reaction at 16°C for 2 h, resuspending the beads every 15 min. Remove the buffer.
2. Blunt-end the cDNA: Add 50 μL of T4 DNA polymerase reaction mix to the beads, and incubate at 16°C for 15 min. Inactivate the enzyme by heating the reaction at 74°C for 10 min. Remove the buffer.
3. To ensure the cDNA has a 5'-phosphate, add 50 μL polynucleotide kinase reaction mix to the beads, and incubate at 37°C for 15 min. Remove the buffer. This step may not be necessary (*see* Note 4).
4. Adapter ligation: Add 20 μL of adapter ligation mix, and incubate at 16°C overnight. Add 50 μL TE to the ligation mix, bind the beads, and remove the buffer.
5. Extend the 3'-end of the first-strand cDNA: Add 50 μL PCR reaction mix, and heat the reaction at 74°C for 10 min to melt off the 12-mer strand of AL adapter (Table 1) and to extend the 3'-end of the cDNA. Heat at 95°C for 2 min to denature the double-stranded cDNA. Remove and discard the supernatant containing the second-strand cDNA.
6. Resynthesis of second-strand cDNA (*see* Note 5): Add 50 μL of PCR reaction mix containing 50 pmol L-primer to the beads, and heat at 72°C for 5 min. Incubate the tube at 95°C for 2 min to denature the cDNA. Bind the beads, and transfer the reaction mix containing the second-strand cDNA to a new tube. Save the beads (Note 6).
7. Amplification of cDNA: Add overlay with 50 μL mineral oil (0.5 pmol) of T-primer to the reaction mix with the second-strand cDNA, and incubate at 30°C for 3 min, 40°C for 3 min, and 72°C for 5 min to resynthesize the antisense strand (*see* Note 7). Both ends of the cDNA now carry the L-primer sequence. Amplify for 15 cycles (95°C for 1 min and 72°C for 5 min), and then incubate at 72°C for 30 min.

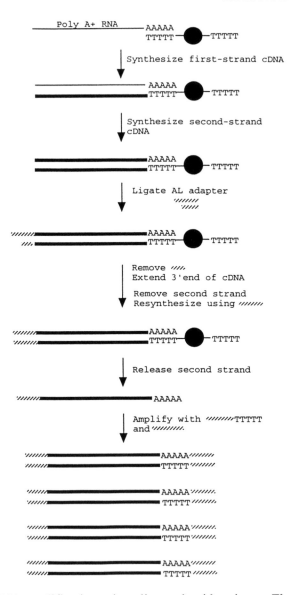

Fig. 1. cDNA amplification using oligonucleotide primers. The thin line represents RNA, the thick line represents DNA, and the crosshatched lines represent primer and adapter sequences.

8. Assessing amplification: Remove 5 µL of amplified cDNA and fractionate on a 2% agarose gel. If the cDNA is not visible after staining with ethidium bromide, reamplify 5 µL of PCR product for 15 additional cycles using 50 pmol of L-primer. The size range of the PCR products is likely to be slightly smaller than

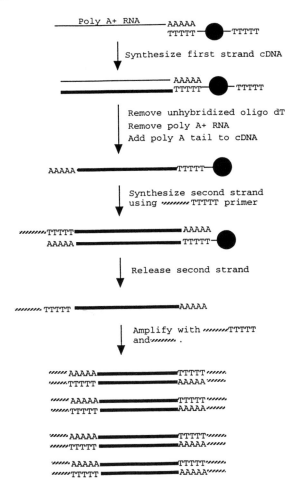

Fig. 2. cDNA amplification using terminal transferase tailing. This figure is adapted from Lambert and Williamson *(15)* by permission of Oxford University Press.

the average size of the starting polyA$^+$ RNA, reflecting selective amplification of smaller cDNAs. The representation of control genes in the cDNA can be determined by Southern hybridization or PCR amplification of the cDNA with appropriate primers. Control genes that differ in abundance and transcript size are most useful to test for bias for abundantly expressed cDNAs or short transcripts.

3.3. cDNA Amplification by Terminal Transferase Method

In this protocol (outlined in Fig. 2), the first-strand cDNA is A-tailed using terminal transferase, and a T-tailed primer is used to initiate the second-strand cDNA. The second-strand is amplified for the first few rounds with T-primer and then with the more stringent L-primer.

1. Removal of the unhybridized oligo (dT) from the beads that contain covalently linked cDNA: This step is important because unbound oligo (dT) can be tailed and amplified. Heat the reaction mixture at 65°C for 10 min to inactivate the reverse transcriptase. Remove the buffer, and add 20 μL of T4 DNA polymerase reaction mix, and incubate at 16°C for 1 h. The oligo dT is removed by the exonuclease activity of the T4 DNA polymerase. Inactivate the enzyme by heating at 74°C for 10 min. Remove the buffer.

2. Removal of polyA$^+$ RNA: Add 20 μL of RNaseH reaction mix, and incubate at 37°C for 1 h. Remove the buffer. Add 50 μL of 1 mM EDTA, and heat the mixture at 75°C for 5 min. Remove the EDTA solution.

3. Addition of polyA tail to first-strand cDNA (*see* Note 8): Add 20 μL of terminal transferase mix, incubate the beads at 37°C for 15 min, and then stop the reaction with 2 μL of 500 mM EDTA. Remove the buffer.

4. Second-strand cDNA synthesis: Add 50 μL of *Taq* polymerase reaction mix containing 29 pmol of T-primer. Extend the primer at 30°C for 3 min, 40°C for 3 min, and then 72°C for 5 min (Note 7). Bind the beads, and discard the supernatant.

5. cDNA amplification: Add 50 μL of fresh *Taq* polymerase reaction mix containing 50 pmol L-primer and 1 pmol T-primer. Heat the reaction at 95°C for 2 min to release the second-strand cDNA. Save the beads for future use (Note 6). Transfer the supernatant to a new tube, and add 50 μL of mineral oil. Incubate at 30°C for 15 min, 40°C for 15 min, and at 72°C for 15 min to extend the T-primer and synthesize a new first-strand cDNA. Heat at 95°C for 2 min, and amplify the cDNA for 15 cycles (95°C for 1 min and 72°C for 5 min) and then incubate at 72°C for 30 min.

6. Assessing amplification (*see* Section 3.2., step 7).

3.4. Cloning Amplified cDNA

Vector and insert ends are modified for efficient ligation (Fig. 3). A 5'-TT overhang is produced on each end of the insert by the 3' to 5'-exonuclease activity of T4 DNA polymerase *(20)*. 5'-AA overhangs are produced on the vector by digesting with *Eco*RI and partially filling in the 5'-overhang using T4 DNA polymerase *(21)*.

1. Digest 5 μg of vector to completion with *Eco*RI. Add $^1/_{10}$ vol 3M sodium acetate and 2.5 vol cold 100% ethanol, and then precipitate at −20°C. Spin in a microcentrifuge, and decant the supernatant. Wash the pellet with 70% ethanol, decant, and vacuum dry. Resuspend the *Eco*RI-digested vector in 50 μL of T4 DNA polymerase reaction mix lacking dTTP, and incubate at 16°C for 1 h. Heat the reaction to 65°C for 15 min to inactivate the enzyme. Gel-purify the cut plasmid to remove traces of uncut vector DNA.

2. Separate the 50 μL of PCR-amplified cDNA from unused primers and small cDNAs on a Chromaspin-100 spin column as recommended by manufacturer. Dilute the size-selected cDNA to 1 mL with TE, and measure the A$_{260}$ on a UV

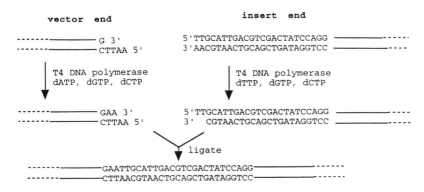

Fig. 3. Cloning the cDNA. Generation and ligation of 2-nucleotide 5'-tails at vector and insert ends. This figure is from Lambert and Williamson *(15)* by permission of Oxford University Press.

spectrophotometer. Ethanol-precipitate 400 ng of amplified cDNA, wash with 70% ethanol, and vacuum dry. Resuspend in 50 μL of T4 DNA polymerase reaction mix lacking dATP, and incubate at 16°C for 1 h and then at 75°C for 10 min (*see* Note 9).

3. Add 1 μL (100 ng) of *Eco*RI-digested vector to 400 ng of amplified cDNA, and concentrate using GeneClean as recommended by the manufacturer; elute in 10 μL of TE. Add 10 μL of ligation mix, and incubate at 16°C overnight. Transform competent *E. coli* cells with the ligated DNA.

4. Assessing the library: Blue/white color selection will determine the percentage of insert-containing clones. In our hands, the fraction of clones that contain inserts falls between 50 and 90%. The average cDNA size can be measured by PCR amplification of individual clones with L-primer. In our hands, the average insert size was approx 600 bp.

4. Notes

1. Addition of TMAC *(22)*, a chemical that causes an oligonucleotide to hybridize based on length and not GC content, results in an improvement in the quality of the PCR products *(23)*. TMAC is included in all PCR amplifications of cDNA. Its effectiveness should be determined empirically for each new primer set.

2. The total amount of nucleic acid used should not contain more than 200 ng of polyA$^+$ RNA, which is the carrying capacity of the beads. The amount of beads can be scaled up proportionally, but should not be reduced because there is some nonspecific binding of beads to the pipet tips and microcentrifuge tubes. The nonspecific sticking of the beads can be reduced by using siliconized microcentrifuge tubes and pipet tips. The lower limit of mRNA is not known, but we have used as little as 5 ng of polyA$^+$ RNA to construct libraries. Karrer et al. *(24)* used a similar protocol to construct a cDNA library from the contents of a single

plant cell. The method of nucleic acid extraction is dependent on the biological system under study. We found that a stainless-steel tissue pulverizer (Fisher Scientific, Pittsburgh, PA) cooled in liquid nitrogen was particularly useful for reducing small amounts of frozen tissue to a fine powder. A number of nucleic acid extraction protocols should be satisfactory. We used a simple phenol/chloroform extraction method *(25)*.

3. The AL adapter (Table 1) is made up of a 12- and a 24-mer. The blunt end of the adapter lacks a 5'-phosphate and the 5'-overhang is nonpalandromic, so that adapters cannot form concatemers *(26,27)* and the 12-mer cannot form a covalent bond with the cDNA. This adapter design maximizes the efficiency of ligation by minimizing competing ligation reactions *(6,11)*. A *Sal*I site is included in the primer to give an alternative cDNA library construction method.

4. To obtain efficient ligation of the adapter, the cDNA must have a 5'-phosphate. The polynucleotide kinase step is included to ensure all cDNAs are phosphorylated *(27,28)*. However, this may be an unnecessary step, since many protocols for adapter ligation do not include a kinase step.

5. The second strand is resynthesized to increase the likelihood that it is full length. RNaseH partially degrades the polyA$^+$ RNA in the RNA/DNA hybrid, whereas *E. coli* DNA polymerase I extends and removes the RNA primers and *E. coli* DNA ligase seals the nicks. However, some nicks or gaps may remain in the second-strand cDNA owing to incomplete action of DNA polymerase I or DNA ligase. This will result in incomplete second-stand cDNA, which will not amplify during PCR. Small RNAs will be less likely to have nicks, and thus, may skew the library toward small inserts. Resynthesis of the second-strand cDNA using only L-primer eliminates the nicking problem and results in the formation of full-length second-strand cDNA.

6. Beads with the first strand attached can be saved to generate additional libraries. We have also found them to be useful for obtaining full-length cDNAs *(4)*.

7. The first cycle annealing is started at 30°C and slowly increased to the 72°C extension temperature, so the $(T)_{15}$-tail of the T-primer can anneal to the polyA-tail to prime a new first-strand cDNA. In subsequent amplification cycles in which primer annealing and extensions are carried out at 72°C, only the L-primer contributes to the amplification.

8. The first-strand cDNA is A-tailed by terminal transferase using conditions that should produce a tail of >500 residues. The absolute tail length is not critical, because in the next step, a large excess of the T_{15}-tailed primer is used to synthesize the second-strand cDNA. The excess primer causes the final A-tail of the second-strand cDNA to be about 15–20 residues.

9. Reaction temperatures of 11–16°C will ensure that a large fraction of the cDNAs have the correct overhang. Higher temperatures can partly denature the ends of the DNA and allow the polymerase to remove more than the terminal two bases *(29)*.

References

1. Gasser, C. S., Budelier, K. A., Smith, A. G., Shah, D. M., and Fraley, R. T. (1989) Isolation of tissue-specific cDNAs from tomato pistils. *Plant Cell* **1**, 15–24.
2. Saiki, R. K., Gelfand, D. H., Stoffel, S., Scharf, S. J., Higuchi, R., Horn, G. T., Mullis, K. B., and Erlich, H. A. (1988) Primer-directed enzymatic amplification of DNA with a thermostable DNA polymerase. *Science* **239**, 487–491.
3. Arnheim, N., Li, H., and Cui, X. (1990) PCR analysis of DNA sequences in single cells: single sperm gene mapping and genetic disease diagnosis. *Genomics* **8**, 415–419.
4. Frohman, M. A., Dush, M. K., and Martin, G. R. (1988) Rapid production of full-length cDNAs from rare transcripts: amplification using a single gene-specific oligonucleotide primer. *Proc. Natl. Acad. Sci. USA* **85**, 8998–9002.
5. Ohara, O., Dorit, R. L., and Gilbert, W. (1989) One-sided polymerase chain reaction: the amplification of cDNA. *Proc. Natl. Acad. Sci. USA* **86**, 5673–5677.
6. Akowitz, A. and Manuelidis, L. (1989) A novel cDNA/PCR strategy for efficient cloning of small amounts of undefined RNA. *Gene* **81**, 295–306.
7. Welsh, J., Liu, J.-P., and Efstratiadis, A. (1990) Cloning of PCR-amplified total cDNA: construction of a mouse oocyte library. *Genet. Anal. Technol. Appl.* **7**, 5–17.
8. Domec, C., Garbay, B., Fournier, M., and Bonnet, J. (1990) cDNA library construction from small amounts of unfractionated RNA: association of cDNA synthesis with polymerase chain reaction amplification. *Anal. Biochem.* **188**, 422–426.
9. Jepson, I., Bray, J., Jenkins, G., Schuch, W., and Edwards, K. (1991) A rapid procedure for the construction of PCR cDNA libraries from small amounts of plant tissue. *Plant Mol. Biol. Reporter* **9**, 131–138.
10. Gurr, S. J., McPherson, M. J., Scollan, C., Atkinson, H. J., and Bowles, D. J. (1991) Gene expression in nematode-infected plant roots. *Mol. Gen. Genet.* **226**, 361–366.
11. Ko, M. S. H., Ko, S. B. H., Takahashi, N., Nishiguchi, K., and Abe, K. (1990) Unbiased amplification of a highly complex mixture of DNA fragments by "lone linker"-tagged PCR. *Nucleic Acids Res.* **18**, 4293,4294.
12. Froussard, P. (1992) A random-PCR method (rPCR) to construct whole cDNA library from low amounts of RNA. *Nucleic Acids Res.* **20**, 2900.
13. Jakobsen, K. S., Breivold, E., and Hornes, E. (1990) Purification of mRNA directly from crude plant tissues in 15 minutes using magnetic oligo dT microspheres. *Nucleic Acids Res.* **18**, 3669.
14. Raineri, I., Moroni, C., and Senn, H. P. (1991) Improved efficiency for single-sided PCR by creating a reusable pool of first-strand cDNA coupled to a solid phase. *Nucleic Acids Res.* **19**, 4010.
15. Lambert, K. N. and Williamson, V. M. (1993) cDNA library construction from small amounts of RNA using paramagnetic beads and PCR. *Nucleic Acids Res.* **21**, 775,776.
16. Kaufman, D. L. and Evans, G. A. (1990) Restriction endonuclease cleavage at the termini of PCR products. *BioTechniques* **9**, 304–306.

17. Clark, J. M. (1988) Novel non-template nucleotide addition reactions catalysed by procaryotic and eucaryotic DNA polymerases. *Nucleic Acids Res.* **20,** 9677–9686.

18. Gubler, U. and Hoffman, B. J. (1983) A simple and very efficient method for generating cDNA libraries. *Gene* **25,** 263–269.

19. Fernandez, J. M., Mc Atee, C., and Herrnstadt, C. (1990) Advances in cDNA technology. *Am. Biotechnol. Lab.* **8,** 46,47.

20. Stoker, A. W. (1990) Cloning of PCR products after defined cohesive termini are created with T4 DNA polymerase. *Nucleic Acids Res.* **18,** 4290.

21. Kripamoy, A., Kusano, T., Suzuki, N., and Kitagawa, Y. (1990) An improved method for the construction of high efficiency cDNA library in plasmid or lambda vector. *Nucleic Acids Res.* **18,** 1071.

22. Hung, T., Mak, K., and Fong, K. (1990) A specificity enhancer for polymerase chain reaction. *Nucleic Acids Res.* **18,** 4953.

23. Wood, W. I., Gitschier, J., Lasky, L. A., and Lawn, R. M. (1985) Base composition-independent hybridization in tetramethylammonium chloride: a method for oligonucleotide screening of highly complex gene libraries. *Proc. Natl. Acad. Sci. USA* **82,** 1585–1588.

24. Karrer, E. E., Lincoln, J. E., Hogenhout, S., Bennett, A. B., Bostock, R. M., Martineau, B., Lucas, W. J., Gilchrist, D. G., and Alexander, D. (1995) In situ isolation of mRNA from individual plant cells: creation of cell-specific cDNA libraries. *Proc. Natl. Acad. Sci. USA* **92,** 3814–3818.

25. Rochester, D. E., Winer, J. A., and Shah, D. M. (1986) The structure and expression of maize genes encoding the major heat shock protein, hsp70. *EMBO J.* **5,** 451–458.

26. Haymerle, H., Herz, J., Bressan, G. M., Frank, R., and Stanley, K. K. (1986) Efficient construction of cDNA libraries in plasmid expression vectors using an adapter strategy. *Nucleic Acids Res.* **14,** 8615–8624.

27. D'Souza, C. R., Deugau, K. V., and Spencer, J. H. (1989) A simplified procedure for cDNA and genomic library construction using nonpalindromic oligonucleotide adaptors. *Biochem. Cell Biol.* **67,** 205–209.

28. Bhat, G. J., Lodes, M. J., Myler, P. J., and Stuart, K. D. (1990) A simple method for cloning blunt ended DNA fragments. *Nucleic Acids Res.* **19,** 398.

29. Challberg, M. D. and Englund, P. T. (1980) Specific labeling of 3' termini with T4 DNA polymerase, in *Methods in Enzymology*, vol. 65, *Nucleic Acids,* Part I (Grossman, L. and Moldave, K., eds.), Academic, New York, pp. 39–43.

2

Increasing the Average Abundance of Low-Abundance cDNAs by Ordered Subdivision of cDNA Populations

David R. Sibson and Michael P. Starkey

1. Introduction

It is estimated that 50,000–100,000 genes are expressed across the cell types of higher eukaryotes. As a consequence of differential expression (either regionally, temporally, or environmentally specific), a large proportion of all transcripts (approx 40–45%) represent low-abundance mRNAs, present at 1–20 molecules/cell *(1)*. In a given cell type, "low-abundance" mRNAs are likely to represent >95% of the different mRNAs expressed. The lower the abundance of a given transcript, the larger the number of clones in a representative cDNA library which must be screened in order to have a reasonable chance of isolating that message; for example, employing a statistical calculation *(2)*, it can be estimated that approx 500,000 clones need be screened to have a 99% probability of finding one representative of a gene expressed at the level of 0.001% (a low-abundance message) of total cellular mRNA. This is of practical significance in all cases, except where high-abundance mRNAs (accumulating to a few percent of total cellular mRNA) are sought. The implications are particularly severe for Human Genome Project expressed sequence tag (EST) studies, which on the basis of automated sequencing, aim to identify all the unknown genes expressed in a particular cell or tissue. Utilizing an unmodified cDNA library, well-resourced laboratories are forced to employ approaches incorporating a 10- to 100-fold redundancy in screening in order to isolate rare mRNAs. However, such approaches are not an option available to the majority of investigators.

"Normalization" is a means of reducing the number of clones in a cDNA library that must be screened in order to detect rare transcripts. This is achieved

From: *Methods in Molecular Biology, Vol. 69: cDNA Library Protocols*
Edited by: I. G. Cowell and C. A. Austin Humana Press Inc., Totowa, NJ

by a process in which abundant messages are reduced in number and the relative proportion of rare messages is increased. Differential hybridization has provided a general strategy for normalization via several approaches. Prescreening of cDNA libraries, employing abundant sequence hybridization probes, is a means of reducing the redundancy of sequencing in EST studies.

Subtractive hybridization has conventionally been directed toward the isolation of differentially expressed genes. In particular, messages (predominantly abundant) common to a related cell type (driver) are subtracted from the cell type-derived mRNA population of interest (target), by hybridizing driver polyA$^+$ RNA at a 10-fold excess with target cDNA. RNA–DNA hybrids representing sequences present in both cell types can be removed *(3–5)*, and the remaining unhybridized cDNA can be used to generate a subtracted library (or be used as a subtracted cell type-specific probe to identify clones of interest).

An alternative means of enriching the proportion of specific low-abundance cDNAs cloned in a given library is based on hybridization with genomic DNA, such that the relative abundance of cDNAs is rendered proportional to the abundance of complementary genes in the genomic DNA. Low-abundance cDNAs, encoded by immobilized genomic clones, can be isolated, amplified by polymerase chain reaction (PCR), and subcloned *(6–8)*.

A further approach to constructing a cDNA library containing an approximately equal representation of all the transcripts present in an initial polyA$^+$ RNA preparation relies on the differential rates at which denatured double-stranded (ds) cDNAs of varying abundance reanneal in solution. Single-stranded (ss) cDNAs (corresponding to sequences of relatively lower abundance) can be cloned, following separation from abundant cDNAs, which reanneal more rapidly to form double-stranded molecules *(9)*. Utilizing a refinement of this procedure *(10)*, the range of abundance of mRNAs represented in a human infant brain cDNA library (constructed in a phagemid vector) has been reduced from 4 to 1 orders of magnitude. In this model, ss phagemid circles were reassociated with short complementary strands, derived by controlled primer extension from the essentially unique noncoding 3'-ends of cDNA inserts cloned in the ss circles.

Hybridization-based approaches to reducing the complexity of cDNA libraries are, however, beset by a common problem. Repetitive sequences (e.g., Alu repeats) shared by nonhomologous cDNAs may be responsible for the elimination of low-abundance cDNAs, and/or the selection of abundant cDNAs. A further problem associated with solution-phase hybridization is the thermal degradation of DNA; the sizes of ss and ds DNA decrease progressively with increasing reassociation time. PCR amplification (utilizing cDNA insert-flanking vector primers), although able to increase the concentration of separated ss

DNAs *(9)*, introduces the element of length-dependent differential amplification of "normalized" sequences.

In this chapter we describe a method of "normalization" based on the sequence-specific sorting of cDNA restriction fragments. Restriction fragment sorting provides a means of subdividing complex cDNA mixtures into distinct subpopulations. In concept, a given cDNA restriction fragment can only be sorted into a single subset. Since an individual subpopulation is of relatively low complexity (as compared to the original population), the concentration of any given cDNA will be higher than in the original population. The subpopulations combined represent the entire original cDNA population.

The restriction fragment sorting of cDNAs is based on the partitioning of cohesive-ended cDNA restriction fragments according to the sequence of their ends. This technique is thus unaffected by the presence of internal repeat sequences common to nonhomologous cDNAs. Restriction fragment sorting necessitates the use of an enzyme that produces staggered cuts outside of its recognition sequence, so that any combination of bases is possible in the cohesive ends produced.

cDNA fragments can be sorted into different subsets by successive base-specific adaptering and base-specific PCR. cDNA fragments separated into different subsets are amplified by PCR. This serves to enrich further the abundance of rare mRNAs, since although the relative proportion of different cDNAs within a given subset remains constant, the absolute abundance of a particular cDNA in a subset is significantly increased above that in the original population. This facilitates the subdivision of a complex cDNA population into subsets, which combined will provide access to a greater repertoire of genes than would be accessible in the original unmodified cDNA population.

We have utilized this technology to reduce the redundancy of sequencing in a human EST program at the UK Human Genome Mapping Project Resource Centre. The elimination of serum albumin (present at the level of a few percent of cellular mRNA in fetal liver) from sorted subpopulations of fetal liver cDNA *(11)*, readily provides an illustration of the effectiveness of the subsetting procedure.

1.1. Sorting of cDNA Restriction Fragments

A population of cDNA molecules is digested with a type-IIS restriction endonuclease, generating fragments with nonidentical cohesive ends. The number of different end sequences is 4^n, where n is the length of the overhang. *Fok*I has an asymmetric pentanucleotide recognition sequence (cutting DNA every 512-bp on average) and generates fragments with 4-base 5'-overhangs (Fig. 1). If the two ends of a *Fok*I fragment (independent of each other) are considered, there are potentially 4^8 (65,536) fragment classes, each with a different pair of

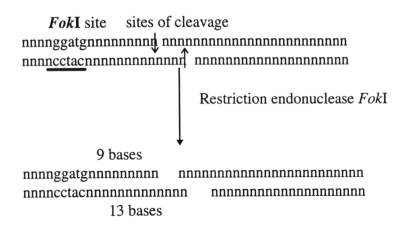

*Fok*I site   sites of cleavage

n - can be any base, but is always the same base at a given
position in a given sequence of DNA

Fig. 1. Nature of the cleavage produced by the restriction endonuclease *Fok*I.

cohesive ends. Each base of a given *Fok*I cohesive end represents an identifier
for the cDNA fragment possessing that overhang. If 4 bases divided between
the two ends of a *Fok*I fragment are considered, 4^4 (256) different classes of
fragment can be identified. The procedure by which 256 *Fok*I cDNA restriction fragment classes can be recognized is as follows.

1.1.1. Sorting by Base-Specific Adaptering

The specificity of the T_4 DNA ligase reaction is employed to select for
cDNA fragments with particular cohesive termini. *Fok*I cDNA fragments are
ligated to two types of adapters (Fig. 2). A given specific adapter (Fig. 2, Inset
A), with specified bases at two positions within the 4-base 5'-cohesive end, is
capable of specific ligation to the $1/16$ of cDNA fragments that have complementary cohesive ends. Solid-phase capture of cDNA fragments containing at
least a single specific adapter (Fig. 2) provides a means of selecting $1/16$ of all
*Fok*I cDNA fragments. Independent application of 16 different specific
adapters would enable a heterogenous mixture of *Fok*I cDNA fragments to be
fractionated into 16 "primary" subsets.

1.1.2. Sorting by Base-Specific PCR

The specificity of primer annealing and extension, as part of the PCR reaction, can be utilized to subdivide adapted *Fok*I cDNA fragment primary
subsets further. Asymmetric PCR, employing a primer with a specified base

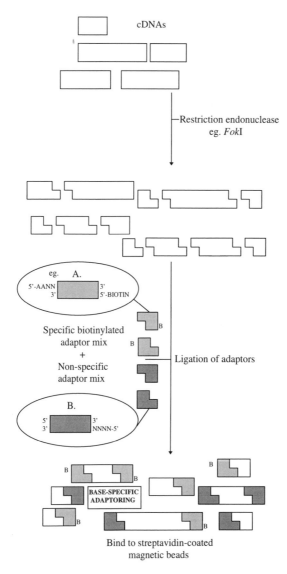

Fig. 2. Base-specific selection of cDNA restriction fragments *(continued on next page).*

at its 3'-terminus, will theoretically discriminate in favor of solid-phase captured ss cDNAs containing a complementary base at position four of the 5'-cohesive terminus of a nonspecific adapter sequence (Fig. 2). Four primers (5'-...A-3', 5'-...C-3', 5'-...G-3', and 5'-...T-3'), used independently, would partition each primary subset into four "secondary" subsets. A second selective round of amplification of specific solution-phase ss cDNAs can be

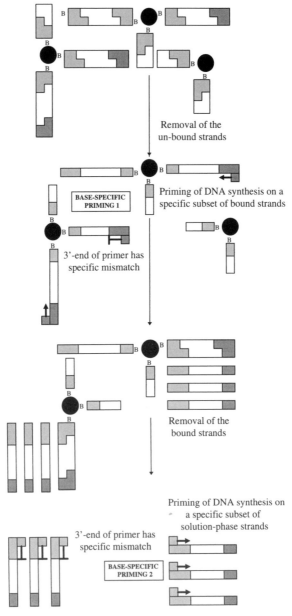

Fig. 2. *(continued)*

achieved employing a primer 100% complementary to cDNAs with a given base at position four of the 5'-cohesive end of a specific adapter sequence (Fig. 2). Amplification of each of the 64 secondary subsets with each of four specific primers (5'-...A-3', 5'-...C-3', 5'-...G-3', or 5'-...T-3') would generate 256 "tertiary" subsets.

In this chapter, we describe an approach by which a population of *Fok*I cDNA restriction fragments can be sorted into 256 subsets, based on the combined effects of 2-base selection imposed by ligation and 2-base selection exacted by PCR. However, in principle, utilizing an enzyme that generates 4-base overhangs (e.g., *Fok*I), it is possible to achieve a subdivision of between 4^1- and 4^8-fold by 1–8 base selection, effected either by any combination of adaptering and PCR, or by one or the other procedure employed singly.

2. Materials
2.1. Isolation of mRNA

1. RNA extraction kit, including extraction buffer: buffered aqueous solution containing guanidium thiocyanate, *N*-lauryl sarcosine, and EDTA; $2M$ potassium acetate, pH 5.0 (Pharmacia Biotech, St. Albans, Hertfordshire, UK).
2. 2.0 ± 0.05 g/mL cesium trifluoroacetate (CsTFA) (Pharmacia Biotech).
3. Sterile 50-mL graduated conical centrifuge tubes with caps (Falcon, London, UK).
4. Sorvall RT6000B refrigerated centrifuge, H1000B swinging-bucket rotor, and $4 \times 4 \times 50$ mL tube buckets (00830) (Dupont, Stevenage, Hertfordshire, UK).
5. Quick seal polyallomer 16×76 mm 13.5-mL ultracentrifuge tubes, and open-top polyallomer 13.2-mL 14×89 mm centrifuge tubes (Beckman, High Wycombe, Buckinghamshire, UK).
6. L7-65 ultracentrifuge, Ti50 fixed angle, and SW41Ti swinging-bucket rotors (Beckman).
7. 0.2% (v/v) Diethylpyrocarbonate (DEPC)-treated water (DTW) (DEPC is a powerful protein denaturant and denatures ribonuclease irreversibly). DEPC is added to the water, which is left to stand for 20 min, and then autoclaved twice at 121°C for 20 min to decompose the DEPC.
8. 0.2% (v/v) DEPC-treated $3M$ lithium chloride (LiCl).
9. Disposable polypropylene micropipet tips and microcentrifuge tubes (Sarstedt, Leicester, UK) baked at 150°C for 1 h.
10. Ultra-Turrax T25 homogenizer probe washed successively in 100% (v/v) ethanol and three times in DEPC-treated water, and baked for 4 h at 250°C.
11. Dynal MPC-E-1 magnet and Dynabeads mRNA purification kit, incorporating 5 mg/mL Dynabeads Oligo (dT)$_{25}$; 2X binding buffer: 20 mM Tris-HCl, pH 7.5, 1M LiCl, 2 mM EDTA; 1X washing buffer: 10 mM Tris-HCl, pH 7.5, 0.15M LiCl, 1 mM EDTA; and elution buffer: 2 mM EDTA diluted 1:1 with DTW (Dynal UK, Bromborough, Merseyside, UK).

2.2. cDNA Synthesis

1. cDNA synthesis kit (Pharmacia Biotech).
2. T$_4$ DNA polymerase (Boehringer Mannheim, Lewes, East Sussex, UK).

3. SizeSep 400 spun columns (for rapid intermediate purification of cDNAs >400 bp in length), prepacked with Sephacryl S-400 (Pharmacia Biotech).
4. Sterile 50-mL graduated conical centrifuge tubes with caps (Falcon).
5. Sorvall RT6000B refrigerated centrifuge, H1000B swinging-bucket rotor, and 4 × 4 × 50 mL tube buckets (00830) (Dupont).

2.3. Restriction of cDNA

1. *Fok*I (Boehringer Mannheim).
2. 10X *Fok*I reaction buffer: 100 m*M* Tris-HCl, 500 m*M* NaCl, 100 m*M* MgCl$_2$, 10 m*M* dithioerythritol, pH 7.5 (incubation buffer M, Boehringer Mannheim).
3. Phenol/chloroform (Sigma, Poole, Dorset, UK).
4. SizeSep 400 spun columns (for rapid intermediate purification of cDNAs >400 bp in length), prepacked with Sephacryl S-400 (Pharmacia Biotech).
5. Sterile 50-mL graduated conical centrifuge tubes with caps (Falcon).
6. Sorvall RT6000B refrigerated centrifuge, H1000B swinging-bucket rotor, and 4 × 4 × 50 mL tube buckets (00830) (Dupont).

2.4. Sorting of cDNA Restriction Fragments

1. Nonphosphorylated oligonucleotides (1 µ*M* syntheses) produced, for example, using an Applied Biosystems 380B synthesizer, and purified by reverse-phase high-performance liquid chromatography (HPLC) (Beckman System Gold):
   a. Nonspecific θ adapter comprising:
      i. Universal θ oligonucleotide: 5'-TGTCTGTCGCAGGAGAAGGA-3'.
      ii. Variable θ oligonucleotide mix: 5'-NNNNTCCTTCTCCTGCGACAG ACA-3', where N = A, C, G, or T (i.e., the variable θ oligonucleotide mix is a mixture of 4^4 different species, and is the sum of 16 oligonucleotide syntheses; e.g., 5'-NNNA..., 5'-NNNC..., 5'-NNNG..., 5'-NNNT..., 5'-NNAN..., etc.).
   b. Base-specific π adapter comprising:
      i. Universal π oligonucleotide: 5'-biotin-GTTCTCGGAGCACTGTCCG AGA-3'.
      ii. Specific π oligonucleotide mix: 5'-XYNNTCTCGGACAGTGCTCCGAG AAC-3', where X and Y are specified bases, and N = A, C, G, or T (i.e., the specific π oligonucleotide is a mixture of 4^2 molecules, and in addition, there are 4^2 possible specific π oligonucleotide mixes).
   c. Base-specific θ PCR primer: 5'-TGTCTGTCGCAGGAGAAGGAX-3', where X is a specified base (i.e., there are four possible base-specific θ PCR primers).
   d. Base-specific π PCR primer: 5'-GTTCTCGGAGCACTGTCCGAGAY-3', where Y is a specified base (i.e., there are four possible base-specific π PCR primers).
2. T$_4$ DNA ligase (Boehringer Mannheim).
3. 10X T$_4$ DNA ligase buffer: 5*M* Tris-HCl, 500 m*M* NaCl, 1*M* MgCl$_2$, 1*M* dithio-threitol, 10 m*M* spermine, 100 m*M* ATP, pH 7.4 (Amersham International, Little Chalfont, Buckinghamshire, UK).

4. Streptavidin-coated magnetic beads (10 mg/mL Dynabeads M-280 Streptavidin, Dynal, UK).
5. Dynabeads buffers: washing: 1M NaCl, 10 mM Tris-HCl, pH 8.3; binding: 50 mM NaCl, 10 mM Tris-HCl, pH 8.3; elution: 0.15M NaOH (freshly prepared).
6. Techne PHC-2 thermocycler.
7. Reagents for PCR: Ampli*Taq* DNA polymerase (Perkin-Elmer, Warrington, Cheshire, UK); 10X *Taq* polymerase buffer: 100 mM Tris-HCl, pH 8.3, 500 mM KCl, 25 mM MgCl$_2$; 10X dNTPs: 4X 2 mM dNTPs (Ultrapure dNTPs; Pharmacia Biotech).
8. NuSieve GTG agarose (FMC BioProducts, Vallensback Strand, Denmark).
9. SizeSep 400 spun columns, prepacked with Sephacryl S-400 (Pharmacia Biotech).
10. Sterile 50-mL graduated conical centrifuge tubes with caps (Falcon).
11. Sorvall RT6000B refrigerated centrifuge, H1000B swinging-bucket rotor, and 4 × 4 × 50 mL tube buckets (00830) (Dupont).
12. T$_4$ DNA polymerase (Boehringer Mannheim).
13. 10X T$_4$ DNA polymerase resection buffer: 100 mM Tris-HCl, 500 mM NaCl, 100 mM MgCl$_2$, 10 mM dithioerythritol, pH 7.5 (incubation buffer M, Boehringer Mannheim).
14. cDNA spun columns (for rapid intermediate purification of cDNAs without significant size selection), prepacked with Sephacryl S-300 (Pharmacia Biotech).

2.5. Preparation of Cloning Vector

1. pBluescript II KS+ (Stratagene, Cambridge, UK).
2. *Hind*III and *Bam*HI (Boehringer Mannheim).
3. 10X *Hind*III + *Bam*HI restriction buffer: 100 mM Tris-HCl, 1M NaCl, 50 mM MgCl$_2$, 10 mM 2-mercaptoethanol, pH 8.0 (incubation buffer B, Boehringer Mannheim).
4. Nonphosphorylated oligonucleotides:
   a. *Hind*III/θ adapter comprising:
      i. *Hind*III compatible oligonucleotide: 5'-AGCTCGGCTCGAGTCTG-3'.
      ii. θ compatible oligonucleotide: 5'-GCGACAGACAGCAGACTCGAG CCG-3'.
   b. *Bam*HI/π adapter comprising:
      i. *Bam*HI compatible oligonucleotide: 5'-GATCCGGCTCGAGT-3'.
      ii. π compatible oligonucleotide: 5'-CCGAGAACACTCGAGCCG-3'.
5. T$_4$ DNA ligase (Boehringer Mannheim).
6. 10X T$_4$ DNA ligase buffer (Amersham International).
7. T$_4$ polynucleotide kinase (Amersham International).
8. 10X T$_4$ polynucleotide kinase buffer: 5M Tris-HCl, 500 mM NaCl, 1M MgCl$_2$, 1M dithiothreitol, 10 mM spermine, 100 mM ATP, pH 7.4 (Amersham International).
9. SizeSep 400 spun columns, prepacked with Sephacryl S-400 (Pharmacia Biotech).
10. Sterile 50-mL graduated conical centrifuge tubes with caps (Falcon).
11. Sorvall RT6000B refrigerated centrifuge, H1000B swinging-bucket rotor, and 4 × 4 × 50 mL tube buckets (00830) (Dupont).
12. cDNA spun columns, prepacked with Sephacryl S-300 (Pharmacia Biotech).

2.6. Cloning and Transformation

1. T_4 DNA ligase (Boehringer Mannheim).
2. 10X T_4 DNA ligase buffer (Amersham International).
3. *Escherichia coli* XL1-Blue recA1, endA1, gyrA96, thi1, hsdR17, supE44, relA1, lac, F' [proAB+, lacI^q, lacZΔM15, Tn10(tet^r)] (Stratagene).
4. LB broth: 1% (w/v) bacto-tryptone, 0.5% (w/v) bacto-yeast extract, 1% (w/v) NaCl (LB agar: LB broth + 1.5% [w/v] agar).
5. "Standard transformation buffer" (STB): 10 mM 2-N-morpholinoethanesulfonic acid-KOH, pH 6.2, 100 mM RbCl, 45 mM MnCl$_2$, 10 mM CaCl$_2$, 3 mM hexamine cobalt (III) chloride (filter-sterilize and store at 4°C).
6. Dimethyl sulfoxide (Sigma) (store in small aliquots at –20°C).
7. 2.25M Dithiothreitol in 40 mM potassium acetate, pH 6.0 (filter-sterilize and store at –20°C).
8. SOC broth: 2% (w/v) bacto-tryptone, 0.5% (w/v) bacto-yeast extract, 10 mM NaCl, 2.5 mM KCl, pH 7.5; sterilize by autoclaving, and supplement with Mg^{2+} (MgCl$_2$ + MgSO$_4$) to 20 mM, and glucose to 20 mM, prior to use.
9. Sterile 50- and 15-mL graduated conical centrifuge tubes with caps (Falcon).
10. Sorvall RT6000B refrigerated centrifuge, H1000B swinging-bucket rotor, 4 × 4 × 50 mL tube buckets (00830), and 4 × 10 × 15 mL tube buckets (00884) (Dupont).
11. 50 mg/mL Ampicillin (sodium salt; Sigma) in water (store in small aliquots at –20°C).
12. 5 mg/mL Tetracycline (tetracycline-HCl; Sigma) in 70% (v/v) ethanol (store in the dark at –20°C).
13. IPTG: 200 mg/mL isopropyl β-D-thiogalactopyranoside in water (store in 1-mL aliquots at –20°C).
14. Xgal: 20 mg/mL 5-bromo-4-chloro-3-indolyl-β-D-galactopyranoside in dimethyl-formamide (store in the dark at –20°C).

3. Methods

3.1. Extraction of Total RNA and Isolation of mRNA

This section describes one method for the extraction of total RNA from human tissues and the purification of polyA$^+$ RNA. The extraction of total RNA is based on the use of the chaotropic salt guanidium thiocyanate to destroy ribonuclease activity and deproteinase nucleic acids *(12)*, and the purification of RNA by isopycnic banding in CsTFA *(13)*.

1. Homogenize (Ultra-Turrax T25, 30s) approx 1 g of tissue in 18 mL of extraction buffer (density = 1.51 g/mL) in a 50-mL centrifuge tube. Remove cell debris from the homogenate by centrifugation at 1349g for 20 min at 15°C.
2. Layer 6.5-mL aliquots of the homogenate on top of 6-mL volumes of 2.0 g/mL CsTFA (the average density of tube contents is therefore 1.75 g/mL) in quick-seal polyallomer centrifuge tubes (3 tubes/1 g tissue). Centrifuge (fixed angle rotor) at 120,000g for 36–48 h at 15°C.

3. Collect successive gradient aliquots of 0.5, 1.5, 1.5, 0.5, and 0.5mL from the bottom of each tube, and identify the fractions containing RNA by analysis of 10 µL of each fraction by 2% (w/v) TBE-agarose gel electrophoresis.

4. Pool the RNA-containing fractions from each of the three tubes (~9–10 mL combined volume), and reload 4.5–5 mL aliquots onto 6 mL of 2.0 g/mL CsTFA in open-top polyallomer centrifuge tubes (2 tubes/1 g tissue). Recentrifuge (swinging-bucket rotor) at 120,000g for 16 h at 15°C.

5. Moving from the top of each tube toward the bottom, remove the top 6 mL of gradient, and then collect 7–10 successive ~0.5-mL gradient aliquots. Identify the fractions containing RNA by analysis of 10 µL of each fraction by 2% (w/v) TBE-agarose gel electrophoresis.

6. Add $^1/_{10}$ volume of 2M potassium acetate, pH 5.0, and 2.5 vol of 100% (v/v) ethanol to each RNA-containing fraction to precipitate the nucleic acid. Incubate at –70°C for 1 h, and centrifuge at 13,800g for 30 min at 4°C. Wash the pelleted RNA in 500 µL of 70% (v/v) ethanol (recentrifuging at 13,800g for 30 min at 4°C). Air-dry the RNA, and resuspend each pellet in 500 µL of DTW.

7. Add an equal volume of 3M LiCl to each aliquot of RNA, mix by vortexing, and leave at 4°C for 10–16 h. Centrifuge at 13,800g for 30 min at 4°C to precipitate the high-mol-wt RNA. Wash each pellet in 500 µL of 70% (v/v) ethanol, air-dry, and resuspend each pellet in 500 µL of DTW. (Confirm selective precipitation of mRNA and rRNA by analysis of a 10 µL of each sample by 2% [w/v] TBE-agarose gel electrophoresis.)

8. Add 490 µL of phenol/chloroform to each aliquot of RNA, mix by vortexing, and centrifuge at 13,800g for 5 min at room temperature. Collect the upper aqueous layer in each case. Add 490 µL of chloroform to each aliquot collected, mix, and recentrifuge. Collect each aqueous layer.

9. Precipitate each aliquot of RNA with $^1/_{10}$ vol of 2M potassium acetate, pH 5.0, and 2.5 vol of 100% (v/v) ethanol. Incubate at –70°C for 1 h, and centrifuge at 13,800g for 30 min at 4°C. Wash, and store each pellet in 500 µL of 70% (v/v) ethanol at –70°C, until required.

10. When necessary, precipitate the RNA samples, air-dry, and resuspend in a combined volume of 200 µL of DTW. Estimate the concentration of the RNA solution by measuring the optical density (OD) at 260 nm of a 10-µL aliquot (OD_{260nm} of 1.0 ≈ 40 µg/mL RNA), and the purity of the RNA by measuring the OD at 280 nm and calculating OD_{260nm}/OD_{280nm} (RNA giving a ratio of 1.8–2.0 is satisfactory). Remove a 300-µg aliquot of RNA, and adjust the volume with DTW to produce a 0.75 µg/µL solution.

11. Heat the 0.75 µg/µL RNA solution to 65°C for 2 min to disrupt secondary structure, and add 100 µL to 1.0 mg of Dynabeads Oligo (dT)$_{25}$ resuspended (after washing in 100 µL of 2X binding buffer) in 100 µL of 2X binding buffer. Mix gently, and allow hybridization to proceed for 5 min at room temperature.

12. Pellet the Dynabeads Oligo Dynabeads (dT)$_{25}$ using a magnet and remove the supernatant. Wash the Dynabeads twice with 200 µL of washing buffer, taking care to remove all the supernatant.

13. Resuspend the Dynabeads in 10 μL of elution buffer, heat at 65°C for 2 min, and immediately remove the supernatant containing the eluted mRNA.

14. Repeat steps 11–14 for an additional 3 × 75 μg of total RNA. Pool the four eluted mRNA samples, and examine the integrity (and quantity) of a 5-μL aliquot by 2% (w/v) TBE-agarose gel electrophoresis. Store the mRNA at −70°C if not used immediately.

3.2 cDNA Synthesis

This section describes the synthesis of cDNA utilizing a kit featuring oligo (dT)$_{12-18}$-primed first-strand synthesis by cloned Moloney Murine Leukemia Virus reverse transcriptase (reduced RNase H activity as compared to Avian Retrovirus reverse transcriptase) and second-strand synthesis via the extension of multiple RNase H-generated RNA fragments *(14)*. The kit protocol is modified by employing T$_4$ DNA polymerase to fill in the recessed 3'-ends of cDNA fragments following second-strand synthesis.

1. Dilute 1–5 μg of polyA$^+$ RNA (≈10 μL of the mRNA prepared as described in Section 3.1.) to 20 μL with DTW. Heat at 65°C for 10 min to denature any secondary structure, and then chill on ice.

2. Add the denatured mRNA to a first-strand reaction mixture, mix by recycling through a micropipet tip, and incubate at 37°C for 1 h.

3. Add the first-strand reaction mixture (32 μL) to a second-strand reaction mixture (68 μL). Mix and incubate at 12°C for 1 h, and then at 22°C for 1 h.

4. Heat at 70°C for 1 min, and cool on ice. Add T$_4$ DNA polymerase (2 U/μg of original mRNA), mix briefly, and incubate at 37°C for 30 min.

5. Add 100 μL of phenol/chloroform, mix by vortexing, and centrifuge at 13,800g for 5 min at room temperature. Collect the upper aqueous layer.

6. Purify the cDNA by loading onto a SizeSep 400 column (equilibrated in 50 m*M* NaCl, 10 m*M* Tris-HCl, pH 7.5), standing in a 1.5-mL microcentrifuge tube placed inside a 50-mL centrifuge tube, and centrifuging at 423g for 2 min and 40 s in a swinging-bucket rotor. Collect the column filtrate.

7. Precipitate the cDNA with $^1/_{10}$ vol of 3*M* sodium acetate and 2.5 vol of 100% (v/v) ethanol (centrifuging at 13,800g for 30 min at 4°C). Wash with 70% (v/v) ethanol, air-dry, and resuspend the cDNA in 30 μL of sterile distilled water. Store the cDNA at −20°C until required.

3.3. Restriction Digestion of cDNA

This section describes the restriction of cDNA with *Fok*I. In principle, any type-IIS restriction enzyme whose cutting site is displaced from its recognition sequence and that generates nonidentical 4-base cohesive overhangs may be utilized.

1. Mix 30 μL of cDNA (prepared from ~1 to 5 μg of mRNA, as described in Section 3.2.) with 10 μL of 10X incubation buffer M, and 16 U of *Fok*I (<$^1/_{10}$ the volume of the reaction mixture), in a total volume of 100 μL. Digest at 37°C for 2 h.

2. Purify the cDNA fragments by two successive 100-µL phenol/chloroform extractions, and by gel filtration through a SizeSep 400 column (equilibrated in 1X T_4 DNA ligase buffer) (*see* Section 3.2.). Store the cDNA at –20°C until required.

3.4. Sorting of cDNA Restriction Fragments

3.4.1. Sequence-Specific Ligation of Adapters

This section details the ligation of adapters to cDNA fragments with *Fok*I-generated 4-base 5'-cohesive ends. In theory, the adapters are in at least a 100- to 200-fold molar excess to the cDNA ends in order to drive the adaptering reaction. In concept, a given base-specific π adapter is capable of annealing to the $1/4^2$ of *Fok*I cDNA ends, which have a complementary 4-base cohesive end. In principle, the aim is to give to the $1/16$ of the cDNA ends, capable of annealing to a given base-specific π adapter, an equal probability of annealing to a nonspecific θ adapter with an identical 4-base overhang. A given specific π oligonucleotide, e.g., 5'-AACG...-3', is $1/4^2$ of the relevant specific π oligonucleotide mix, i.e., 5'-AANN...-3'. However, since the variable θ oligonucleotide with an equivalent 4-base 5'-overhang is $1/4^4$ of the variable θ oligonucleotide mix (5'-NNNN...3'), it is necessary to employ 16 times more variable θ oligonucleotide mix than specific π oligonucleotide mix.

1. Mix 25% of the cDNA fragments (prepared in Section 3.3.) with 94.1 pmol of 10 µM universal θ oligonucleotide, 94.1 pmol of 10 µM variable θ oligonucleotide mix, 5.9 pmol of 10 µM universal π oligonucleotide, and 5.9 pmol of a 10 µM specific π oligonucleotide mix. Heat at 65°C for 3 min. Allow to cool to ambient temperature, and add 9 µL of 10X T_4 DNA ligase buffer (heat to 37°C and vortex to dissolve precipitates formed on freezing), sterile distilled water (to adjust final volume to 90 µL), and 2.4 U of T_4 DNA ligase. Incubate at 12°C for 16 h.
2. Remove excess adapters and DNA ligase, and size select ≥400-bp cDNA fragments, by two successive phenol/chloroform extractions and SizeSep 400 column (equilibrated in 1X *Taq* polymerase buffer) chromatography.
3. Add 20 µL of 10X *Taq* polymerase buffer, 20 µL of 10X dNTPs, and sterile distilled water (to adjust the final reaction volume to 200 µL) to the adapted cDNA fragments. Heat to 78°C, and add 5 U of *Taq* polymerase. Maintain at 78°C for 5 min to remove the non-5'-phosphorylated adapter oligonucleotides, which are not ligated to the recessed 3'-hydroxyl-bearing termini of cDNA fragments. Incubate at 72°C for 10 min to allow *Taq* polymerase to extend the 3'-hydroxyl termini of the cDNA fragments, thereby filling the gaps in the adapters. (Store the adapted cDNA fragments at –20°C if not required immediately.)
4. Add the adapted cDNA fragments to 0.8 mg (80 µL) of streptavidin-coated magnetic beads, resuspended in 200 µL of binding buffer (after washing three times in 160 µL of washing buffer). Bind cDNA fragments adapted with at least a single biotinylated base-specific π adapter to the beads by incubating at 28°C for 30 min and mixing by recycling through a micropipet tip every 10 min.

5. Pellet the beads using a magnet, and wash twice in 200 μL of binding buffer. Remove the nonbiotinylated adapted ss cDNAs from the beads by washing four times in 200 μL of elution buffer (incubating at 28°C for 5 min on each occasion). Wash the beads twice in 200 μL of sterile distilled water and finally in 200 μL of 1X *Taq* polymerase buffer. Resuspend the beads in 230 μL of 1X *Taq* polymerase buffer, 1X dNTPs.

3.4.2. Sequence-Specific PCR

This section describes the initial selection of a subset of θ adapter-containing cDNAs, which had a specific base at the most 5'-position of the 4-base cohesive end ligated to a θ adapter, and the secondary selection of a further subset of cDNAs, which also had a specific base at the most 5'-position of the 4-base cohesive end ligated to the π adapter. In the presence of dTTP, the 3'→5' exonuclease activity of T$_4$ DNA polymerase is utilized to degrade the amplified adapted cDNAs generating specific 3'-recessed termini compatible with directional cloning into an adapted pBluescript II KS+ vector (*see* Section 3.5.).

1. Divide the bead-bound ss cDNAs into four. Add 2 pmol of a different 1-μM base-specific θ PCR primer (i.e., 5'-...A-3', 5'-...C-3', 5'-...G-3', or 5'-...T-3'), and 2.5 U (in 2 μL) of *Taq* polymerase to each 57.5-μL aliquot. Under mineral oil, perform 16 cycles of asymmetric PCR: 95°C, 30 s; 65°C, 2 min; 72°C, 3 min.

2. Add 0.3 mg (30 μL) of magnetic beads resuspended in 30 μL of binding buffer (after washing three times in 60 μL of washing buffer) to each PCR mixture (60 μL), and incubate at 28°C for 30 min (mixing every 10 min). Remove the bead-bound, biotinylated adapted ss cDNAs with a magnet.

3. Purify the nonbiotinylated adapted ss cDNAs present in each of the four supernatants by 2× phenol/chloroform extractions and filtration through a SizeSep 400 column (equilibrated in 1X *Taq* polymerase buffer). Store the ss cDNAs at –20°C until required.

4. Remove 4 × 10 μL aliquots from each of the four column filtrates (products of the θ PCR primers "A," "C," "G," and "T," respectively; each ~90 μL). To each aliquot of ss PCR product, add 2 pmol of the 1 μM base-specific θ PCR primer originally used to generate the product, and 2 pmol of a given 1 μM base-specific π PCR primer (i.e., to the 4 × 10 μL ss DNA aliquots produced by the "A" θ PCR primer, add 2 pmol of the 5'-...A-3' θ PCR primer, and 2 pmol of either the 5'-...A-3', 5'-...C-3', 5'-...G-3', or 5'-...T-3' base-specific π PCR primer, and so forth).

5. To each of the 16 DNA + primer mixes, add 4 μL of 10X *Taq* polymerase buffer, 4 μL of 10X dNTPs, sterile distilled water (to ensure a final reaction volume of 40 μL), and 2.5 U of *Taq* polymerase. Perform five cycles of 95°C, 30 s; 65°C, 2 min; 72°C, 3 min.

6. After five cycles, add an additional 20 pmol of both the appropriate 10 μM base-specific θ PCR primer and the appropriate 10 μM base-specific π PCR primer to each of the 16 PCR reaction mixtures. Perform an additional 16–32 cycles (*see* Note 5) of 95°C, 30 s; 65°C, 2 min; 72°C, 3 min, followed by 1 × 10 min at 72°C.

7. Adjust the volume of each PCR reaction mixture to 100 µL with sterile distilled water. Purify the amplification products of each reaction by 2× phenol/chloroform extractions and filtration through a SizeSep 400 column (equilibrated in 1X T_4 DNA polymerase resection buffer).
8. To 75% (~75 µL) of each of the 16 column filtrates, add 10 µL of 10X T_4 DNA polymerase resection buffer, 10 µL of 0.5 mM dTTP, sterile distilled water (to adjust final volume to 100 µL), and 16 U of T_4 DNA polymerase. Incubate at 37°C for 30 min.
9. Purify each of the 16 PCR products by 2× phenol/chloroform extractions and filtration (centrifuge at 423g for 2 min and 40 s in a swinging-bucket rotor) through a cDNA spun column, equilibrated in 1X T_4 DNA ligase buffer. Store the resected reaction products (each ~100 µL) at –20°C until required.

3.5. Preparation of Cloning Vector

In this section, the manipulation of pBluescript II KS+ to permit the directional cloning of the θ and π adapted cDNAs is described. In this case, the vector is restricted and adapted to permit the ligation of the θ adapter end of a cDNA molecule adjacent to the *Hind*III site (and hence KS sequencing primer site) in pBluescript II KS+, and the π adapter end of a cDNA adjacent to the *Bam*HI site (and hence M13 Universal sequencing primer site) in pBluescipt II KS+. However, it is possible to construct a range of adapted vectors to permit directional cloning of θ and π cDNA ends in either orientation.

1. Mix 10 µg of pBluescript II KS+ with 10 µL of 10X *Hind*III + *Bam*HI restriction buffer, sterile distilled water (to adjust final volume to 100 µL), 75 U of *Hind*III and 45 U of *Bam*HI. Incubate at 37°C for 60 min.
2. Remove the excised polylinker fragment from the plasmid vector by 2× phenol/chloroform extractions and SizeSep 400 column (equilibrated in 1X T_4 DNA ligase buffer) chromatography. Store the restricted plasmid at –20°C until required.
3. To 10 µL (~1 µg) of restricted plasmid, add 200 pmol of each of 20 µM *Hind*III-compatible oligonucleotide, 20 µM π-compatible oligonucleotide, 20 µM *Bam*HI-compatible oligonucleotide, and 20 µM π-compatible oligonucleotide. Heat at 65°C for 3 min. Allow to cool to ambient temperature, and add 6 µL of 10X T_4 DNA ligase buffer, sterile distilled water (to adjust final volume to 60 µL), and 2.4 U of T_4 DNA ligase. Incubate at 12°C for 16 h. In parallel, perform five ¹/₅X full-scale control reactions, omitting either T_4 DNA ligase or one of the four adapter oligonucleotides. Evaluate the success (perceptible 31-bp increase in the size of the linearized plasmid) of the adaptering reaction by analyzing a 10-µL aliquot of each reaction by 1.5% (w/v) TBE-agarose gel electrophoresis.
4. Remove the excess adapters by 2X phenol/chloroform extractions and SizeSep 400 column (equilibrated in 1X T_4 polynucleotide kinase buffer) chromatography.
5. Add 1 U of T_4 polynucleotide kinase to the column filtrate and incubate at 37°C for 30 min, in order to phosphorylate the protruding 5'-hydroxyl termini of the adapted plasmid. Purify the adapted vector by 2X phenol/chloroform extrac-

tions and cDNA spun column (equilibrated in 1X T_4 DNA ligase buffer) gel filtration (*see* Section 3.4.).

3.6. Cloning and Transformation

The procedure described in this section can be applied to each of the resected cDNA products generated by the 4 × 4 combinations of θ and π PCR primers. Aseptic technique should be followed through the preparation of competent cells and the transformation procedure *(15)*. Each of the 16 resected cDNA products will potentially yield a maximum of 10 × 9 cm diameter Petri dishes of bacterial colonies. Transformation of *E. coli* XL1-Blue (200 μL of competent cells) with 40 ng of pBluescript II KS+ should be performed in parallel, in order to permit estimation of the efficiency (no. of transformants/μg of DNA) with which the competent cells may be transformed.

1. Mix 2 μL (~30 ng) of adaptered pBluescript II KS+ with 10 μL of resected cDNA, 2 μL of 10X T_4 DNA ligase buffer, sterile distilled water (to adjust final volume to 20 μL), and 1.2 U of T_4 DNA ligase. Incubate at 12°C for 16 h.
2. Pick a single colony from an LB agar + 12.5 μg/mL tetracycline dilution streak plate of *E. coli* XL1-Blue into 5 mL of LB broth + 12.5 μg/mL tetracycline, and culture overnight at 37°C. Inoculate 100 mL of prewarmed LB broth + 12.5 μg/mL tetracycline (in a 1000-mL conical flask) with 1 mL of the overnight culture. Incubate with agitation (275 rpm, approx 1.437g) at 37°C until an $OD_{550 \, nm}$ = 0.5 is achieved (~2–2.5 h).
3. Chill the culture on ice for 10–15 min. Transfer 50-mL aliquots (50 mL of cells are sufficient for 12× discrete *E. coli* transformations) into precooled 50-mL centrifuge tubes and centrifuge at 1349g for 12 min at 4°C. On ice, carefully resuspend (by gentle swirling) the bacterial pellet in 17 mL of STB.
4. Recentrifuge at 1349g for 10 min at 4°C, and (on ice) carefully resuspend the bacterial pellet in 4 mL of STB. Add 145 μL of dimethyl sulfoxide (DMSO), swirl gently, and leave on ice for 5 min. Add 138 μL of dithiothreitol, mix, and leave on ice for 10 min. Add a further 138 μL of DMSO, mix gently, and chill on ice for 5 min.
5. Dispense 210 μL of bacterial cells into precooled 15-mL centrifuge tubes. Add 20 μL of ligated vector:cDNA to a 210-μL cell aliquot, mix, and incubate on ice for 30 min. Heat shock (without agitation) at 42°C for 90 s, and place on ice for 2 min.
6. Add 800 μL of SOC broth, and incubate with agitation (225 rpm, approx 0.962g) at 37°C for 1 h (in order to allow expression of the ampicillin resistance gene carried by the plasmid vector). In order to concentrate the bacterial cells, centrifugate at 485g for 5 min at 20°C.
7. Remove 800 μL of supernatant, and carefully resuspend (by recycling through a micropipet tip) the bacterial cells in the remaining fluid (~200 μL). Using an L-shaped glass spreader, spread the cell suspension over the surface of a LB agar plate (9-cm diameter Petri dish), supplemented with 12.5 μg/mL tetracycline,

50 µg/mL ampicillin, 0.24 mg/mL IPTG, and 0.2 mg/mL Xgal. Incubate at 37°C overnight.

8. Incubate the bacterial colonies for a further 4 h at 4°C in order to allow additional time for the nonrecombinant clones to exhibit β-galactosidase activity, and develop an unambiguous blue coloration. Analyze (and archive) the likely insert-containing white colonies.

4. Notes

1. In effect, in order to screen an entire cDNA population, it is necessary to study all the sorted individual subpopulations (in this case 256), which together constitute the original population. In order to optimize the efficiency of the screening procedure, a plausible approach would be to produce and analyze the minimal number of cDNA clones from a given subset that are likely to offer a comprehensive representation of mRNAs, before proceeding to analyze a different subset, and so on. This is particularly pertinent to EST programs, in which the basis of screening is DNA sequencing, and for which identification of all the unknown genes (the majority, by definition, encoding rare messages) expressed in a given source material is sought. Consequently, some degree of expectation regarding the likely complexity of a given subset would be of value. The evidence available to date *(11)* implies a variation in the number of different transcripts represented in each subset. However, some indication can be gained from the subset (fetal liver) studied in most depth, in which after sequencing 529 cDNAs, 60.5% of cDNAs sequenced represented transcripts already encountered within the subset.

2. The fidelity of the T_4 DNA ligase-catalyzed reaction is the basis for the sorting of cDNA fragments into distinct subsets by 2-base specific adaptering. Specific primer annealing and extension are the prerequisites for the partitioning of cDNA fragments by PCR employing two single base-specific primers. Restriction fragment sorting yields distinct subpopulations of cDNA fragments *(11,16)*, although overlap in transcript composition is observed between certain subsets. This is likely to be attributable to misligation, and the extension of PCR primers with a 3'-end mismatch. Since misligation may be expected to occur at a rate at least 1000-fold less than specific ligation under "standard" conditions (16–37°C) *(17)*, the amplification of adaptered fragments may exaggerate the otherwise undetectable effects of misligation. The contribution of mismatched primer extension to overlap between subsets may be negated by utilizing an enzyme, such as Stoffel Fragment (Perkin-Elmer), which is reported to have a mismatched primer extension rate 1000-fold lower than that of Ampli*Taq*. However, given that each PCR primer/primer pair is employed under a single regime of thermocycling parameters, it is conceivable that the specificity of selection achieved by PCR is liable to be less than that obtained by adaptering. A refined approach would be to achieve the desired level of subdivision (e.g., sorting into 256 subsets by 4-base selection) entirely by adaptering, and perform nonselective amplification of all the cDNA fragments in each subset utilizing a standard primer pair under established conditions.

3. Since higher eukaryotic cell types express a nonidentical catalog of genes (10,000–15,000 different transcripts in a "typical" cell), the complexity of a given subset (and hence the number of different rare transcripts) can be enlarged by increasing the heterogeneity of the source material from which mRNA is isolated, i.e., fetal liver mRNA combined with mRNA extracted from five other fetal organs was found to generate a cDNA subset of greater complexity than that produced from fetal liver mRNA alone *(11)*.

4. The preparation of mRNA free from contaminating genomic DNA is essential in a cDNA cloning procedure featuring the amplification power of PCR. Hence, a tripartite RNA purification scheme is utilized.

5. It is not possible to deduce *a priori* the number of cycles required by a particular combination of θ and π base-specific PCR primers to generate sufficient specific amplification products for cloning. Since even a single base difference in either one of the PCR primers can make orders of magnitude difference to the overall amplification efficiency, it is necessary to determine the number of cycles required by each primer pair. This is achieved by analyzing 4-μL aliquots of a 40-μL PCR reaction (prepared as described in Section 3.4.2.), removed after five cycles (prior to the addition of an extra 20 pmol of each primer), and then at four-cycle intervals up to 32 cycles. (Add each 4-μL reaction aliquot to an equal volume of 10 mM EDTA, and analyze by 2% [w/v] TBE-NuSieve GTG agarose gel electrophoresis.) For each primer combination, select the number of cycles that yields all the observable products, plus an additional four cycles.

6. In order to encourage the accumulation of specific amplification products at the expense of dimers formed between primers with complementary sequences at their 3'-termini, the concentration of the PCR primers is maintained at a low level (0.05 $\mu$$M$) during the first five cycles of PCR. An alternative approach is to employ the "hot-start" technique *(18)* to prevent nonspecific amplification occurring as a consequence of pre-PCR mis-priming and primer dimerization. In this procedure, at least one component of the PCR reaction essential for extension (typically *Taq* polymerase, but possibly the primers, or dNTPs) is omitted from the PCR reaction until the temperature of the reaction mixture has exceeded the T_m of specific duplexes (i.e., 75–80°C) formed by the PCR primers.

7. A problem encountered during the amplification of adapted cDNA fragments is the amplification of fragments of high molecular weight (larger than the expected ~1500-bp average size of even a full-length cDNA), which appear close to the loading slots of agarose gels. These molecules, which tend to appear between 12 and 20 cycles of PCR, are likely to represent concatemers formed between cDNAs (possibly nonhomologous), which anneal to form short overlaps as a consequence of regions of complementarity. Extension from the 3'-ends (either part of an overlap or sufficiently close to base pair with an adjacent strand) of the ss cDNAs would generate a hybrid product. These hybrids often appear to be exponentially amplified at the expense of specific target molecules. Given the apparent stability (at the requisite primer annealing temperature) of overlaps formed between certain cDNAs, gel purification of specific amplification products formed may represent the viable alternative to the elimination of these concatemers.

8. The involvement of PCR in this procedure avails both advantages and disadvantages. The inefficiencies of standard cDNA synthesis procedures mean that possibly only about 10% of input mRNA is converted to ds cDNA. The inefficiencies of bacterial transformation (or in vitro phage packaging) further reduce the amount of ligated DNA that may be introduced into cells. Consequently, the construction of a representative library when the amount of RNA is limiting (e.g., studies involving small cell numbers) may not be possible. The ability of PCR to amplify exponentially (by at least 10^5-fold) small numbers of molecules provides a means of generating credible amounts of DNA for cloning from small amounts of source material. However, the constraints imposed by a PCR-based approach to cDNA cloning are, in particular, a variability in the efficiency with which different templates are amplified (i.e., factors affecting amplification efficiency include target size, G/C content, and the presence of secondary structure), and also the issue of fidelity (*Taq* polymerase has an error rate of 0.2–0.3%).

9. The capability of the restriction fragment sorting technique to increase the abundance of rare mRNAs in a given population means that the utility of the approach is not limited to the construction of cDNA libraries. For example, the procedure may effectively be employed for the generation of representative specific probes (cell-type-specific), for use either in the identification of differentially expressed mRNAs or in cDNA library subtraction strategies.

References

1. Davidson, E. H. and Britten, R. J. (1979) Regulation of gene expression: possible role of repetitive sequences. *Science* **204,** 1052–1059.
2. Sargent, T. D. (1987) Isolation of differentially expressed genes. *Methods Enzymol.* **152,** 423–432.
3. Davis, M. M., Cohen, D. I., Nielson, E. A., Steinmetz, M., Paul, W. E., and Hood, L. (1984) Cell-type-specific cDNA probes and the murine-I region—the localisation and orientation of A-α-d. *Proc. Natl. Acad. Sci. USA* **81,** 2194–2198.
4. Sive, H. L. and St. John, T. (1988) A simple subtractive hybridization technique employing photoreactivatable biotin and phenol extraction. *Nucleic Acids Res.* **16,** 10,937.
5. Batra, S. K., Metzgar, R. S., and Hollingsworth, M. A. (1991) A simple, effective method for the construction of subtracted cDNA libraries. *Gene Anal. Technol.* **8,** 129–133.
6. Lovett, M., Kere, J., and Hinton, L. M. (1991) Direct selection: a method for the isolation of cDNAs encoded by large genomic regions. *Proc. Natl. Acad. Sci. USA* **88,** 9628–9632.
7. Parimoo, S., Patanjali, S. R., Shukla, H., Cahplin, D. D., and Weissman, S. M. (1991) cDNA selection: efficient PCR approach for the selection of cDNAs encoded in large chromosomal DNA fragments. *Proc. Natl. Acad. Sci. USA* **88,** 9623–9627.
8. Morgan, J. G., Dolganov, G. M., Robbins, S. E., Hinton, L. M., and Lovett, M. (1992) The selective isolation of novel cDNAs encoded by the regions surrounding the human interleukin-4 and interleukin-5 genes. *Nucleic Acids Res.* **20,** 5173–5179.

9. Patanjali, S. R., Parimoo, S., and Weissman, S. M. (1991) Construction of a uniform-abundance (normalised) cDNA library. *Proc. Natl. Acad. Sci. USA* **88,** 1943–1947.

10. Soares, M. B., Bonaldo, de Fatima, M., Jelene, P., Su, L., Lawton, L., and Efstratiadis, A. (1994) Construction and characterization of a normalized cDNA library. *Proc. Natl. Acad. Sci. USA* **91,** 9228–9232.

11. Dearlove, A. M., Discala, C., Gross, J., Parsons, J., Starkey, M. P., Umrania, Y., and Sibson, D. R. (1994) Restriction Fragment Sorting of cDNA Populations. Poster presented at the Human Genome 1994 meeting, "The genes and beyond," October 2–5, 1994, Washington, DC.

12. Okayama, H., Kawaichi, M., Brownstein, M., Lee, F., Yokota, T., and Arai, K. (1987) High-efficiency cloning of full-length cDNA; construction and screening of cDNA expression libraries for mammalian cells. *Methods Enzymol.* **154,** 3–28.

13. Carter, C., Britton, V. J., and Haff, L. (1983) CsTFA: a centrifugation medium for nucleic acid isolation and purification. *Biotechniques* **1,** 142–147.

14. Gubler, U. and Hoffman, B. J. (1983) A simple and very efficient method for generating cDNA libraries. *Gene* **25,** 263–269.

15. Hanahan, D. (1983) Studies on transformation of *E . coli* with plasmids. *J. Mol. Biol.* **166,** 557–580.

16. Sibson, D. R., Attan, J., Dearlove, A. M., Howells, D. D., Jain, Y. P., Kelly, M., Parsons, J., Smith, S., and Starkey, M. P. (1993) Subsets of Sorted cDNA Restriction Fragments Comprise a Useful Resource for Gene Identification. Human Genome Mapping Workshop 93 (HGM'93), November 14–17, 1993, Kobe, Japan.

17. Unrau, P. and Deugau, K. V. (1994) Non-cloning amplification of specific DNA fragments from whole genomic DNA digests using DNA "indexers." *Gene* **145,** 163–169.

18. Chou, Q., Russell, M., Birch, D. E., Raymond, J., and Bloch, W. (1992) Prevention of pre-PCR mis-priming and primer dimerisation improves low-copy-number amplifications. *Nucleic Acids Res.* **20,** 1717–1723.

3

Isolation of Messenger RNA from Plant Tissues

Alison Dunn

1. Introduction

The starting material for any representative plant cDNA library is a supply of good-quality messenger RNA from the plant tissue of choice. Extraction of RNA can be made from several grams of tissue or as little as 50 mg. However, large samples are generally more representative of the genes expressed in a population of plants in response to environmental cues or at a defined stage of development. Therefore, large-scale extraction of RNA is the method of choice for preparations to be used for cDNA libraries.

Several published protocols describe the rapid extraction of RNA from small quantities of plant tissue *(1,2)*, and a number of small scale RNA extraction kits are commercially available (Gibco-BRL Life Technologies, Middlesex, UK; Dynal, Oslo, Norway; Qiagen, Germany). Small-scale methods can be used to extract RNA for the preparation of cDNA libraries when quantities of suitable plant material are severely limited. However, for reasons already given, they are not ideal and are generally more useful for the analysis of expression of large numbers of individual plants. They are not further considered here.

The greatest obstacle to obtaining good-quality RNA is the ubiquitous and persistent nature of highly active RNases in plant tissues. All RNA extraction techniques are therefore based on initial inactivation or inhibition of RNases by chemical or physical means, such as suboptimal pH and temperature, followed by separation of nucleic acids from proteins, usually by extraction with phenol, thereby separating the RNA from RNases. The extraction buffer described here contains $4M$ guanidinium thiocyanate and β-mercaptoethanol, both of which irreversibly inactivate RNases *(3)*, and subsequent steps in the protocol employ high-pH buffers (pH 9.0) and a temperature of 50°C, which

From: *Methods in Molecular Biology, Vol. 69: cDNA Library Protocols*
Edited by: I. G. Cowell and C. A. Austin Humana Press Inc., Totowa, NJ

inhibits RNase activity. Proteins are subsequently removed by phenol/chloroform extraction. Extraction of RNA is inevitably accompanied by extraction of cellular DNA, and enrichment for total cellular RNA is desirable, although not essential. Only approx 10% of total cellular RNA is messenger RNA; the rest is mainly ribosomal RNA. Since mRNA is the substrate for cDNA synthesis, enrichment for this fraction of total RNA is highly desirable and is achieved by affinity chromatography on oligo (dT)-cellulose. Polyadenylated mRNA (polyA$^+$ RNA) binds to oligo (dT)-cellulose at high salt concentrations and is eluted in a salt-free buffer. This is the basis of the method described here. It is worth noting that a kit for polyA$^+$ RNA purification is commercially available (Pharmacia, Uppsala, Sweden), which has the advantage of prepacked RNase-free columns and reagents reducing sources of RNase contamination.

Since RNases are ubiquitous and are not inactivated by autoclaving, there is a danger of contamination from laboratory glassware, reagents, and handling. Steps should be taken:

1. To decontaminate, as far as possible, all equipment to be used in the extraction,
2. To ensure inactivation of RNases in laboratory-prepared solutions; and,
3. To avoid recontamination.

All work should be carried out wearing clean, disposable plastic gloves and, if possible, equipment and chemicals kept only for RNA extraction and analysis should be used, e.g., pestles, mortars, spatulas, Corex® tubes (DuPont, Wilmington, DE), reagent bottles. Sterile, disposable plasticware, which has not been handled, is RNase-free. RNA preparation requires a degree of paranoia!

2. Materials

2.1. RNA Extraction

2.1.1. Equipment for RNA Extraction

1. Corex tubes (12 × 40 mL): Siliconized by soaking in dichlorodimethylsilane (5% [v:v]) in chloroform for 15 min, rinsing four times with deionized water and once with 100% ethanol. The siliconizing solution can be stored for further use.
2. Glassware and nondisposable plasticware: glass rod, measuring cylinder, 10-mL pipets, bottles (4 × 100 mL), filter funnel, sterile universal. Pretreat by soaking overnight in a 2% (v/v) solution of Absolve (DuPont; see Note 1) and thoroughly rinse before sterilization, preferably by baking at 180°C for several hours or by autoclaving for 15 min at 15 psi.
3. Pestle (not wooden handles) and mortar: pretreat as for glassware (see Notes 1 and 2).
4. Sterile Miracloth (Calbiochem, La Jolla, CA).

2.1.2. Chemicals for RNA Extraction

1. Guanidinium thiocyanate (GT) (Fluka, Buchs, Switzerland).
2. β-Mercaptoethanol.
3. Sodium citrate.
4. Sodium lauryl sarcosine.
5. Oligo (dT)-cellulose (Pharmacia).
6. Diethylpyrocarbonate (DEPC).
7. Liquid nitrogen.
8. Phenol (equilibrated in 0.1M Tris-HCl, pH 8.0).

2.1.3. Solutions for RNA Extraction (see Notes 3 and 4)

Except where noted, prepare 100 mL of each of the following:

1. 200 mM Sodium citrate.
2. 10% (w/v) Sodium lauryl sarcosine.
3. 1M Acetic acid.
4. 50 mM Tris-HCl, pH 9.0, 0.2M NaCl, 5 mM EDTA, 1% (w/v) sodium dodecyl sulfate (SDS).
5. 3M Sodium acetate, pH 5.5.
6. 0.2M Sodium acetate, pH 5.5.
7. Chloroform:pentan-2-ol (24:1 [v/v]).
8. TE Buffer 10 mM Tris-HCl, pH 7.4, 1 mM EDTA.
9. 1 L DEPC water.
10. 4M GT medium: Dissolve 50 g guanidinium thiocyanate in 50 mL sterile deionized water and heat to 60 or 70°C to dissolve if necessary. Add 10.6 mL 200 mM sodium citrate solution and 10.6 mL 10% sodium lauryl sarcosine, make up to 106 mL with sterile water. Filter if necessary (only if sediment is visible) and divide between two sterile dark bottles (wide-necked 100-mL are best so that ground plant tissue can be added directly to the bottle). Store at 4°C until required (stable indefinitely). Add β-mercaptoethanol to 0.1M just before use (0.43/50 mL GT medium).

2.2. Purification of PolyA+ RNA

2.2.1. Equipment for PolyA+ Purification

1. Two plastic mini-columns (5 mL).
2. 15-mL Corex centrifuge tubes.

2.2.2. Solutions for PolyA+ Purification

Prepare 20 mL of each of the following from stock solutions.

1. 10 mM Tris-HCl, pH 7.5, 1M KCl.
2. 10 mM Tris-HCl, pH 7.5, 0.5M KCl (loading buffer).
3. 10 mM Tris-HCl, pH 7.5, 0.1M KCl (wash buffer).
4. 10 mM Tris-HCl, pH 7.5, preheated to 60°C (elution buffer).
5. 3M Sodium acetate, pH 5.5.

3. Methods

3.1. RNA Extraction

This method is a modification of that described in ref. *4*.

1. Add 4–5 g plant tissue to liquid nitrogen in precooled mortar and pestle, when N_2 has almost evaporated, grind vigorously—keep cool and add more N_2 if necessary. The ground tissue should be a very fine powder (*see* Note 5).
2. Scrape frozen powder with a sterile spatula into 55-mL GT medium in a sterile bottle and shake for 2 min. Filter through two layers of sterile Miracloth into a clean bottle or directly into Corex tubes.
3. Dispense into Corex tubes, and centrifuge for 10 min at 12,000*g* at 10°C.
4. Collect supernatant, add 0.025 vol 1*M* acetic acid and 0.75 vol absolute ethanol at room temperature, invert to mix (cover with Nescofilm while mixing) and centrifuge for 15 min at 12,000*g* at 10°C.
5. Remove supernatant, dry pellet, and redissolve in 4 mL 50 m*M* Tris-HCl, pH 9.0, 0.2*M* NaCl, 5 m*M* EDTA, 1% SDS (*see* Note 6).
6. Transfer to a sterile universal tube, warm to 50°C (pellet from previous step may not fully dissolve until now, but make sure it is dissolved before proceeding). Add an equal volume of phenol and shake vigorously.
7. Transfer to a fresh Corex tube, centrifuge for 10 min at 4500*g*. Transfer the aqueous phase to a fresh Corex tube, extract with an equal volume of chloroform:pentan-2-ol, recentrifuge as previously, and remove the aqueous phase to a fresh Corex tube.
8. Add 0.1 vol 3*M* sodium acetate, pH 5.5 and 2.5 vol absolute ethanol, mix, and leave at 20°C overnight to precipitate nucleic acids.
9. Centrifuge for 30 min at 12,000*g*, decant supernatant, and dry the pellet. Add 2 mL 0.2*M* sodium acetate, pH 5.5, and allow to stand for 1 h with gentle agitation occasionally (*see* Note 7).
10. Remove solution to a fresh Corex tube (leaving behind any undissolved nucleic acid), add 2.5 vol absolute ethanol, and precipitate overnight at –20°C.
11. Centrifuge for 30 min at 12,000*g*, decant supernatant and dry the RNA pellet. Redissolve in either TE buffer or loading buffer for polyA$^+$ purification (*see* Notes 8 and 9).

3.2. Preparation of PolyA$^+$ RNA

1. Equilibrate 0.3 g oligo(dT)-cellulose with loading buffer (*see* Section 2.2.2.) and load onto a minicolumn. Allow to drain.
2. Heat 0.5 mg total RNA (*see* Note 10) dissolved in 1 mL loading buffer at 65°C for 5 min, cool on ice, and then load onto the column. Collect the eluant, and reload it onto the column. Allow to drain and discard eluant.
3. Wash column with 5 mL wash buffer (*see* Section 2.2.2.) and discard eluant.
4. Have ready elution buffer (*see* Section 2.2.2.) at 60°C. Elute polyA$^+$ RNA with 5 mL elution buffer at 60°C.

5. Collect eluant in Corex tube, add 0.1 vol 3M sodium acetate, pH 5.5, 2.5 vol absolute ethanol, and precipitate for 2 h or overnight at –20°C.
6. Centrifuge for 30 min at 12,000g at 10°C, to recover polyA$^+$ RNA. Dissolve in TE buffer or DEPC-treated sterile water.
7. To prepare RNA that is 90% polyA$^+$, steps 1–6 should be repeated with a fresh oligo(dT) cellulose column. The final pellet should be washed in 2 mL 70% ethanol and recentrifuged for 5 min at 12,000g at 10°C, dried, and redissolved in 50 μL DEPC-treated water.

4. Notes

1. Absolve is an alkaline detergent that is effective in removing RNases, but it can cause degradation of RNA if glassware is not thoroughly rinsed.
2. Pestles and mortars should be precooled at –70°C following sterilization.
3. Common laboratory chemicals are not listed.
4. DEPC is a powerful RNase inhibitor that can be used to treat water and solutions for RNA preparation. DEPC should be added to water and solutions (except those containing Tris buffers) in the ratio 1 mL:1 L (0.1%). Solutions should be thoroughly shaken, then left overnight in a fume hood, and autoclaved at 15 psi for 15 min to remove residual DEPC. DEPC carboxymethylates purine residues in RNA, which affects efficiency of in vitro translation, but not hybridization. DEPC is volatile and a suspected carcinogen. It should be handled with care in a fume hood. Where DEPC treatment is inadvisable, chemicals kept separate from general laboratory stocks should be used.
5. This is probably the most important step in the procedure. Well-ground plant tissue gives high yields of RNA.
6. This step can be slow, and it may be necessary to disrupt the pellet gently by pipeting.
7. This preferentially redissolves RNA leaving most of the DNA in the pellet. It may be necessary to recentrifuge briefly (5 min at 12,000g) if the residual pellet is not sticking to the tube.
8. Use TE buffer or manufacturer's recommendation if using a kit for polyA$^+$ purification.
9. All ethanol precipitations are incubated overnight at –20°C, but 2 h (minimum) at –20°C can be substituted if you are in a hurry.
10. RNA concentrations can be determined using a spectrophotometer (1 A_{260} U = 40 μg/mL RNA) or by visualization of a diluted aliquot on an ethidium bromide-impregnated agarose gel against tRNA standards. (Use 1 μL of each dilution.)
11. The described RNA extraction has been used to make good-quality RNA to prepare cDNA libraries from barley (*Hordeum vulgare*) *(5)*, white clover (*Trifolium repens*) (unpublished), and cassava (*Manihot esculenta*) *(6)*, and typically yields 1–2 mg total RNA/extraction (0.5 mg/g of starting material). More recently, polyA$^+$ RNA has been prepared using an oligo(dT) cellulose kit (Pharmacia) as an alternative to the method described here.

References

1. Jakobsen, K. S., Breiwold, E., and Hornes, E. (1990) Purification of mRNA directly from crude plant tissues in 15 minutes using magnetic oligo dT microspheres. *Nucleic Acids Res.* **18,** 3669.
2. Verwoerd, T. C., Dekker, B. M. M, and Hoekema, A. (1989) A small scale procedure for the rapid isolation of plant RNAs. *Nucleic Acids Res.* **17,** 2362.
3. Chirgwin, J. M., Przbyla, A. E., McDonald, R. J., and Rutler, W. J. (1979) Isolation of biologically active ribonucleic acid from sources enriched in ribonuclease. *Biochemistry* **18,** 5294–5299.
4. Broglie, R., Coruzzi, G., Keith, B., and Chua, N.-H. (1986) Molecular biology of C4 photosynthesis in Zea mays: differential localisation of proteins and mRNAs in two leaf cell types. *Plant Mol. Biol.* **3,** 431–444.
5. Dunn, M. A., Hughes, M. A., Pearce, R. S., and Jack, P. L. (1990) Molecular characterisation of a barley gene induced by cold treatment. *J. Exp. Bot.* **41,** 1405–1413.
6. Hughes, M. A., Brown, K., Pancoro, A., Murray, B. S., Oxtoby, E., and Hughes, J. (1992) A molecular and biochemical analysis of the structure of cyanogenic β-glucosidase (linamarase) from cassava (*Manihot esculenta* Crantz). *Arch. Biochem. Biophys.* **295,** 273–279.

4

cDNA Library Construction
for the Lambda ZAP®-Based Vectors

Marjory A. Snead, Michelle A. Alting-Mees, and Jay M. Short

1. Introduction

Each organism and tissue type has a unique population of messenger RNA (mRNA) molecules. These mRNA populations are difficult to maintain, clone, and amplify; therefore, they must be converted to more stable DNA molecules (cDNA). Successful cDNA synthesis should yield full-length copies of the original population of mRNA molecules. Hence, the quality of the cDNA library can be only as good as the quality of the mRNA. Pure, undegraded mRNA is essential for the construction of large, representative cDNA libraries *(1)*. Secondary structure of mRNA molecules can cause the synthesis of truncated cDNA fragments. In this case, treatment of the mRNA with a denaturant, such as methyl-mercuric hydroxide, prior to synthesis may be necessary *(2)*. Other potential difficulties include DNA molecules contaminating the mRNA sample. DNA can clone efficiently, and their introns can confuse results. RNase-free DNase treatment of the sample is recommended.

After synthesis, the cDNA is inserted into an *Escherichia coli*-based vector (plasmid or λ), and the library is screened for clones of interest. Since 1980, lambda has been the vector system of choice for cDNA cloning *(3–10)*. The fundamental reasons are that in vitro packaging of λ generally has a higher efficiency than plasmid transformation, and λ libraries are easier to handle (amplify, plate, screen, and store) than plasmid libraries. However, most λ vectors have the disadvantage of being poorer templates for DNA sequencing, site-specific mutagenesis, and restriction fragment shuffling, although this trend is reversing to some degree with the continued development of polymerase chain reaction (PCR) techniques.

From: *Methods in Molecular Biology, Vol. 69: cDNA Library Protocols*
Edited by: I. G. Cowell and C. A. Austin Humana Press Inc., Totowa, NJ

5'   GAGAGAGAGAGAGAGAGAGAGAGA**CTCGAG**TTTTTTTTTTTTTTTTTT   3'

      Protective Sequence   *Xho* I   Poly(dT)

Fig. 1. Forty-eight base pair oligonucleotide hybrid oligo(dT) linker-primer.

The development of excisable λ vectors, such as those based on restriction enzyme digestion *(11)*, site-specific recombination *(12)*, or filamentous phage replication *(13)*, has increased the flexibility of DNA cloning. Now it is possible to clone and screen libraries with the efficiency and ease of λ systems, and be able to analyze positive clones with the ease and versatility of a plasmid. The vectors that are compatible with the cDNA synthesis protocol described in this chapter are based on the Lambda ZAP® excision system (Stratagene Cloning Systems) *(13,14,* manuscript in preparation for SeqZAP). These vectors use an excision mechanism that is based on filamentous helper phage replication (e.g., M13). The choice of vector (Lambda ZAP, ZAP Express, or SeqZAP) depends on whether one requires such features as prokaryotic expression, eukaryotic expression, in vitro transcription, in vitro translation, directional cloning, single-strand replication, automated sequencer compatibility, and special antibiotic resistance selection.

Several cloning procedures for constructing cDNA libraries exist *(15–19)*. Here we describe a modification of a directional cDNA cloning protocol *(16)*. This procedure has been successfully used for generating hundreds of directional cDNA libraries representing a vast number of plant and animal species containing polyA+ mRNA.

A hybrid oligo(dT) linker-primer containing an *Xho*I site is used to make directional cDNA. This 48-base oligonucleotide was designed with a protective sequence to prevent the *Xho*I restriction enzyme recognition site from being damaged in subsequent steps and an 18-base poly(dT) sequence, which binds to the 3' polyA region of the mRNA template (*see* Fig. 1).

First-strand synthesis is primed with the linker-primer and is transcribed by reverse transcriptase in the presence of nucleotides and buffer. An RNase H-deficient reverse transcriptase may produce larger yields of longer cDNA transcripts *(20,21)*. The use of 5-methyl dCTP in the nucleotide mix during first-strand synthesis "hemi-methylates" the cDNA, protecting it from digestion during a subsequent restriction endonuclease reaction used to cleave the internal *Xho*I site in the linker-primer.

The cDNA/mRNA hybrid is treated with RNase H in the second-strand synthesis reaction. The mRNA is nicked to produce fragments that serve as primers for DNA polymerase I, synthesizing second-strand cDNA. The second-strand nucleotide mixture is supplemented with dCTP to dilute the 5-methyl dCTP, reducing the probability of methylating the second-strand, since the

*Xho*I restriction site in the linker-primer must be susceptible to restriction enzyme digestion for subsequent ligation into the vector.

The uneven termini of the double-stranded cDNA must be polished with cloned *Pfu* DNA polymerase to allow efficient ligation of adapters *(22,23)*. Adapters are complementary oligonucleotides which, when annealed, create a phosphorylated blunt end and a dephosphorylated cohesive end. This double-stranded adapter will ligate to other blunt termini on the cDNA fragments and to other adapters. Since the cohesive end is dephosphorylated, ligation to other cohesive ends is prevented. After the adapter ligation reaction is complete and the ligase has been inactivated, the molecules are phosphorylated to allow ligation to the dephosphorylated vector.

An *Xho*I digestion releases the adapter and protective sequence on the linker-primer from the 3'-end of the cDNA. These fragments are separated from the cDNA on a size fractionation column. The purified cDNA is then precipitated and ligated to the vector. This strategy is illustrated in Fig. 2.

2. Materials

2.1. First-Strand Synthesis

1. 10X First-strand buffer: 500 m*M* Tris-HCl, pH 7.6, 700 m*M* KCl, 100 m*M* MgCl$_2$.
2. First-strand methyl-nucleotide mixture: 10 m*M* dATP, dGTP, dTTP, and 5 m*M* 5-methyl dCTP.
3. Linker-primer (3.0 μg at 1.5 μg/μL).
4. Diethylpyrocarbonate (DEPC)-treated water.
5. Ribonuclease inhibitor (40 U).
6. PolyA$^+$ mRNA (5.0 μg in ≤36 μL DEPC-treated water; *see* Notes 1 and 2).
7. [α-^{32}P]-Labeled deoxynucleotide (800 Ci/mmol) [α-^{32}P]dATP, [α-^{32}P]dGTP, or [α-^{32}P]dTTP. Do not use [α-^{32}P]dCTP (*see* Note 3).
8. Reverse transcriptase (250 U) (RNase H-deficient is recommended *[20,21]*).

2.2. Second-Strand Synthesis

1. 10X Second-strand buffer: 700 m*M* Tris-HCl, pH 7.4, 100 m*M* (NH$_4$)$_2$SO$_4$, 50 m*M* MgCl$_2$.
2. Second-strand dNTP mixture: 10 m*M* dATP, dGTP, dTTP, and 26 m*M* dCTP.
3. *E. coli* RNase H (4.0 U).
4. *E. coli* DNA polymerase I (100 U).

2.3. Blunting the cDNA Termini

1. Blunting dNTP mixture (2.5 m*M* dATP, dGTP, dTTP, and dCTP).
2. Cloned *Pfu* DNA polymerase (5 U).
3. Phenol-chloroform: 1:1 (v/v), pH 7.0–8.0 (*see* Note 5).
4. Chloroform.
5. 3*M* Sodium acetate.
6. 100% (v/v) Ethanol.

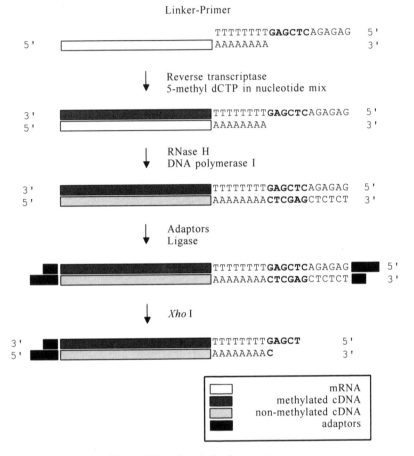

Fig. 2. Directional cloning strategy.

2.4. Ligating the Adapters

1. 70% (v/v) Ethanol.
2. Adapters: 4 µg at 0.4 µg/µL.
3. 5% Nondenaturing acrylamide gel.
4. 10X Ligation buffer: 500 m*M* Tris-HCl, pH 7.4, 100 m*M* MgCl₂, 10 m*M* dithiothreitol (DTT).
5. 10 m*M* rATP.
6. T4 DNA ligase (4 Weiss U).

2.5. Phosphorylating the Adapters

1. 10X Ligation buffer (*see* Section 2.4.).
2. 10 m*M* rATP.
3. T4 Polynucleotide kinase (10 U).

2.6. XhoI Digestion

1. *Xho*I reaction buffer: 200 m*M* NaCl, 15 m*M* MgCl$_2$.
2. *Xho*I restriction endonuclease (120 U).
3. 10X STE buffer: 1*M* NaCl, 100 m*M* Tris-HCl, pH 8.0, 10 m*M* EDTA.

2.7. Size Fractionation

1. 1X STE buffer: 100 m*M* NaCl, 10 m*M* Tris-HCl, pH 8.0, 1 m*M* EDTA.
2. Sephacryl® S-500 column filtration medium (Pharmacia).
3. 5% Nondenaturing acrylamide gel.
4. Phenol-chloroform (1:1 [v/v], pH 7.0–8.0) (*see* Note 5).
5. Chloroform.
6. 100% (v/v) Ethanol.

2.8. Quantitating the cDNA

1. 70% (v/v) Ethanol.
2. TE buffer: 10 m*M* Tris-HCl, pH 8.0, 1 m*M* EDTA.
3. 0.8% Agarose.
4. 10 mg/mL Ethidium bromide.

2.9. Ligating the cDNA to Prepared Vector

1. λ vector (such as Lambda ZAP, ZAP Express, SeqZAP) double-digested and dephosphorylated. Vectors are digested with *Xho*I and a second restriction enzyme that leaves ends compatible with the adapters.
2. 10X Ligation buffer (*see* Section 2.4.).
3. 10 m*M* rATP.
4. T4 DNA ligase (4 Weiss U).

2.10. Packaging and Plating

1. NZY medium, plates and top agarose: 5 g NaCl, 2 g MgSO$_4$·7H$_2$O, 5 g yeast extract, 10 g NZ amine (casein hydrolysate)/L. Add 15 g agar for plates, or add 0.7% (w/v) agarose for top agarose. Adjust the pH to 7.5 with NaOH, and sterilize by autoclaving.
2. Appropriate *E. coli* host strains (such as XL1-Blue MRF' or DH5αMCR) freshly streaked on an LB agar plate containing the appropriate antibiotic (*see* Note 8).
3. 10 m*M* MgSO$_4$.
4. Packaging extract (such as Gigapack® II λ packaging extract [Stratagene] *[23,24]*).
5. SM buffer: 5.8 g NaCl, 2.0 g MgSO$_4$ · 7H$_2$O, 50.0 mL 1*M* Tris-HCl, pH 7.5, 5.0 mL 2% (w/v) gelatin/L. Autoclave.
6. Chloroform.
7. LB agar plates: 10 g NaCl, 10 g bacto-tryptone, 5 g bacto-yeast extract, 15 g agar/L. Adjust the pH to 7.5 with NaOH, and sterilize by autoclaving.
8. Isopropyl-β-D-thio-galactopyranoside (IPTG), 0.5*M* in water, and 5-bromo-4-chloro-3-indolyl-β-D-galactopyranoside (X-gal), 250 mg/mL in dimethyl formamide (*see* Note 10).

2.11. Amplification of the Primary Library

1. Packaged and titered primary library.
2. Prepared, appropriate *E. coli* host strains.
3. NZY medium, plates and top agarose (*see* Section 2.10.).
4. SM buffer (*see* Section 2.10.).
5. Chloroform.
6. Dimethylsulfoxide (DMSO).

3. Methods

3.1. First-Strand Synthesis

The final volume of the first-strand synthesis reaction should be 50 μL. Take this into account when determining the volumes necessary.

1. In an RNase-free microcentrifuge tube, add the reagents in the following order: 5.0 μL 10X first-strand buffer, 3.0 μL methyl-nucleotide mixture, 2.0 μL linker-primer (1.5 μg/μL), X μL DEPC-treated water, 40 U ribonuclease inhibitor.
2. Mix the reagents well. Add X μL of polyA$^+$ mRNA (5 μg), and gently vortex (*see* Notes 1 and 2).
3. Allow the mRNA template and linker-primer to anneal for 10 min at room temperature.
4. Add 0.5 μL of [α-^{32}P]-labeled deoxynucleotide (800 Ci/mmol). Do not use [α-^{32}P]dCTP (*see* Note 3).
5. Add 250 U of reverse transcriptase. The final volume of the reaction should now be 50 μL.
6. Gently mix the sample, and briefly spin down the contents in a microcentrifuge.
7. Incubate at 37°C for 1 h.
8. After the 1-h incubation, place on ice.

3.2. Second-Strand Synthesis

The final volume of the second-strand synthesis reaction should be 200 μL. Take this into account when determining the necessary volumes.

1. To the first strand reaction (50 μL), add the following components in the following order: 20.0 μL 10X second-strand buffer, 6.0 μL second-strand dNTP mixture, X μL sterile distilled water (DEPC-treated water is not required), 4 U *E. coli* RNase H, 100 U *E. coli* DNA polymerase I.
2. The final volume of the reaction should now be 200 μL. Quickly vortex and spin down the reaction in a microcentrifuge. Incubate for 2.5 h at 16°C.
3. After the 2.5-h incubation, place on ice.

3.3. Blunting the cDNA Termini

1. Add the following reagents to the synthesized cDNA: 23.0 μL blunting dNTP mixture, 2.0 μL cloned *Pfu* DNA polymerase (2.5 U/μL).
2. Mix well, and incubate at 70°C for 30 min. Do not exceed 30 min.

3. Phenol-chloroform/chloroform extract (*see* Note 5).
4. Precipitate the cDNA by adding the following: 20 μL 3*M* sodium acetate, 400 μL 100% (v/v) ethanol.
5. Mix by gently vortexing, and incubate on ice for 10 min or overnight at –20°C.

3.4. Ligating the Adapters

1. Microcentrifuge the precipitated cDNA sample at maximum speed and at 4°C for 1 h.
2. A large white pellet will form at the bottom of the microcentrifuge tube. Carefully remove the radioactive ethanol, and properly discard. Counts left in this supernatant are unincorporated nucleotides.
3. Wash the pellet by gently adding 500 μL of 70% (v/v) ethanol, and microcentrifuge for 2 min.
4. Aspirate the ethanol wash, and lyophilize the pellet until dry.
5. Resuspend the pellet in 9.0 μL of adapters (0.4 μg/μL) by gentle pipeting. Use a Geiger counter to confirm that the cDNA is in solution.
6. Remove 1.0 μL for analysis of cDNA synthesis on a 5% nondenaturing acrylamide gel. This aliquot may be frozen at –20°C (*see* Notes 1–4).
7. Add the following components to the tube containing the 8.0 μL of blunted DNA and adapters: 1.0 μL 10X ligation buffer, 1.0 μL 10 m*M* rATP, 1.0 μL T4 DNA ligase (4 U/μL).
8. Mix well and briefly spin in a microcentrifuge. Incubate overnight at 8°C or for 2 d at 4°C.

3.5. Phosphorylating the Adaptors

The final volume of the phosphorylation reaction will be 25 μL. Take this into account when determining the necessary volumes.

1. After ligation, heat-inactivate the ligase by incubating at 70°C for 30 min.
2. Spin down, and allow the reaction to cool at room temperature for 5 min. Add 1.5 μL 10X ligation buffer, 2.0 μL 10 m*M* rATP, *X* μL sterile distilled water, and 7 U T4 polynucleotide kinase.
3. Incubate at 37°C for 30 min.
4. Heat-inactivate the kinase by incubating at 70°C for 30 min.
5. Spin down, and allow the reaction to cool at room temperature for 5 min.

3.6. XhoI Digestion

The final volume of the digestion reaction will be 60 μL.

1. Add the following components to the phosphorylation reaction (25 μL): 30.0 μL *Xho*I reaction buffer, *X* μL sterile distilled water, 120 U *Xho*I restriction endonuclease. Be sure the volume of enzyme is ≤10% of the reaction volume.
2. Incubate for 1.5 h at 37°C.
3. Cool the reaction to room temperature, and add 15 μL of 10X STE buffer and 75 μL water.

3.7. Size Fractionation

There are many types of filtration media used to separate DNA molecules. Sephacryl S-500 medium separates efficiently in the 2-kb size range. Drip columns made with Sephacryl S-500 medium separate by size, the larger cDNA molecules eluting from the column first and the small unligated adapters and unincorporated nucleotides eluting later. The cDNA will not have a high number of counts, but will be detectable by a handheld monitor at ≤250 cps.

3.7.1. Drip-Column Preparation

1. Discard the plunger from a 1-mL plastic syringe, and insert a small cotton plug. Push the cotton to the bottom of the syringe.
2. Fill the syringe to the top with Sephacryl S-500 filtration medium.
3. Place the syringe in a rack and allow the column to drip "dry."
4. Fill the syringe up to ~0.5 cm from the top with medium, and drip through as in step 3.
5. Rinse the column with four aliquots of 300 μL of 1X STE buffer (total wash volume of 1200 μL). Drip dry after each addition of buffer.

3.7.2. Collecting Fractions

1. Pipet the cDNA into the washed Sephacryl S-500 drip column, and allow to drip through. This is fraction 1. The recovery volume is ~150 μL and does NOT contain cDNA (*see* Note 6).
2. Load two more aliquots of 150 μL of 1X STE buffer on the column and drip through. These are fractions 2 and 3.
3. Collect fraction 4 in a fresh tube. Load 150 μL of 1X STE buffer and drip as before.
4. Collect fraction 5 as in step 3. Two fractions are usually adequate. The size of the cDNA decreases. Two fractions are usually adequate. The size of the cDNA decreases in each additional fraction. Most of the radioactivity will remain in the column owing to unincorporated nucleotides. Discard the radioactive drip column appropriately.
5. Remove 5 μL from each fraction (or up to $^1/_{10}$ of the fraction volume) for analysis of cDNA size on a 5% nondenaturing acrylamide gel. These aliquots can be frozen at –20°C.
6. To remove any residual enzyme from previous reactions, phenol-chloroform/chloroform extract (*see* Note 5).
7. Add twice the volume of 100% (v/v) ethanol to precipitate the cDNA.
8. Place on ice for 1 h or at –20°C overnight.

3.8. Quantitating the cDNA

1. Microcentrifuge the fractionated cDNA at maximum speed for 30–60 min at 4°C. Carefully transfer the ethanol to another tube, and monitor with a Geiger counter. Most of the counts should be present in the pellet. Discard the ethanol appropriately.

2. Wash the cDNA pellet with 200 µL of 70% (v/v) ethanol, and microcentrifuge for 2 min.
3. Carefully remove the ethanol wash, and vacuum evaporate until the cDNA pellet is dry.
4. Each fraction can contain 0–250 cps. If the pellet contains 0–10 cps, resuspend the cDNA in 5.0 µL of sterile water. If the pellet contains >10 cps, resuspend the cDNA in 12.0 µL of sterile water.
5. Quantitate the cDNA by UV visualization of samples spotted on ethidium bromide agarose plates (*see* Note 7). The cDNA can be stored at –20°C.

3.9. Ligating the cDNA to Prepared Vector

The cloning vector should be double-digested with *Xho*I and an enzyme which leaves ends compatible with the adapters. The vector should also be dephosphorylated to prevent vector-to-vector ligations. The final ligation reaction volume is 5 µL.

1. To a 0.5-mL microcentrifuge tube, add in order: X µL water, 0.5 µL 10X ligation buffer, 0.5 µL 10 mM rATP, 1 µg prepared λ arms, 100 ng cDNA, 0.5 µL T4 DNA ligase (4 Weiss U/µL).
2. Incubate overnight at 4°C.

3.10. Packaging and Plating

The ligation is packaged and transfected into an appropriate *E. coli* host strain.

3.10.1. Preparation of Plating Cells

1. Inoculate 50 mL of NZY medium with a single colony of the appropriate *E. coli* host. Do not add antibiotic.
2. Grow at 30°C with gentle shaking overnight (*see* Note 9).
3. Spin the culture at 1000g for 10 min.
4. Gently resuspend the cells in 20 mL sterile 10 mM MgSO$_4$.
5. Determine the concentration of the cells by reading OD$_{600}$ on a spectrophotometer. Store this cell stock at 4°C for no more than 1 wk. To use, dilute cells to OD$_{600}$ = 1.0 in 10 mM MgSO$_4$.

3.10.2. Packaging

Package the ligation reaction following manufacturer's instructions. Stop the reaction by adding 500 µL SM buffer and 20 µL chloroform.

3.10.3. Plating

1. Mix the following components in a Falcon 2059 polypropylene tube: 200 µL appropriate diluted host cells (*see* Section 3.10.1.), and 1 µL final packaged reaction.
2. Incubate the phage and the bacteria at 37°C for 15 min to allow the phage to attach to the cells.

3. Add 2–3 mL of NZY top agarose (48°C) containing IPTG and X-gal (*see* Note 10). Plate onto NZY agar plates, and place the plates upside down in a 37°C incubator.
4. Plaques should be visible after 6–8 h. Background plaques are blue. Recombinant plaques are clear and should be 10- to 100-fold above the background.
5. Count the plaques and calculate the titer. Primary libraries can be unstable. Immediate amplification of at least a portion of the library is recommended to produce a large, stable quantity of a high-titer stock of the library.

3.11. Amplification of the Primary Library

After amplification, the library is suitable for screening by a variety of techniques *(2)*. More than one round of amplification is not recommended, since slower growing clones may be significantly underrepresented.

1. Prepare the host strains (*see* Section 3.10.1.).
2. Mix aliquots of the packaged library containing ~50,000 PFU (≤300 µL vol) with 600 µL of host cells in Falcon 2059 polypropylene tubes. Usually, 1×10^6 plaques are amplified (20 tubes).
3. Incubate the tubes containing the phage and host cells for 15 min at 37°C.
4. Mix 8.0 mL of melted NZY top agarose (48°C) with each aliquot of infected bacteria, and spread evenly onto a freshly poured 150-mm NZY plate.
5. Incubate the plates at 37°C for 6–8 h. Do not allow the plaques to grow larger than 1–2 mm.
6. Overlay the plates with 8–10 mL of SM buffer. Store the plates at 4°C overnight with gentle rocking. The phage will diffuse into the SM buffer.
7. Recover the SM buffer containing the bacteriophage from each plate, and pool it in a sterile polypropylene container. Add chloroform to a 5% final concentration and mix well.
8. Incubate for 15 min at room temperature.
9. Remove the cell debris by centrifugation for 10 min at 500*g*.
10. Recover the supernatant, and transfer it to a sterile polypropylene container. Add chloroform to a 0.3% final concentration, and store at 4°C.
11. Check the titer of the amplified library by making serial dilutions in SM buffer and plating on host cells (*see* Section 3.10.1.). The average titer is usually 10^9–10^{12} PFU/mL.
12. Frozen stocks can be made by adding DMSO to a final concentration of 7%, mixing well, and freezing at –80°C.

4. Notes

1. The mRNA sample must be highly purified for efficient cDNA synthesis. The mRNA sample may contain inhibitors that can be removed by phenol-chloroform extractions. The presence of DNA or rRNA will give an inaccurate concentration of mRNA leading to an insufficient amount of sample used. Treat the mRNA with RNase-free DNase, or use more mRNA sample.

2. Some populations of mRNA molecules may have tight secondary structures. Methyl-mercuric hydroxide treatment of the RNA sample may be necessary. Perform the following protocol under a fume hood. Resuspend the mRNA in 20 µL of DEPC-treated water, and incubate at 65°C for 5 min. Cool to room temperature, and add 2 µL of 100 mM methyl-mercuric hydroxide. Incubate at room temperature for 1 min, add 4 µL of 700 mM β-mercaptoethanol (dilute stock in DEPC-treated water), and incubate at room temperature for 5 min. The final volume is 26 µL. This denatured mRNA is ready for first-strand synthesis.

3. Do not use [α-^{32}P]dCTP. The 5-methyl dCTP present in the nucleotide mixture will be diluted, and the synthesized cDNA will not be protected from the subsequent restriction digest. Gel analysis may show a false-negative result if the [α-^{32}P]dNTP is degraded, since it may not incorporate into the cDNA even though synthesis is occurring.

4. Gel analysis may show hairpinning of the cDNA, which is caused by a number of factors: an insufficient amount of mRNA was used in the first-strand reaction (*see* Note 1), the mRNA population had tight secondary structure (*see* Note 2), the second-strand incubation temperature was higher than 16°C (cool the first-strand reaction by placing it on ice before adding the second-strand synthesis reaction components), or an excessive amount of DNA polymerase was used in the second-strand reaction.

5. Phenol-chloroform (1:1 [v/v], pH 7.0–8.0) is recommended. Do not use low-pH phenol routinely used for RNA isolation *(1,2)*. To extract the cDNA sample, add an equal volume of phenol-chloroform (1:1 [v/v], pH 7.0–8.0) and vortex. Microcentrifuge at maximum speed for 2 min. Transfer the upper aqueous layer, which contains the cDNA, to a new sterile tube. Avoid removing any interface. Add an equal volume of chloroform and vortex. Microcentrifuge for 2 min at maximum speed. Save the upper aqueous layer, and transfer it to a new tube.

6. Sephacryl S-500 drip columns can be run "dry." A reservoir at the top of the column is not required. Each 150 µL wash yields ~150 µL fraction volume. Fractions 1–3 can be collected in one tube since these fractions do not contain cDNA. The cDNA elutes in fractions 4 (containing fragments ≥1.5 kb) and 5 (containing fragments >500 bp).

7. Ethidium bromide agarose plate quantitation is performed as follows. Using a DNA sample of known concentration (such as a plasmid), make serial dilutions (200, 150, 100, 75, 50, 25, and 10 µL) in TE buffer. Melt 10 mL of 0.8% (w/v) agarose in TE buffer, and cool to 50°C. Under a hood, add 10 µL of 10 mg/mL ethidium bromide, swirl to mix, and pour into a 100-mm Petri dish. Allow the plate to harden. Label the bottom of the Petri dish with a marker to indicate where the sample and standards will be spotted. Carefully spot 0.5 µL of each standard onto the surface of the plate. Do not puncture the agarose. Allow capillary action to pull the small volume from the pipet tip to the surface. Spot 0.5 µL of the cDNA sample onto the plate adjacent to the standards. Allow the spots to absorb into the agarose for 10–15 min at room temperature. Invert the plate and visualize on a UV light box. Compare the spotted sample of unknown concentration with the standards to determine the concentration of the cDNA.

8. Since the cDNA is heavily methylated, introduction into a host with an McrA, McrCB, hsdSMR, Mrr phenotype would be subject to digestion by these restriction systems. Therefore, the choice of packaging extract and an *E. coli* host strain is crucial *(24–29)*.

9. Since λ phage can adhere to dead as well as to viable cells, the lower temperature prevents the bacteria from overgrowing.

10. Most cDNA vectors have color selection by IPTG and X-gal. These components can be added to the top agarose before plating to produce the background blue color. Use 15 µL of 0.5*M* IPTG (in water) and 50 µL of X-gal at 250 mg/mL in dimethylformamide.

References

1. Chomczynski, P. and Sacchi, N. (1987) Single-step method of RNA isolation by acid guanidinium thiocyanate-phenol-chloroform extraction. *Anal. Biochem.* **162,** 156–159.

2. Sambrook, J., Fritsch, E. F., and Maniatis, T. (eds.) (1989) *Molecular Cloning: A Laboratory Manual*, 2nd ed. Cold Spring Harbor Laboratory Press, Cold Spring Harbor, NY.

3. Han, J. H. and Rutter, W. J. (1987) Lambda gt22, an improved lambda vector for the directional cloning of full-length cDNA. *Nucleic Acids Res.* **15,** 6304.

4. Huynh, T. V., Young, R. A., and Davis, R. W. (1985) *DNA Cloning*, vol. I (Glover, D. A., ed.), IRL, Washington, DC, pp. 49–78.

5. Meissner, P. S., Sisk, W. P., and Berman, M. L. (1987) Bacteriophage lambda cloning system for the construction of directional cDNA libraries. *Proc. Natl. Acad. Sci. USA* **84,** 438–447.

6. Murphy, A. J. M. and Efstratiadis, A. (1987) Cloning vectors for expression of cDNA libraries in mammalian cells. *Proc. Natl. Acad. Sci. USA* **84,** 8277–8281.

7. Palazzolo, M. J. and Meyerowitz, E. M. (1987) A family of lambda phage cDNA cloning vectors, lambda SWAJ, allowing the amplification of RNA sequences. *Gene* **52,** 197–206.

8. Scherer, G., Telford, J., Baldari, C., and Pirrotta, V. (1981) Isolation of cloned genes differentially expressed at early and late stages of Drosophila embryonic development. *Dev. Biol.* **86,** 438–447.

9. Young, R. A. and Davis, R. W. (1983) Efficient isolation of genes by using antibody probes. *Proc. Natl. Acad. Sci. USA* **80,** 1194–1198.

10. Young, R. A. and Davis, R. W. (1983) Yeast RNA polymerase II genes: isolation with antibody probes. *Science* **222,** 778–782.

11. Swaroop, A. and Weissman, S. M. (1988) Charon BS (+) and (–), versatile lambda phage vectors for constructing directional cDNA libraries and their efficient transfer to plasmids. *Nucleic Acids Res.* **16,** 8739.

12. Palazzolo, M. J., Hamilton, B. A., Ding, D. L., Martin, C. H., Mead, D. A., Mierendorf, R. C., Raghavan, K. V., Meyerowitz, E. M., and Lipshitz, H. D. (1990) Phage lambda cDNA cloning vectors for subtractive hybridization, fusion-protein synthesis and Cre-loxP automatic plasmid subcloning. *Gene* **88,** 25–36.

13. Short, J. M., Fernandez, J. M., Sorge, J. A., and Huse, W. D. (1988) Lambda ZAP: a bacteriophage lambda expression vector with in vivo excision properties. *Nucleic Acids Res.* **16,** 7583–7600.

14. Alting-Mees, M., Hoener, P., Ardourel, D., Sorge, J. A., and Short, J. M. (1992) New lambda and phagemid vectors for prokaryotic and eukaryotic expression. *Strategies Mol. Biol.* **5(3),** 58–61.

15. Gubler, U. and Hoffman, B. J. (1983) A simple and very efficient method for generating cDNA libraries. *Gene* **25,** 263–269.

16. Huse, W. D. and Hansen, C. (1988) cDNA cloning redefined: a rapid, efficient, directional method. *Strategies Mol. Biol.* **1(1),** 1–3.

17. Kimmel, A. R. and Berger, S. L. (1989) Preparation of cDNA and the generation of cDNA libraries: overview. *Methods Enzymol.* **152,** 307–316.

18. Krug, M. S. and Berger, S. L. (1989) First strand cDNA synthesis primed with oligo (dT). *Methods Enzymol.* **152,** 316–325.

19. Okayama, H. and Berg, P. (1982) High-efficiency cloning of full-length cDNA. *Mol. Cell. Biol.* **2,** 161–170.

20. Gerard, G. (1989) cDNA synthesis by cloned Moloney Murine Leukemia Virus reverse transcriptase lacking RNaseH activity. *Focus* **11,** 66.

21. Nielson, K., Simcox, T. G., Schoettlin, W., Buchner, R., Scott, B., and Mathur, E. (1993) StratascriptTM RNaseH-reverse transcriptase for larger yields of full-length cDNA transcripts. *Strategies Mol. Biol.* **6(2),** 45.

22. Costa, L., Grafsky, A., and Weiner, M. P. (1994) Cloning and analysis of PCR-generated DNA fragments. *PCR Methods Applic.* **3,** 338–345.

23. Hu, G. (1993) DNA polymerase-catalyzed addition of nontemplated extra nucleotides to the 3' end of a DNA fragment. *DNA Cell Biol.* **12(8),** 763–770.

24. Kretz, P. L., Reid, C. H., Greener, A., and Short, J. M. (1989) Effect of lambda packaging extract mcr restriction activity on DNA cloning. *Nucleic Acids Res.* **17,** 5409.

25. Kretz, P. L. and Short, J. M. (1989) GigapackTMII: restriction free (hsd-, mcrA-, mcrB-, mrr-) lambda packaging extracts. *Strategies Mol. Biol.* **2(2),** 25,26.

26. Bullock, W., Fernandez, J. M., and Short, J. M. (1987) XL1-blue: a high efficiency plasmid transforming recA *Escherichia coli* strain with beta-galactosidase selection. *Biotechniques* **5(4),** 376–379.

27. Kohler, S. W., Provost, G. S., Kretz, P. L., Dycaico, M. J., Sorge, J. A., and Short, J. M. (1990) Development of short-term in vivo mutagenesis assay. The effects of methylation on the recovery of a lambda phage shuttle vector from transgenic mice. *Nucleic Acids Res.* **18,** 3007–3013.

28. Kretz, P. L., Kohler, S. W., and Short, J. M. (1991) Gigapack® III high efficiency lambda packaging extract with single-tube convenience. *Strategies Mol. Biol.* **7(2),** 44,45.

29. Kretz, P. L., Kohler, S. W., and Short, J. M. (1991). Identification and characterization of a gene responsible for inhibiting propagation of methylated DNA sequences in mcrA and mcrB *Escherichia coli* strains. *J. Bacteriol.* **173,** 4707–4716.

5

Clone Excision Methods
for the Lambda ZAP®-Based Vectors

Marjory A. Snead, Michelle A. Alting-Mees, and Jay M. Short

1. Introduction

The Lambda ZAP® vectors have been designed to allow in vivo excision and recircularization of the cloned insert and phagemid sequences contained within the λ vector *(1)*. Once a Lambda ZAP library is constructed and amplified, putative clones or the λ library itself may be excised into the phagemid form. Two versions of the excision protocol for Lambda ZAP-based vectors are included here.

1.1. ExAssist Procedure

The ExAssist protocol is suitable for excision of a few clones or mass excision of amplified libraries *(2,3)*. Mass excision can be used to generate subtraction libraries and subtraction DNA probes *(4)*. Converting the library to phagemid form also allows screening of the phage library in eukaryotic cells by transformation of eukaryotic cells with supercoiled plasmid DNA *(1,5)*. Excision is performed by coinfecting an excision strain with both λ-phage and helper phage. This in vivo excision is dependent on DNA sequences placed in the λ-phage genome, which normally serve as the f1 bacteriophage "origin of replication" for positive strand synthesis, and on the presence of a variety of proteins, including f1 bacteriophage-derived proteins *(2,6)*. The origin of the plus strand replication within the λ vector is divided into two overlying parts: the site of initiation and the site of termination for DNA synthesis. The λ-phage is made accessible to the f1-derived proteins by simultaneously infecting a strain of *Escherichia coli* with both the λ vector and the f1 "helper" bacteriophage. Inside *E. coli*, the helper proteins (i.e., proteins from f1 or M13

From: *Methods in Molecular Biology, Vol. 69: cDNA Library Protocols*
Edited by: I. G. Cowell and C. A. Austin Humana Press Inc., Totowa, NJ

phage) recognize the initiator DNA that is within the λ vector. These proteins nick one of the two DNA strands. At the site of this nick, new DNA synthesis begins and duplicates whatever DNA exists in the λ vector "downstream" of the nicking site. DNA synthesis of a new single strand of DNA continues through the cloned insert until the termination signal, positioned 3' of the initiator signal in the λ vector, is encountered. The single-stranded DNA molecule is circularized by the gene II product from the f1 phage forming a circular DNA molecule. This circular DNA molecule contains everything between the initiator and terminator. In the case of the Lambda ZAP vector, this includes all sequences of the phagemid, pBluescript® SK(–), and the insert, if one is present. This conversion is the "excision" step, since all sequences associated with normal λ vectors are positioned outside of the initiator and terminator signals, and are not contained within the circularized DNA. In addition, the circularizing of the DNA automatically recreates a functional f1 origin as found in the f1 bacteriophage or phagemids.

Signals for "packaging" the newly created phagemid are contained within the f1 terminator origin DNA sequence. They permit the circularized DNA to be "packaged" and secreted from the *E. coli*. Once the phagemid is secreted, the *E. coli* cells used for in vivo excision of the cloned DNA can be removed from the supernatant by heating at 70°C. The heat treatment kills the *E. coli* cells and λ-phage particles, whereas the phagemid and helper phage particles remain resistant to the heat. For production of double-stranded DNA, the packaged phagemid DNA is mixed with fresh *E. coli* cells and is spread on LB plates containing an appropriate antibiotic to select for the colonies containing the excised phagemid DNA. The helper phage contains an amber mutation that prevents replication of the helper phage genome in a nonsuppressing *E. coli* strain. The cells are also engineered to be resistant to λ infection. This allows only the excised phagemid to replicate in the host, removing the possibility of counterproductive coinfection from the helper phage and λ-phage after the excision process is finished. DNA from minipreps of these colonies can be used for analysis of insert DNA, including DNA sequencing, in vitro transcription, in vitro translation, mapping, and expression.

1.2. Rapid Excision Procedure

The Rapid Excision protocol is useful for excising clones from a primary library (prior to library amplification), excision of single or multiple clones, or as a faster alternative to the ExAssist method *(7)*. The rapid excision procedure uses the same in vivo mechanism; however, excision is performed on NZY agar plates instead of in solution. In addition to requiring less hands-on time, this modified protocol limits representation bias occurring in liquid-phase excision and can be monitored at the clonal level, which makes it suitable for excising primary libraries for sequencing.

In this system, the λ-phage, 704 helper phage, and *E. coli* XPORT and XLOLR host strains are plated together in top agarose, allowing the excision process to occur during plaque development on the plates. The plaques that develop overnight are called "excision plaques" to differentiate them from plaques generated using only λ-phage particles. As the λ-phage and 704 helper phage coreplicate in the XPORT cells, the pII protein encoded by the helper phage mediates single-stranded replication (excision) of the phagemid sequences within the λ DNA *(1)*. The excised phagemid single-stranded DNA (ssDNA) is then packaged as filamentous phage particles and is released within the excision plaque. These released phagemid particles, in turn, infect the XLOLR cells, that are also present in the top agarose. Helper phage do not propagate in the XLOLR cells because the 704 helper phage contain amber stop codons, which the XLOLR cells do not suppress. Also, λ-phage do not propagate in the XLOLR cells, because these cells are resistant to λ infection. Therefore, each excision plaque contains λ-phage particles, helper phage, packaged excised phagemid particles, lysed XPORT cells, and phagemid-infected XLOLR cells.

To miniprep the phagemid DNA, XLOLR cells infected with the excised phagemid can be cultured by touching the excision plaque with a toothpick, placing it into LB medium with the appropriate antibiotics, and growing overnight at 37°C in an air shaker. Tetracycline added to the growth medium kills the XPORT cells used for the excision, but allows survival of the XLOLR cells. Concomitantly, addition of the phagemid-encoded antibiotic directly selects for XLOLR cells harboring the phagemid DNA. For example, tetracycline plus ampicillin selects for XLOLR cells containing pBluescript phagemid in Lambda ZAP vector excisions, and tetracycline plus kanamycin selects for XLOLR cells containing the pBK-CMV vector in ZAP Express and SeqZAP vector excisions. After overnight growth, plasmid DNA can be prepared. In addition, by coring the excision plaque, it is possible to maintain a stock of λ-phage particles and phagemid particles, in case it is necessary to return to the λ-phage after excision.

2. Materials
2.1. ExAssist Excision

1. ExAssist helper phage at ~1 × 10^{10} PFU/mL (Stratagene, La Jolla, CA). ExAssist helper phage is M13 virus containing amber codons in gene I and gene II and α-complementing β-galactosidase sequences. These sequences may interfere with sequencing or site-directed mutagenesis where oligonucleotide primers (e.g., M13–20 and reverse primers) hybridize to β-galactosidase sequences *(3)*.

2. XLOLR *E. coli* host strain: {Δ(*mcrA*)*183* Δ(*mcrCB-hsdSMR-mrr*)*173 endA1 thi-1 recA1 gyrA96 relA1 lac* [F' *proAB lacIqZΔM15* Tn*10* (Tet^r)] Su– (nonsuppressing), λ^r (λ resistant)} freshly streaked on an LB agar plate containing 25 μg/mL tetracycline.

3. XL1-Blue MRF' *E. coli* host strain: {Δ(*mcrA*)*183* Δ(*mcrCB-hsdSMR-mrr*)*173 endA1 supE44 thi-1 recA1 gyrA96 relA1 lac* [F' *proAB lacIqZΔM15* Tn*10* (Tet^r)]} freshly streaked on an LB agar plate containing 25 μg/mL tetracycline.

4. SM buffer: 5.8 g NaCl, 2.0 g MgSO$_4$·7H$_2$O, 50.0 mL 1*M* Tris-HCl, pH 7.5, 5.0 mL 2% (w/v) gelatin/L. Autoclave.

5. LB medium and plates: 10 g NaCl, 10 g bacto-tryptone, 5 g bacto-yeast extract/L. Add 15 g agar for plates. Adjust the pH to 7.5 with NaOH, and sterilize by autoclaving.

6. 2% (v/v) Maltose.

7. 100 m*M* MgSO$_4$.

8. 10 m*M* MgSO$_4$.

9. Chloroform.

2.2. Rapid Excision

1. 704 Helper phage at ~1 × 10^8 PFU/mL (Stratagene). 704 Helper phage is a *lac Z* deletion mutant of the Phagescript vector *(8)*.

2. XLOLR *E. coli* host strain: {Δ(*mcrA*)*183* Δ(*mcrCB-hsdSMR-mrr*)*173 endA1 thi-1 recA1 gyrA96 relA1 lac* [F' *proAB lacIqZΔM15* Tn*10* (Tet^r)] Su– (nonsuppressing), λ^r (λ resistant)} freshly streaked on an NZY plate.

3. XPORT *E. coli* host strain: [Δ(*mcrA*)*183* Δ(*mcrCB-hsdSMR-mrr*)*173 endA1 supE44 thi-1 recA1 gyrA96 relA1 lac* (F' *proAB lacIqZΔM15*)] freshly streaked on an NZY plate.

4. SM buffer (*see* Section 2.1., item 4).

5. Chloroform.

6. NZY medium, plates, and top agarose: 5 g NaCl, 2 g MgSO$_4$·7H$_2$O, 5 g yeast extract, 10 g NZ amine (casein hydrolysate)/L. Add 15 g agar for plates or add 0.7% (w/v) agarose for top agarose. Adjust the pH to 7.5 with NaOH and sterilize by autoclaving.

7. LB medium (*see* Section 2.1., item 5).

3. Methods
3.1. ExAssist Excision

This procedure can be performed either on single clones or as a mass excision on amplified libraries. Colonies appearing on the plate contain the double-stranded phagemid with the cloned DNA insert. Helper phage will not grow, since helper phage is unable to replicate in Su– (nonsuppressing) XLOLR strains and does not contain kanamycin resistance genes. XLOLR cells are also resistant to λ-phage infection, thus preventing λ-phage contamination after excision.

3.1.1. Preparation of Plating Cells

1. Inoculate 50 mL of LB medium (supplemented with 0.2% [v/v] maltose and 10 mM MgSO$_4$ final concentration) in a sterile flask with a single colony of the XL1-Blue MRF' cells. In addition, inoculate 50 mL of LB medium (without supplements) with a single colony of XLOLR cells. Grow overnight cultures of the XL1-Blue MRF' and XLOLR cells with shaking at 30°C. The cells should not overgrow at this lower temperature.
2. Gently spin down the XL1-Blue-MRF' and XLOLR cells (1000g). Resuspend the cells in ~15 mL of 10 mM MgSO$_4$. Do not vortex.
3. Dilute the XL1-Blue MRF' and XLOLR cells to an OD$_{600}$ of 1.0 in 10 mM MgSO$_4$.

3.1.2. Excision

1. After library screening, core the plaque of interest from the agar plate, and transfer the plaque to a sterile microcentrifuge tube containing 500 µL of SM buffer and 20 µL of chloroform. Alternatively, an aliquot of an amplified Lambda ZAP library may be "mass excised."
2. Vortex to release the phage particles into the SM buffer. Incubate the tube for 1–2 h at room temperature or overnight at 4°C. This phage stock is stable for up to 6 mo at 4°C.
3. In a 50-mL conical tube, combine the following: 200 µL XL1-Blue MRF' cells, 250 µL λ-phage stock (containing >1 × 10^5 phage particles), 1 µL ExAssist helper phage (>1 × 10^6 PFU/ µL) (*see* Note 1).
4. Incubate the mixture at 37°C for 15 min.
5. Add 3 mL of LB medium (25 mL of LB medium for mass excision), and incubate for 2–2.5 h at 37°C with shaking (*see* Note 2).
6. Heat the tube at 70°C for 15 min, and then spin at 4000g for 15 min.
7. Decant the supernatant into a sterile tube. This stock contains the excised phagemid packaged as filamentous phage particles, and may be stored at 4°C for 1–2 mo.
8. To plate the excised phagemids, add to a 1.5-mL tube: 200 µL freshly grown XLOLR cells (*see* Section 3.1.1.) and 100-µL phage supernatant (1 µL of the phage supernatant for mass excision) (*see* step 7). To another 1.5-mL tube add: 200 µL freshly grown XLOLR cells (*see* Section 3.1.1.) and 10-µL phage supernatant (*see* step 7).
9. Incubate the tubes at 37°C for 15 min.
10. Add 300 µL of LB medium, and incubate at 37°C for 45 min to allow expression of the antibiotic resistance gene.
11. Plate 200 µL from each tube on LB agar plates containing the appropriate antibiotic for the phagemid vector (*see* Note 3).
12. Incubate overnight at 37°C or until colonies form (*see* Note 4).

3.2. Rapid Excision

The rapid excision system is particularly useful for excision of single or multiple cored plaques and multiple clones within a primary (unamplified) library.

3.2.1. Preparation of Plating Cells

Grow cultures of the XPORT and XLOLR *E. coli* strains using colonies from fresh NZY agar plates or a 1:10 dilution of overnight cultures in 10 mL of NZY broth in 15-mL Falcon® 2059 polypropylene tubes. Grow the cells to midlog phase ($OD_{600} = 1.0$).

3.2.2. Excision

1. After library screening, core a positive λ plaque from the agar plate and transfer the plaque to a sterile microcentrifuge tube containing 500 μL of SM buffer and 20 μL of chloroform. Alternatively, an aliquot of an unamplified library can be used.
2. Vortex the microcentrifuge tube to release the phage particles into the SM buffer. This phage stock may be used directly or may be stored for up to 6 mo at 4°C.
3. In a 15-mL Falcon 2059 polypropylene tube, combine the following components in order: 100 μL XPORT cells (*see* Section 3.2.1., step 1), 10 μL 704 helper phage (10^8 PFU/mL), 1 μL cored λ plaque (10–1000 PFU) or an unamplified library aliquot (10–1000 PFU).
4. Add 10 μL of XLOLR cells (*see* Section 3.2.1.) to each sample.
5. Add 3–5 mL of NZY top agarose to each sample, and plate each entire sample separately on NZY agar plates (*see* Note 5).
6. Incubate the plates overnight at 37°C or until plaques are visible (*see* Note 6).
7. Remove the plates from the 37°C incubator. Visible plaques on the sample excision plates contain λ-phage particles, 704 helper phage particles, and excised, packaged phagemid particles. Also contained within the plaque are XLOLR cells that have been infected by the phagemid particles.
8. Touch a toothpick to an isolated plaque on the overnight plates, and inoculate in a Falcon 2059 polypropylene tube 3 mL of LB medium containing the appropriate antibiotic (*see* Note 7).
9. Grow the cultures overnight at 37°C in an air shaker and purify the DNA using a standard miniprep protocol *(9)* (*see* Note 8).

4. Notes

1. When excising an entire library, 10- to 100-fold more of the amplified λ-phage should be excised than is found in the primary library to ensure statistical representation of the excised clones. Cells should be added at a 10:1 cells-to-amplified λ-phage ratio and ExAssist helper phage should be added at a 10:1 phage-to-cell ratio. For example, use: 10^8 cells (1 OD_{600} = 8.0 × 10^8 cells/mL), 10^9 ExAssist helper phage, 10^7 PFU of amplified λ library.
2. Incubation times for mass excision in excess of 3 h may alter the clonal representation. Single-clone excision reactions can be safely performed overnight, since clonal representation is not relevant.
3. Use LB-ampicillin (100 μg/mL) agar plates for the Lambda ZAP II vectors and LB-kanamycin (50 μg/mL) agar plates for ZAP Express and SeqZAP vectors.

4. If the number of colonies is low, be sure that the appropriate antibiotic was used in the LB agar plates (*see* Note 3). Vortex and allow the cores to sit overnight to elute completely. The number of excised colonies is dependent on the Lambda ZAP phage titer. It may be necessary to make a high-titer stock of the phage for the excision procedure. Increase the excision time (*see* Section 3.1.2., step 5) or the incubation time (*see* Section 3.1.2., step 10) to increase the number of colonies. If the number of colonies is high, owing to the high efficiency of the excision process, it may be necessary to titrate the supernatant to achieve single-colony isolation.

5. Negative control: Prepare an excision reaction as outlined using all components (the XPORT cells, the 704 helper phage and the XLOLR cells) (*see* Section 3.2.2., steps 3–6), except the cored plaque or λ-phage library aliquot. Since every XPORT cell should be infected with helper phage, plaques should form a confluent lawn instead of individual plaques. If individual helper phage plaques are observed on the negative control, check the titer of the helper phage (10^6 PFU/100-mm plate) to be sure a sufficient amount of helper phage was used. Also check the buffers and the host cells for λ-phage contamination.

6. If a clone is difficult to excise, the insert may be toxic to the cells. The probability of success can be increased using the following variation of the rapid excision protocol. Mix 5000 PFU Lambda ZAP clone, 10^6 PFU 704 helper phage, 100 µL XPORT cells (*see* Section 3.2.1.), 3 mL NZY top agarose. Plate this mixture, which does not include the XLOLR cells, on NZY agar plates. Incubate at 37°C overnight to allow the λ plaques almost to reach confluency. Elute the phagemid particles from the plate by overlaying the lawn with 3 mL of SM buffer and rock at room temperature for 1 h. Recover the bacteriophage suspension from the plate, and remove the cell debris by centrifugation for 10 min at 2000*g*. This supernatant contains λ-phage, 704 helper phage, and excised phagemid particles. Transfer the supernatant to a sterile 15-mL Falcon polypropylene tube, and use an aliquot of the recovered phagemid suspension to infect 100 µL of fresh XLOLR cells (*see* Section 3.2.1.). Plate on agar plates containing the appropriate antibiotic (*see* Note 3). The resulting colonies can be tested for functional inserts.

7. Use tetracycline (25 µg/mL) and either carbenicillin or ampicillin (100 µg/mL) for the Lambda ZAP II vector and kanamycin (50 µg/mL) for the ZAP-Express or SeqZAP vectors.

8. If there is an absence of growth in the miniprep, be sure the correct amounts of cells and antibiotic were used. XLOLR cells should be freshly grown to log phase (*see* Section 3.2.1.). If no DNA is present when miniprepped, check for cell growth and be sure the appropriate antibiotic was used (*see* Note 7).

References

1. Short, J. M., Fernandez, J. M., Sorge, J. A., and Huse, W. D. (1988) Lambda ZAP: a bacteriophage lambda expression vector with in vivo excision properties. *Nucleic Acids Res.* **16,** 7583–7600.

2. Short, J. M. and Sorge, J. A. (1992) In vivo excision properties of bacteriophage Lambda ZAP® expression vectors. *Methods Enzymol.* **216**, 495–508.
3. Hay, B. and Short, J. M. (1992) ExAssistTM helper phage and SOLRTM cells for lambda ZAP®II excisions. *Strategies* **5(1)**, 16–18.
4. Schweinfest, C. W., Henderson, K. W., Gu, J. R., Kottaridis, S. D., Besbeas, S., Panotopoulou, E., and Papas, T. S. (1990) Subtraction hybridization cDNA libraries from colon carcinoma and hepatic cancer. *Genetic Anal. and Technol. Appl.* **7**, 64–70.
5. Alting-Mees, M., Hoener, P., Ardourel, D., Sorge, J. A., and Short, J. M. (1992) New lambda and phagemid vectors for prokaryotic and eukaryotic expression. *Strategies* **5(3)**, 58–61.
6. Dotto, G. P., Horiuchi, K., and Zinder, N. D. (1984) The functional origin of bacteriophage f1 DNA replication. Its signals and domains. *J. Mol. Biol.* **172**, 507–521.
7. Alting-Mees, M. A. and Short, J. M. (1994) New rapid excision kit for the lambda ZAP® and ZAPExpressTM vectors. *Strategies* **7(3)**, 70–72.
8. Alting-Mees, M. A. and Short, J. M. (1993) Polycos vectors: a system for packaging filamentous phage and phagemid vectors using lambda phage packaging extracts. *Gene* **137**, 93–100.
9. Sambrook, J., Fritsch, E. F., and Maniatis, T. (eds.) (1989) *Molecular Cloning: A Laboratory Manual*, 2nd ed. Cold Spring Harbor Laboratory Press, Cold Spring Harbor, NY.

6

Using Rapid Amplification of cDNA Ends (RACE) to Obtain Full-Length cDNAs

Yue Zhang and Michael A. Frohman

1. Introduction

Most attempts to identify and isolate a novel cDNA result in the acquisition of clones that represent only a part of the mRNA's complete sequence (Fig. 1). The approach described here to clone the missing sequence (cDNA ends) employs polymerase chain reaction (PCR). Since the initial reports of rapid amplification of cDNA ends (RACE) *(1)* or related techniques *(2,3)*, many labs have developed significant improvements on the basic approach *(4–18)*. The most recent hybrid version of the relatively simple classic RACE is described here, as well as a more powerful, but technically more challenging "new RACE" protocol, which is adapted from the work of a number of laboratories *(19–26)*. Commercial RACE kits are available from Bethesda Research Laboratories (Gaithersburg, MD) *(11)* and Clontech (Palo Alto, CA) that are convenient, but not as powerful as the most recent versions of classic and new RACE.

1.1. Overview

Why use PCR (RACE) at all instead of screening (additional) cDNA libraries? RACE cloning is advantageous for several reasons: First, it takes weeks to screen cDNA libraries, obtain individual cDNA clones, and analyze the clones to determine if the missing sequence is present; using PCR, such information can be generated within a few days. As a result, it becomes practical to modify RNA preparation and/or reverse transcription conditions until full-length cDNAs are generated and observed. In addition, essentially unlimited numbers of independent clones can be generated using RACE, unlike library screens in which generally a single to a few cDNA clones are recovered. The availability of large numbers of clones provides confirmation of nucleotide sequence, and

From: *Methods in Molecular Biology, Vol. 69: cDNA Library Protocols*
Edited by: I. G. Cowell and C. A. Austin Humana Press Inc., Totowa, NJ

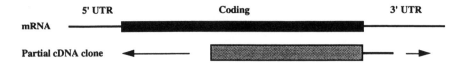

Fig. 1. Schematic representation of the setting in which RACE is useful in cDNA cloning strategies. Depicted is an mRNA for which a cDNA representing only an internal portion of the transcript has been obtained. Such circumstances often arise, for example, when open reading frame fragments are obtained from expression library, two-hybrid, or Genbank Expressed Sequence Tag searches.

allows the isolation of unusual transcripts that are alternately spliced or that begin at infrequently used promoters.

1.2. Principles

1.2.1. Classic RACE

PCR is used to amplify partial cDNAs representing the region between a single point in an mRNA transcript and its 3'- or 5'-end (Fig. 2). A short internal stretch of sequence must already be known from the mRNA of interest. From this sequence, gene-specific primers are chosen that are oriented in the direction of the missing sequence. Extension of the partial cDNAs from the unknown end of the message back to the known region is achieved using primers that anneal to the preexisting polyA tail (3'-end) or an appended homopolymer tail (5'-end). Using RACE, enrichments on the order of 10^6- to 10^7-fold can be obtained. As a result, relatively pure cDNA "ends" are generated that can be easily cloned or rapidly characterized using conventional techniques *(1)*.

To generate "3'-end" partial cDNA clones, mRNA is reverse transcribed using a "hybrid" primer (Q_T) that consists of 17 nt of oligo(dT) followed by a unique 35-base oligonucleotide sequence $(Q_I–Q_O;$ Fig. 2A,C), which in many reports is denoted as an "anchor" primer. Amplification is then performed using a primer containing part of this sequence (Q_O) that now binds to each cDNA at its 3'-end, and using a primer derived from the gene of interest (GSP1). A second set of amplification cycles is then carried out using "nested" primers $(Q_I$ and GSP2) to quench the amplification of nonspecific products. To generate "5'-end" partial cDNA clones, reverse transcription (primer extension) is carried out using a gene-specific primer (GSP-RT; Fig. 2B) to generate first-strand products. Then, a polyA tail is appended using terminal deoxynucleotidyltransferase (TdT) and dATP. Amplification is then achieved using:

1. The hybrid primer Q_T to form the second strand of cDNA;
2. The Q_O primer; and
3. A gene-specific primer upstream of the one used for reverse transcription.

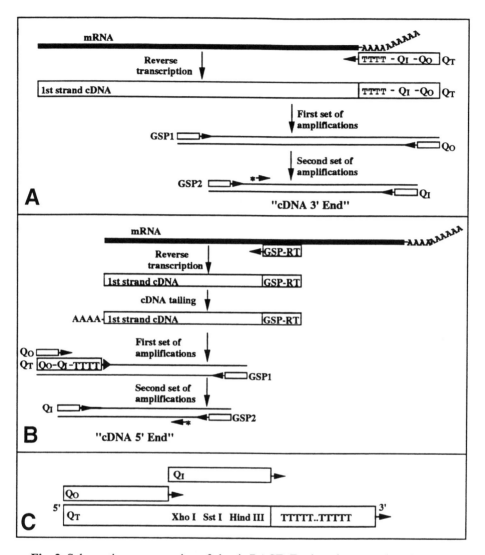

Fig. 2. Schematic representation of classic RACE. Explanations are given in the text. At each step, the diagram is simplified to illustrate only how the new product formed during the previous step is utilized. GSP1, gene-specific primer 1; GSP2, gene-specific primer 2; GSP-RT, gene-specific primer used for reverse transcription; $*\rightarrow$, GSP-Hyb/Seq or gene-specific primer for use in hybridization and sequencing reactions. **(A)** Amplification of 3′ partial cDNA ends. **(B)** Amplification of 5′ partial cDNA ends. **(C)** Schematic representation of the primers used in classic RACE. The 52-nt Q_T primer (5′ Q_O-Q_I-TTTT 3′) contains a 17-nt oligo-(dT) sequence at the 3′-end followed by a 35-nt sequence encoding *Hin*dIII, *Sst*I, and *Xho*I recognition sites. The Q_I and Q_O primers overlap by 1 nt; the Q_I primer contains all three of the recognition sites. Primers: Q_T: 5′-CCAGTGAGCAGAGTGACGAGGACTCGAGCTCAAGCTTTTTTTTTTTTTTTTT-3′, Q_O: 5′-CCAGTGAGCAGAGTGACG-3′, Q_I: 5′-GAGGACTCG AGCTCAAGC-3′.

$$5' \atop \text{XXXXXXXXX-AAAAAAAAA-} \begin{pmatrix} G \\ T \\ C \end{pmatrix}\begin{pmatrix} G \\ A \\ T \\ C \end{pmatrix} 3'$$

Fig. 3. Lock-docking primer. "X" represent one or more restriction sites at the 5'-end of the primer.

Finally, a second set of PCR cycles is carried out using nested primers (Q_I and GSP2) to increase specificity *(5)*.

1.2.2. Classic RACE Variations

In general, as described, the gene-specific primer is derived from a short stretch of sequence that is already known from the mRNA of interest. A frequent question is whether degenerate primers, i.e., ones directed against predicted nucleotide sequence based on known amino acid sequence, can be used instead. Although such primers increase the quantity of spurious amplification, the approach can work if other parameters are favorable (i.e., message abundance, GC composition, and cDNA end size; *see* refs. *14* and *27*).

At the unknown end of the cDNA, the 5'-end can be tailed with Cs instead of As, and then amplified using a hybrid primer with a tail containing Gs *(2)*, or a mixture of Gs and inosines (I) *(11)*. Although the G:I approach entails synthesizing a primer that can be used for 5' RACE only (since a T-tailed primer must be used to anneal to the polyA tail of the 3'-end), there may be sufficient benefits from using a mixed G:I tail to justify the cost, since the G:I region should anneal at temperatures similar to those of other primers normally used in PCR. In contrast, it is believed that homopolymers of either Ts or Gs present problems during PCR owing to the very low and very high annealing temperatures, respectively, required for their optimal usage *(1,11)*. On the other hand, the inosine residues function as degenerate nucleotides and will lead to higher spurious amplification, so the magnitude of the benefit of using a mixed G:I primer is unknown.

To minimize the length of homopolymer tail actually amplified, a "Lock-Docking" primer was developed by Borson et al. *(8)*. In this approach, the final 2 nt on the 3'-end of the primer are degenerate. For example, to amplify cDNAs linked to an A-tail, the lock-docking primer would appear as shown in Fig. 3. The advantage of this approach is that it forces the primer to anneal to the junction of the natural or appended homopolymer tail and the cDNA sequence. The disadvantage is that it is necessary to synthesize four primers, since most synthesizers can only synthesize primers starting from an unambiguous 3' end.

In another variation, the location of the anchor primer is changed from the end of the unknown region of sequence to random points within the unknown region *(7)*. This is accomplished using a primer containing an anchor region followed by six random nucleotides (5' XXXXXXXNNNNNN 3') either for reverse transcription (3' RACE) or for creation of the second strand of cDNA (5' RACE). This approach is valuable when the 3'- or 5'-ends lie so far away from the region of known sequence that the entire unknown region cannot be amplified effectively. Using this approach, cDNA ends of defined sizes are not generated; instead, one obtains a library of randomly sized fragments, all of which initiate at the gene-specific primer. The largest fragments can be cloned and characterized, extending the length of the known sequence, and the process (or standard RACE) repeated until the real unknown end is identified. The development of "long" PCR may make this approach unnecessary.

1.2.3. New RACE

The most technically challenging step in classic 5' RACE is to cajole reverse transcriptase to copy the mRNA of interest in its entirety into first-strand cDNA. Since prematurely terminated first-strand cDNAs are tailed by terminal transferase just as effectively as full-length cDNAs, cDNA populations composed largely of prematurely terminated first strands will result primarily in the amplification and recovery of cDNA ends that are not full length either (Fig. 4A). This problem is encountered routinely with vertebrate genes, which are often quite GC-rich at their 5'-ends and thus frequently contain sequences that hinder reverse transcription. A number of laboratories have developed steps or protocols designed to approach the problem *(19–26)*; the protocol described here and denoted "new RACE" is for the most part a composite adapted from the cited reports.

New RACE departs from classic RACE in that the "anchor" primer is attached to the 5'-end of the mRNA before the reverse transcription step; hence, the anchor sequence becomes incorporated into the first-strand cDNA if, and only if, the reverse transcription proceeds through the entire length of the mRNA of interest (and through the relatively short anchor sequence) as shown in Fig. 4B.

Before beginning new RACE (Fig. 5A), the mRNA is subjected to a dephosphorylation step using calf intestinal phosphatase (CIP). This step actually does nothing to full-length mRNAs, which have methyl-G caps at their termini, but it does dephosphorylate degraded mRNAs, which are uncapped at their termini *(21)*. This makes the degraded RNA biologically inert during the ensuing ligation step, because the phosphate group is required to drive the reaction. The full-length mRNAs are then decapped using tobacco acid pyrophosphatase (TAP), which leaves them with an active and phosphorylated 5'-terminus

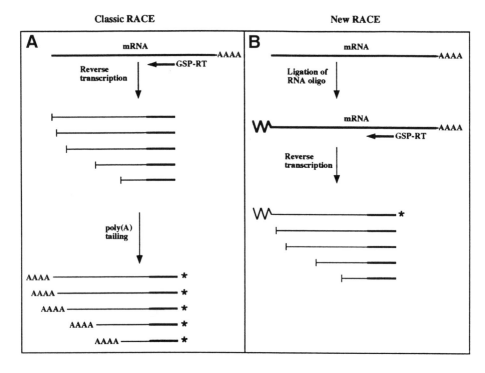

Fig. 4. A depiction of the advantage of using new RACE over classic RACE. **(A)** In Classic RACE, premature termination in the reverse transcription step results in polyadenylation of less-than-full-length first-strand cDNAs, all of which can be amplified using PCR to generate less-than-full-length cDNA 5' ends. *Indicates cDNAs ends created that will be amplified in the subsequent PCR reaction. **(B)** In new RACE, less-than-full-length cDNAs are also created, but are not terminated by the anchor sequence and, hence, cannot be amplified in the subsequent PCR reaction.

(20,24). Using T4 RNA ligase, this mRNA is then ligated to a short synthetic RNA oligo that has been generated by in vitro transcription of a linearized plasmid (Fig. 5B) *(19)*. The RNA oligo–mRNA hybrids are then reverse transcribed using a gene-specific primer or random primers to create first-strand cDNA. Finally, the 5' cDNA ends are amplified in two nested PCR reactions using additional gene-specific primers and primers derived from the sequence of the RNA oligo.

The new RACE approach can also be used to generate 3' cDNA ends *(21; see also* related protocols in *20,22)*, and is useful in particular for nonpoly-adenylated RNAs. In brief, cytoplasmic RNA is dephosphorylated and ligated to a short synthetic RNA oligo as described. Although ligation of the oligo to the 5'-end of the RNA was emphasized, RNA oligos actually ligate to both

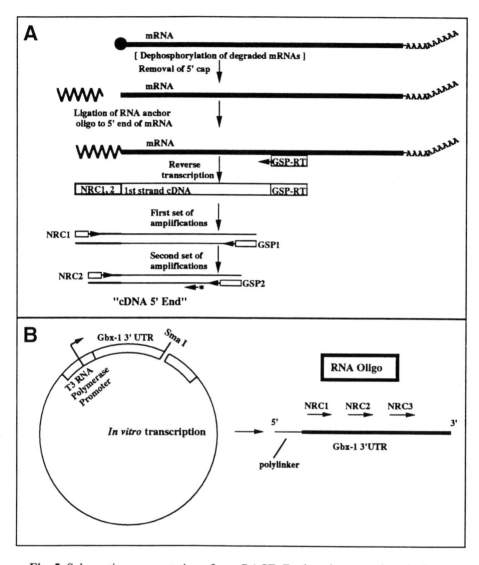

Fig. 5. Schematic representation of new RACE. Explanations are given in the text. At each step, the diagram is simplified to illustrate only how the new product formed during the previous step is utilized. *See* legend to Fig. 2 for description of some primers. **(A)** Amplification of 5' partial cDNA ends. **(B)** In vitro synthesis of the RNA oligo used for ligation in new RACE and schematic representation of the corresponding required primers. A 132-nt RNA oligo is produced by in vitro transcription of the plasmid depicted using T3 RNA polymerase. Primers NRC-1, NRC-2, and NRC-3 are derived from the sequence of the oligo, but do not encode restriction sites. To assist in the cloning of cDNA ends, the sequence "ATCG" is added to the 5'-end of NRC-2, as described in the cloning section of the text.

ends of the cytoplasmic RNAs. For the reverse transcription step, a primer derived from the RNA oligo sequence is used (e.g., the reverse complement of NRC-3; Fig. 5). Reverse transcription of the RNA oligos that happen to be ligated to the 3'-end of the cytoplasmic RNAs results in the creation of cDNAs that have the RNA oligo sequence appended to their 3'-end. Gene-specific primers oriented in the 5'→3' direction and new RACE primers (e.g., the reverse complements of NRC-2 and NRC-1; Fig. 5) can be used in nested PCR reactions to amplify the 3'-ends.

2. Materials

2.1. 3'-End cDNA Amplification Using Classic RACE

2.1.1. Reverse Transcription to Generate cDNA Templates

1. 5X Reverse transcription buffer: 250 mM Tris-HCl, pH 8.3, 375 mM KCl, 15 mM MgCl$_2$.
2. QT primer (*see* Fig. 2).
3. SuperScript II reverse transcriptase and 5X buffer (Bethesda Research Labs; BRL).
4. 1M Dithiothreitol (DTT) in H$_2$O.
5. RNase H (Bethesda Research Labs; BRL).
6. RNasin (Promega Biotech).
7. dNTPs: 10 mM solutions; PL-Biochemicals/Pharmacia or Boehringer Mannheim.
8. TE: 10 mM Tris-HCl, pH 7.5, 1 mM EDTA.

2.1.2. PCR Amplification

1. *Taq* polymerase and 10X buffer (Promega Biotech).
2. *(Optional)* Instead of using the recommended reaction mixture, a 10X buffer consisting of 670 mM Tris-HCl, pH 9.0, 67 mM MgCl$_2$, 1700 μg/mL bovine serum albumin (BSA), and 166 mM (NH$_4$)$_2$SO$_4$ can be substituted, and reaction conditions are altered as further described in Note 3 *(1,16)*.
3. PCR cocktail: 1X *Taq* polymerase buffer (described in item 1 or 2), each dNTP at 1.5 mM, and 10% dimethyl sulfoxide (DMSO).
4. GSP1, GSP2, Q$_O$ and Q$_I$ primers. Oligonucleotide primer sequences are listed in the legend to Fig. 2. Primers can be used "crude," except for Q$_T$, which should be purified to ensure that it is uniformly full-length.

2.2. 5'-End cDNA Amplification Using Classic RACE

2.2.1. Reverse Transcription

1. *See* Section 2.1.1., items 1, 3–8.
2. GSP-RT primer.

2.2.2. Tailing First-Strand cDNA Products

1. TdT (BRL or Boehringer Mannheim).
2. Tailing buffer (5X): 125 mM Tris-HCl, pH 6.6, 1M K Cacodylate, and 1250 μg/mL BSA.

3. Micron-100 spin filters (Amicon, Beverly, MA).
4. 25 mM Cobolt chloride.
5. 1 mM Adenosine triphosphate (ATP).

2.2.3. Amplification

For amplification materials, refer to Section 2.1.2.

2.3. New RACE 5'-End cDNA Amplification

2.3.1. Dephosphorylation of Degraded RNAs

1. CIP and 10X buffer (Boehringer Mannheim).
2. 1 mM DTT in H$_2$O.
3. RNasin (40 U/μL).
4. Proteinase K (Boehringer Mannheim).

2.3.2. Decapping of Intact RNAs

1. TAP and buffer (Epicentre, Madison, WI).
2. RNasin.
3. 100 mM ATP.

2.3.3. Preparation of RNA Oligo

1. RNA transcription kit (Epicentre; Madison, WI).
2. Proteinase K (Boehringer Mannheim).
3. DNase (RNase-free) (Epicentre).
4. Diethyl pyrocarbonate (DEPC)-treated water.

2.3.4. RNA Oligo Ligation

1. T4 RNA ligase (New England Biolabs or Boehringer Mannheim).
2. 10X Buffer: 500 mM Tris-HCl, pH 7.9, 100 mM MgCl$_2$, 20 mM DTT, 1 mg/mL BSA. Note that 10X T4 RNA ligase buffers supplied by some manufacturers contain too much ATP *(19)*. Check the composition of any commercially supplied 10X buffer and make your own if it contains more than 1 mM ATP (final 1X concentration should be 0.1 mM).
3. Micron-100 spin filters (Amicon).

2.3.5. Reverse Transcription

1. See Section 2.1.1.
2. Random hexanucleotides.
3. Antisense specific primer (*see* Fig. 2).

2.3.6. Amplification

For amplification materials, refer to Section 2.1.2.

3. Methods

3.1. 3'-End cDNA Amplification Using Classic RACE

3.1.1. Reverse Transcription to Generate cDNA Templates (see Notes 1 and 9)

1. Assemble reverse transcription components on ice: 4 μL of 5X reverse transcription buffer, 1 μL of dNTPs, 2 μL of 0.1M DTT, 0.5 μL of Q_T primer (100 ng/μL), and 0.25 μL (10 U) of RNasin.
2. Heat 1 μg of polyA$^+$ RNA or 5 μg of total RNA in 13 μL of water at 80°C for 3 min, cool rapidly on ice, and spin for 5 s in a microfuge. Combine with step 1.
3. Add to reverse transcription components 1 μL (200 U) of SuperScript II reverse transcriptase, and incubate for 5 min at room temperature, 1 h at 42°C, and 10 min at 50°C.
4. Incubate at 70°C for 15 min to inactivate reverse transcriptase. Spin for 5 s in a microfuge.
5. Add 0.75 μL (1.5 U) of RNase H to the tube, and incubate at 37°C for 20 min to destroy RNA template.
6. Dilute the reaction mixture to 1 mL with TE, and store at 4°C ("3'-end cDNA pool").

3.1.2. Amplification (see Notes 3 and 11)

First round:

1. Add an aliquot of the cDNA pool (1 μL) and primers (25 pmol each of GSP1 and Q_O) to 50 μL of PCR cocktail in a 0.5-mL microfuge tube.
2. Heat in a DNA thermal cycler for 5 min at 98°C to denature the first-strand products. Cool to 75° C. Add 2.5 U *Taq* polymerase, overlay the mixture with 30 μL of mineral oil (Sigma 400-5; preheat it in the thermal cycler to 75°C), and incubate at the appropriate annealing temperature (52–60°C) for 2 min. Extend the cDNAs at 72°C for 40 min.
3. Carry out 30 cycles of amplification using a step program (94°C, 1 min; 52–60°C, 1 min; 72°C, 3 min), followed by a 15-min final extension at 72°C. Cool to room temperature.

Second round:

1. Dilute 1 μL of the amplification products from the first round into 20 μL of TE.
2. Amplify 1 μL of the diluted material with primers GSP2 and Q_I using the first-round procedure, but eliminate the initial 2-min annealing step and the 72°C, 40-min extension step.

Dialyze and further manipulate the RACE products as described in Notes 7 and 8.

3.2. 5'-End cDNA Amplification Using Classic RACE

3.2.1. Reverse Transcription to Generate cDNA Templates (see Notes 2 and 9)

1. Assemble reverse transcription components on ice: 4 μL of 5X reverse transcription buffer, 1 μL of dNTPs, 2 μL of 0.1M DTT, and 0.25 μL (10 U) of RNasin.
2. Heat 0.5 μL of GSP-RT primer (100 ng/μL) and (1 μg of polyA$^+$ RNA or 5 μg of total RNA) in 13 μL of water at 80°C for 3 min, cool rapidly on ice, and spin for 5 s in a microfuge.
3. Add to reverse transcription components. Add 1 μL (200 U) of SuperScript II reverse transcriptase, and incubate for 1 h at 42°C, and 10 min at 50°C.
4. Incubate at 70°C for 15 min to inactivate reverse transcriptase. Spin for 5 s in a microfuge.
5. Add 0.75 μL (1.5 U) of RNase H to the tube, and incubate at 37°C for 20 min to destroy RNA template.
6. Dilute the reaction mixture to 400 μL with TE, and store at 4°C ("5'-end nontailed cDNA pool").

3.2.2. Appending a PolyA Tail to First-Strand cDNA Products (see Notes 5 and 10)

1. Remove excess primer using Microcon-100 spin filters (Amicon) or an equivalent product, following the manufacturer's instructions. Wash the material by spin filtration twice more using TE. The final volume recovered should not exceed 10 μL. Adjust volume to 10 μL using water.
2. Add 4 μL 5X tailing buffer, 1.2 μL 25 mM CoCl$_2$, 4 μL 1 mM dATP, and 10 U TdT.
3. Incubate for 5 min at 37°C and then 5 min at 65°C.
4. Dilute to 500 μL with TE ("5'-end-tailed cDNA pool").

3.2.3. Amplification (see Notes 4 and 11)

First round:

1. Add an aliquot of the "5'-end-tailed cDNA pool" (1 μL) and primers (25 pmol each of GSP1 and Q$_0$ [shown in Fig. 2B], and 2 pmol of Q$_T$) to 50 μL of PCR cocktail in a 0.5-mL microfuge tube.
2. Heat in a DNA thermal cycler for 5 min at 98°C to denature the first-strand products. Cool to 75°C. Add 2.5 U *Taq* polymerase, overlay the mixture with 30 μL of mineral oil (Sigma 400-5; preheat it in the thermal cycler to 75°C), and incubate at the appropriate annealing temperature (48–52°C) for 2 min. Extend the cDNAs at 72°C for 40 min.
3. Carry out 30 cycles of amplification using a step program (94°C, 1 min; 52–60°C, 1 min; 72°C, 3 min), followed by a 15-min final extension at 72°C. Cool to room temperature.

Second round:

1. Dilute 1 μL of the amplification products from the first round into 20 μL of TE.
2. Amplify 1 μL of the diluted material with primers GSP2 and Q_I using the first-round procedure, but eliminate the initial 2-min annealing step and the 72°C, 40-min extension step.

Notes 7 and 8 describe the analysis and further manipulation of the RACE PCR products.

3.3. New RACE 5'-End cDNA Amplification

The following procedure is described using relatively large amounts of RNA and can be scaled down if RNA quantities are limiting. The advantage of starting with large amounts of RNA is that aliquots can be electrophoresed quickly after each step of the procedure to confirm that detectable degradation of the RNA has not occurred, and the dephosphorylated-decapped-ligated RNA can be stored indefinitely for many future experiments.

3.3.1. Dephosphorylation of Degraded RNAs

In general, follow the manufacturer's recommendations for use of the phosphatase.

1. Prepare a reaction mixture containing 50 μg RNA in 41 μL of water, 5 μL 10X buffer, 0.5 μL 100 mM DTT, 1.25 μL RNasin (40 U/μL), and 3.5 μL CIP (1 U/μL).
2. Incubate the reaction at 50°C for 1 h.
3. Add proteinase K to 50 μg/mL, and incubate at 37°C for 30 min.
4. Extract the reaction with a mixture of phenol/chloroform, extract again with chloroform, and precipitate the RNA using $^1/_{10}$ vol of 3M NaOAc and 2.5 vol of ethanol. Resuspend the RNA in 43.6 μL of water.
5. Electrophorese 2 μg (1.6 μL) on a 1% TAE agarose gel adjacent to a lane containing 2 μg of the original RNA preparation, stain the gel with ethidium bromide, and confirm visually that the RNA remained intact during the dephosphorylation step.

3.3.2. Decapping of Intact RNAs (see Note 6)

1. Prepare a reaction mixture containing 48 μg of RNA in 42 μL of water (this is the RNA recovered from Section 3.3.1., step 1), 5 μL 10X TAP buffer, 1.25 μL RNasin (40 U/μL), 1 μL 100 mM ATP, and 1 μL TAP (5 U/μL).
2. Incubate the reaction at 37°C for 1 h, and then add 200 μL TE.
3. Extract the reaction with a mixture of phenol/chloroform, extract again with chloroform, and precipitate the RNA using $^1/_{10}$ vol of 3M NaOAc and 2.5 vol of ethanol. Resuspend the RNA in 40 μL of water.
4. Electrophorese 2 μg on a TAE 1% agarose gel adjacent to a lane containing 2 μg of the original RNA preparation, stain the gel with ethidium bromide, and confirm visually that the RNA remained intact during the decapping step.

3.3.3. Preparation of RNA Oligo

Choose a plasmid that can be linearized at a site ~100 bp downstream from a T7 or T3 RNA polymerase site (*see* Fig. 5B). Ideally, a plasmid containing some insert cloned into the first polylinker site is optimal, because primers made from palindromic polylinker DNA do not perform well in PCR. For our experiments, we use the 3'UTR of the mouse gene Gbx-1 *(28)*, which is cloned into the *Sst*I site of pBS-SK (Stratagene); we linearize with *Sma*I and transcribe with T3 RNA polymerase to produce a 132 nt RNA oligo, of which all but 17 nt are from Gbx-1. Note that adenosines are the best "acceptors" for the 3'-end of the RNA oligo to ligate to the 5'-end of its target, if an appropriate restriction site can be found. The primers subsequently used for amplification are all derived from the Gbx-1 3' UTR sequence. Interested investigators are welcome to the Gbx-1 NRC primer sequences and plasmid on request.

Carry out a test transcription to make sure that everything is working; then scale up. The oligo can be stored at −80°C indefinitely for many future experiments, and it is important to synthesize enough oligo so that losses owing to purification and spot-checks along the way will leave plenty of material at the end of the procedure.

1. Linearize 25 µg of the plasmid that is to be transcribed (the plasmid should be reasonably free of RNases).
2. Treat the digestion reaction with 50 µg/mL proteinase K for 30 min at 37°C, followed by two phenol/CHCl$_3$ extractions, one CHCl$_3$ extraction, and ethanol precipitation.
3. Resuspend the template DNA in 25 µL TE, pH 8.0, for a final concentration of approx 1 µg/µL.
4. Transcription: Mix at room temperature in the following order:

	Test scale, µL	Prep scale, µL
DEPC water	4	80
5X buffer	2	40
0.1*M* DTT	1	20
10 m*M* UTP	0.5	10
10 m*M* ATP	0.5	10
10 m*M* CTP	0.5	10
10 m*M* GTP	0.5	10
Restricted DNA (1 µg/µL)	0.5	10
RNasin (40 U/µL)	0.25	5
RNA polymerase (20 U/µL)	0.25	5

Incubate at 37°C for 1 h.
5. DNase template: Add 0.5 µL DNase (RNase-free) for every 20 µL of reaction volume and incubate at 37°C for 10 min.
6. Run 5 µL of test or prep reaction on a 1% TAE agarose gel to check. Expect to see a diffuse band at about the right size (or a bit smaller) in addition to some smearing all up and down the gel.

7. Purify the oligo by extracting with phenol/CHCl$_3$ and CHCl$_3$, and then rinse three times using water and a microcon spin filter (prerinsed with water).
8. Run another appropriately sized aliquot on a 1% TAE agarose gel to check integrity and concentration of oligo. Microcon 30 spin filters have a cutoff size of 60 nt, and microcon 100 spin filters have a cutoff size of 300 nt. Microcon 10 spin filters are probably most appropriate if the oligo is smaller than 100 nt, and microcon 30 spin filters for anything larger.

3.3.4. RNA Oligo-Cellular RNA Ligation

1. Set up two tubes—one with TAPped cellular RNA, and the other with unTAPped cellular RNA with: 11.25 μL water, 3 μL 10X buffer, 0.75 μL RNasin (40 U/μL), 2 μL 4 μg RNA oligo (3–6 molar excess over target cellular RNA), 10 μL 10 μg TAPped (or unTAPped) RNA, 1.5 μL 2 mM ATP, and 1.5 μL T4 RNA ligase (20 U/μL).
2. Incubate for 16 h at 17°C.
3. Purify the ligated oligo-RNA using microcon 100 spin filtration (three times in water; prerinse filter with RNase-free water). The volume recovered should not exceed 20 μL.
4. Run ¹/₃ of the ligation on a 1% TAE agarose gel to check integrity of the ligated RNA. It should look about as it did before ligation.

3.3.5. Reverse Transcription (see Note 9)

1. Assemble reverse transcription components on ice: 4 μL of 5X reverse transcription buffer, 1 μL of dNTPs (stock concentration is 10 mM of each dNTP), 2 μL of 0.1M DTT, and 0.25 μL (10 U) of RNasin.
2. Heat 1 μL of antisense specific primer (20 ng/μL) or random hexamers (50 ng/μL) and the remaining RNA (~6.7 μg) in 13 μL of water at 80°C for 3 min, cool rapidly on ice, and spin for 5 s in a microfuge.
3. Add to reverse transcription components 1 μL (200 U) of SuperScript II reverse transcriptase. Incubate for 1 h at 42°C, and 10 min at 50°C. If using random hexamers, insert a room temperature 10-min incubation period after mixing everything together.
4. Incubate at 70°C for 15 min to inactivate reverse transcriptase. Spin for 5 s in a microfuge.
5. Add 0.75 μL (1.5 U) of RNase H to the tube, and incubate at 37°C for 20 min to destroy RNA template.
6. Dilute the reaction mixture to 100 μL with TE and store at 4°C ("5'-end oligo-cDNA pool").

3.3.6. Amplification (see Note 11)

First round:

1. Add an aliquot of the "5'-end oligo-cDNA pool" (1 μL) and primers (25 pmol each of GSP1 and NRC-1) to 50 μL of PCR cocktail (1X *Taq* polymerase buffer

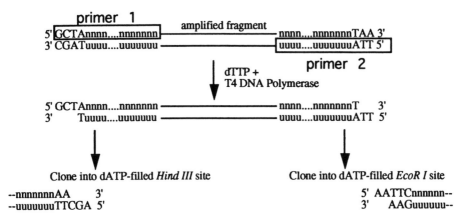

Fig. 6. A safe and easy cloning method.

[described in Section 2.1.2.], each dNTP at 1.5 m*M*, and 10% DMSO) in a 0.5-mL microfuge tube.

2. Heat in a DNA thermal cycler for 5 min at 98°C to denature the first-strand products. Cool to 75°C. Add 2.5 U *Taq* polymerase, overlay the mixture with 30 μL of mineral oil (Sigma 400-5, preheat it in the thermal cycler to 75°C), and incubate at the appropriate annealing temperature (52–60°C) for 2 min. Extend the cDNAs at 72°C for 40 min.

3. Carry out 35 cycles of amplification using a step program (94°C, 1 min; 52–60°C, 1 min; 72°C, 3 min), followed by a 15-min final extension at 72°C. Cool to room temperature.

Second round:

1. Dilute 1 μL of the amplification products from the first round into 20 μL of TE.

2. Amplify 1 μL of the diluted material with primers GSP2 and NRC-2 using the first-round procedure, but eliminate the initial 2-min annealing step and the 72°C, 40-min extension step.

3.4. Safe and Easy Cloning Protocol (see also Note 8)

1. Insert preparation: Select a pair of restriction enzymes for which you can synthesize half sites appended to PCR primers that can be chewed back to form the appropriate overhangs, as shown for *Hin*dIII and *Eco*RI, in Fig. 6. For RACE cloning, add "TTA" to the 5'-end of Q$_I$ or NRC-2, and add "GCTA" to the 5'-end of GSP2. Carry out PCR as usual.

2. After PCR, add proteinase K (10 mg/mL stock) to the PCR reaction to a concentration of 50 μg/mL, and incubate at 37°C for 30 min, to remove sticky *Taq* polymerase from the amplified DNA *(29)*.

3. Extract the PCR products with phenol/CHCl$_3$ and then CHCl$_3$ (but do not precipitate!) to remove proteins.

4. Filter the PCR products through a microcon-100 spin column (or microcon-30 if your product is <150 bp) three times using TE (not water) as the wash buffer to remove unwanted organics, primers, and dNTPs.
5. On ice, add the selected dNTP(s) (e.g., dTTP) to a final concentration of 0.2 mM, $^1/_{10}$ vol of 10X T4 DNA polymerase buffer, and 1–2 U T4 DNA polymerase.
6. Incubate at 12°C for 15 min and then 75°C for 10 min to heat-inactivate the T4 DNA polymerase. (*Optional:* Gel isolate DNA fragment of interest, depending on degree of success of PCR amplification.)
7. Vector preparation: Digest vector (e.g., pGem-7ZF [Promega]) using the selected enzymes (e.g., *Hin*dIII and *Eco*RI) under optimal conditions, in a volume of 10 µL.
8. Add a 10-µL mixture containing the selected dNTP(s) (e.g., dATP) at a final concentration of 0.4 mM, 1 µL of the restriction buffer used for digestion, 0.5 µL Klenow, and 0.25 µL Sequenase.
9. Incubate at 37°C for 15 min and then 75°C for 10 min to heat-inactivate the polymerases.
10. Gel isolate the linearized vector fragment.
11. For ligation, use equal molar amounts of vector and insert.

4. Notes

1. The following comments relate to reverse transcription for 3' RACE. PolyA$^+$ RNA is preferentially used for reverse transcription to decrease background, but it is unnecessary to prepare it if only total RNA is available. An important factor in the generation of full-length 3'-end partial cDNAs concerns the stringency of the reverse transcription reaction. Reverse transcription reactions were historically carried out at relatively low temperatures (37–42°C) using a vast excess of primer (~$^1/_2$ the mass of the mRNA, which represents an ~30:1 molar ratio). Under these low stringency conditions, a stretch of A residues as short as six to eight nucleotides will suffice as a binding site for an oligo(dT)-tailed primer. This may result in cDNA synthesis being initiated at sites upstream of the polyA tail, leading to truncation of the desired amplification product (*see* Fig. 7). One should be suspicious that this has occurred if a canonical polyadenylation signal sequence is not found near the 3'-end of the cDNAs generated. This can be minimized by controlling two parameters: primer concentration and reaction temperature. The primer concentration can be reduced dramatically without decreasing the amount of cDNA synthesized significantly *(30)* and will begin to bind preferentially to the longest A-rich stretches present (i.e., the polyA tail). The quantity recommended above represents a good starting point; it can be reduced fivefold further if significant truncation is observed.

   In the described protocol, the incubation temperature is raised slowly to encourage reverse transcription to proceed through regions of difficult secondary structure. Since the half-life of reverse transcriptase rapidly decreases as the incubation temperature increases, the reaction cannot be carried out at elevated temperatures in its entirety. Alternatively, the problem of difficult secondary structure (and nonspecific reverse transcription) can be approached using heat-

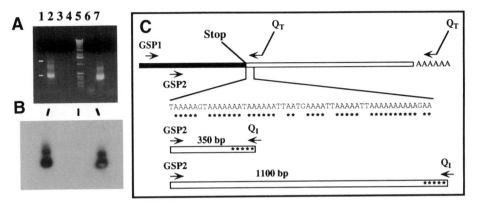

Fig. 7. Amplification of the 3'-end of the mouse type II sodium channel gene—an example of cDNA synthesis being initiated at sites upstream of the polyA tail. (A) Ethidium bromide-stained 1% agarose gel of the PCR product amplified from mouse brain 3' cDNA pool. Lane 1, no template control; lane 2, PCR product from primer GSP1 to Q_T; lanes 3 and 4, blank; lane 5, 1-kb ladder; lane 6, no template control; lane 7, nested PCR product from primer GSP2 to Q_I as reamplified from products of lane 2. The 350- and 1100-bp products were generated and are indicted by "–"s. (B) Southern blot hybridization with a ^{32}P-labeled oligonucleotide primer located 3' of GSP2 confirming the identity of the products. The hybridizing bands correspond to the 350 and 100 bp fragments observed in A. (C) Cloning and sequencing of the products revealed that the 350-bp product was a truncation, resulting from initiation of reverse transcription in the A-rich region just 3' of the stop codon TAA. The sequence of the A-rich region is shown.

stable reverse transcriptases, which are now available from several suppliers (Perkin-Elmer-Cetus, Amersham, Epicentre Technologies, and others). As in PCR reactions, the stringency of reverse transcription can thus be controlled by adjusting the temperature at which the primer is annealed to the mRNA. The optimal temperature depends on the specific reaction buffer and reverse transcriptase used, and should be determined empirically, but will usually be found to be in the range of 48–56°C for a primer terminated by a 17-nt oligo(dT)-tail.

2. Many of the remarks in Note 1 are also relevant for reverse transcription for 5' RACE and should be noted. There is, however, one major difference. The efficiency of cDNA extension is now critically important, since each specific cDNA, no matter how short, is subsequently tailed and becomes a suitable template for amplification (Fig. 2A). Thus, the PCR products generated directly reflect the quality of the reverse transcription reaction. Extension can be maximized by using clean, intact RNA, by selecting the primer for reverse transcription to be near the 5'-end of region of known sequence, and in theory, by using heat-stable reverse transcriptase at elevated temperatures or a combination of SuperScript II and heat-stable reverse transcriptase at multiple temperatures. Syn-

thesis of cDNAs at elevated temperatures should diminish the amount of secondary structure encountered in GC-rich regions of the mRNA. Random hexamers (50 ng) can be substituted for GSP-RT to create a universal 5'-end cDNA pool. A "universal" pool can be used for amplification of the 5'-end of any cDNA created in the reverse transcription. Correspondingly, though, each cDNA is present at a much lower level than if created using the gene-specific reverse transcription described in Section 3.2.1. If using random hexamers, insert a room temperature 10-min incubation period after mixing everything together.

3. For 3'-end amplification, it is important to add the *Taq* polymerase **after** heating the mixture to a temperature above the T_m of the primers ("hot-start" PCR). Addition of the enzyme prior to this point allows one "cycle" to take place at room temperature, promoting the synthesis of nonspecific background products dependent on low-stringency interactions.

An annealing temperature close to the effective T_m of the primers should be used. The Q_I and Q_O primers work well at 60°C under the PCR conditions recommended here, although the actual optimal temperature may depend on the PCR machine used. Gene-specific primers of similar length and GC content should be chosen. Computer programs to assist in the selection of primers are widely available and should be used. An extension time of 1 min/kb expected product should be allowed during the amplification cycles. If the expected length of product is unknown, try 3–4 min initially.

Very little substrate is required for the PCR reaction. One microgram of polyA$^+$ RNA typically contains ~5 × 10^7 copies of each low-abundance transcript. The PCR reaction described here works optimally when 10^3–10^5 templates (of the desired cDNA) are present in the starting mixture; thus, as little as 0.002% of the reverse transcription mixture suffices for the PCR reaction! Addition of too much starting material to the amplification reaction will lead to production of large amounts of nonspecific product and should be avoided. The RACE technique is particularly sensitive to this problem, since every cDNA in the mixture, desired and undesired, contains a binding site for the Q_O and Q_I primers.

It was found empirically that allowing extra extension time (40 min) during the first amplification round (when the second strand of cDNA is created) sometimes resulted in increased yields of the specific product relative to background amplification, and in particular, increased the yields of long cDNAs vs short cDNAs when specific cDNA ends of multiple lengths were present *(1)*. Prior treatment of cDNA templates with RNA hydrolysis or a combination of RNase H and RNase A infrequently improves the efficiency of amplification of specific cDNAs.

For some applications intended for cloned PCR products, such as expressing cDNAs to generate proteins, it is critically important to minimize the rate at which mutations occur during amplification. In other applications, such as using the cloned DNA as a probe in hybridization experiments, the presence of a few mutations is relatively unimportant, and thus, it is most convenient to use PCR conditions that maximize the likelihood of generating the desired product the first time a set of primers is used. Unfortunately, PCR conditions that result in a minimum of mutations are finicky, and often the desired product cannot be generated until the PCR conditions have been optimized, whereas PCR conditions that reliably produce desired products result in a relatively high mutation rate (~1% after 30 rounds). Thus, appropriate conditions must be chosen to generate the PCR products required prior to undertaking cloning steps. PCR conditions that result in a minimum of mutations require the use of nucleotides (dNTPs) at low concentrations (0.2 mM). Using the conditions recommended for *Taq* polymerase by Perkin-Elmer-Cetus results in an error rate of ~0.05% after 30 rounds of amplification. However, the conditions recommended often have to be optimized, meaning that the pH of the buffer and the concentration of magnesium have to be adjusted until the desired product is observed. In addition, inclusion of DMSO or dimethyl formamide may be required. For those who do not wish to prepare their own reagents to carry out optimization experiments, such kits are commercially available (e.g., from Invitrogen and Stratagene). PCR conditions that work much more frequently in the absence of optimization steps require the use of DMSO, ammonium sulfate, and relatively high concentrations (1.5 mM) of dNTPs, as described. It should be noted that the inclusion of DMSO to 10% decreases primer melting temperatures (and thus optimal annealing temperatures) by about 5–6°C.

4. Many of the remarks from Note 3 are also relevant for 5'-end amplification and should be noted. There is, however, one major difference. The annealing temperature in the first step (48–52°C) is lower than that used in successive cycles (52–60°C). This is because cDNA synthesis during the first round depends on the interaction of the appended polyA tail and the oligo(dT)-tailed Q_T primer, whereas in all subsequent rounds, amplification can proceed using the Q_O primer, which is composed of ~60% GC and which can anneal at a much higher temperature to its complementary target.

5. The following comments relate to TdT tailing the first-strand products. To attach a known sequence to the 5'-end of the first-strand cDNA, a homopolymeric tail is appended using TdT. We prefer appending polyA tails rather than polyC tails for several reasons. First, the 3'-end strategy is based on the naturally occurring polyA tail; thus, the same adapter primers can be used for both ends, decreasing variability in the protocol and cost. Second, since A:T binding is weaker than G:C binding, longer stretches of A residues (~2×) are required before the oligo(dT)-tailed Q_T primer will bind to an internal site and truncate the amplification product. Third, vertebrate coding sequences and 5'-untranslated regions tend to be biased toward GC residues; thus, use of a polyA tail further decreases the likelihood of inappropriate truncation.

Unlike many other situations in which homopolymeric tails are appended, the actual length of the tail added here is unimportant, provided it exceeds 17 nt. This is because although the oligo(dT)-tailed primers subsequently bind all along the length of the appended polyA tail, only the innermost one becomes incorporated into the amplification product, and consequently, the remainder of the polyA tail is lost *(1)*. The truncation appears to happen because *Taq* polymerase is unable to resolve branched structures (efficiently). The conditions described in the procedure result in the addition of 30–400 nt.

6. Most protocols call for much more TAP than recommended here. The enzyme is very expensive, and it is not necessary!

7. The following describes the analysis of the quality of the RACE PCR products. The production of specific partial cDNAs by the RACE protocol is assessed using Southern blot hybridization analysis. After the second set of amplification cycles, the first- and second-set reaction products are electrophoresed in a 1% agarose gel, stained with ethidium bromide, denatured, and transferred to a nylon membrane. After hybridization with a labeled oligomer or gene fragment derived from a region contained within the amplified fragment (e.g., GSP-Hyb/Seq in Fig. 2A,B), gene-specific partial cDNA ends should be detected easily. Yields of the desired product relative to nonspecific amplified cDNA in the first-round products should vary from <1% of the amplified material to nearly 100%, depending largely on the stringency of the amplification reaction, the amplification efficiency of the specific cDNA end, and the relative abundance of the specific transcript within the mRNA source. In the second set of amplification cycles, ~100% of the cDNA detected by ethidium bromide staining should represent specific product. If specific hybridization is not observed, then troubleshooting steps should be initiated.

Information gained from this analysis should be used to optimize the RACE procedure. If low yields of specific product are observed because nonspecific products are being amplified efficiently, then annealing temperatures can be raised gradually (~2°C at a time) and sequentially in each stage of the procedure until nonspecific products are no longer observed. Alternatively, some investigators have reported success using the "touchdown PCR" procedure to optimize the annealing temperature without trial and error *(31)*. Optimizing the annealing temperature is also indicated if multiple species of specific products are observed, which could indicate that truncation of specific products is occurring. If multiple species of specific products are observed after the reverse transcription and amplification reactions have been fully optimized, then the possibility should be entertained that alternate splicing or promoter use is occurring.

For classic RACE only: If a nearly continuous smear of specific products is observed up to a specific size limit after 5'-end amplification, this suggests that polymerase pausing occurred during the reverse transcription step. To obtain nearly full-length cDNA ends, the amplification mixture should be electrophoresed, and the longest products recovered by gel isolation. An aliquot of this material can then be reamplified for a limited number of cycles.

For new RACE only: Expect to see one or two extra nucleotides inserted between the RNA oligo 3'-end, and the 5'-end of the gene of interest—these come from the transcription step using T7, T3, or SP6, which can add an extra nucleotide or two to oligo past the end of the template (template-independent transcription).

Compare the results you obtain from unTAPped RNA vs TAPped RNA. Junction sites (where the oligo is connected to the 5'-end of your gene) in common arise from ligation of the oligo to degraded RNA; unique junctions in the TAPped RNA population represent candidate transcription start sites. If you have "RNA degradation" sites (e.g., TTTᵗAAA) in your 5' RNA end, you may have substantial numbers of clones that begin at exactly the same nucleotide, but that arise from ligation of the oligo to degraded RNA molecules, not from ligation of the oligo to the true 5'-end of the RNA. Look for TATA, CCAAT, and initiator element (Inr) sites at or around your candidate transcription site in the genomic DNA sequence if it is available—you should usually be able to find either a TATA or an Inr.

8. RACE PCR products can be further manipulated as described:
   a. Cloning: RACE products can be cloned like any other PCR products.
      i. Option 1: To clone the cDNA ends directly from the amplification reaction (or after gel purification, which is recommended), ligate an aliquot of the products to plasmid vector encoding a one-nucleotide 3'-overhang consisting of a "T" on both strands. Such vector DNA is available commercially (Invitrogen's "TA Kit") or can be easily and cheaply prepared *(16,32–35)*.
      ii. Option 2: The classic RACE Q_I primer encodes *Hin*dIII, *Sst*I, and *Xho*I restriction enzyme sites. Products can be efficiently cloned into vectors that have been double-cut with one of these enzymes and with a blunt-cutting enzyme, such as *Sma*I. (Note—remember to "polish" the amplification products with Klenow enzyme or T4 DNA polymerase, and separate them from residual *Taq* Polymerase and dNTPs before carrying out the restriction enzyme digest.) If clones are not obtained, determine whether the restriction enzyme chosen is cutting the amplified gene fragment a second time, at some internal location in the new and unknown sequence. A somewhat easier strategy is to append a restriction site (not *Hin*dIII, *Sst*I, or *Xho*I) onto the 5'-end of the GSP2 primer to allow for the creation of overhanging strands at both ends of the amplified product.
      iii. Option 3: A safer and very effective approach is to modify the ends of the primers to allow the creation of overhanging ends using T4 DNA polymerase to chew back a few nucleotides from the amplified product in a controlled manner and Klenow enzyme (or Sequenase) to fill in partially restriction-enzyme-digested overhanging ends on the vector, as shown in Fig. 6 (adapted from *36,37*) *(see* Section 3.4.). For another conceptual variation, *see* Rashtchian et al. *(10)*.

   The advantages of this approach are that it eliminates the possibility that the restriction enzymes chosen for the cloning step will cleave the cDNA end in the unknown region, vector dephosphorylation is not required since vector self-ligation is no longer possible, and insert kinasing (and polishing) is not necessary and insert multimerization and fusion clones are not observed either. In addition, the procedure is more reliable than "TA" cloning.

b. Sequencing: RACE products can be sequenced directly on a population level using a variety of protocols, including cycle sequencing, from the end at which the gene-specific primers are located. Note that the classic RACE products cannot be sequenced on a population level using the Q_I primer at the unknown end, since individual cDNAs contain different numbers of A residues in their polyA tails, and as a consequence, the sequencing ladder falls out of register after reading through the tail. 3'-end products can be sequenced from their unknown end using the following set of primers (TTTTTTTTTTTTTTTTTTTA, TTTTTTTTTTTTTTTTTTTG, TTTTTTTTTTTTTTTTTTTC). The non-T nucleotide at the 3'-end of the primer forces the appropriate primer to bind to the inner end of the polyA tail *(38)*. The other two primers do not participate in the sequencing reaction. Individual cDNA ends, once cloned into a plasmid vector, can be sequenced from either end using gene-specific or vector primers.

c. Hybridization probes: RACE products are generally pure enough that they can be used as probes for RNA and DNA blot analyses. It should be kept in mind that small amounts of contaminating nonspecific cDNAs will always be present. It is also possible to include a T7 RNA polymerase promoter in one or both primer sequences and to use the RACE products with in vitro transcription reactions to produce RNA probes *(5)*. Primers encoding the T7 RNA polymerase promoter sequence do not appear to function as amplification primers as efficiently as the others listed in the legend to Fig. 2 (personal observation). Thus, the T7 RNA polymerase promoter sequence should not be incorporated into RACE primers as a general rule.

d. Construction of full-length cDNAs: It is possible to use the RACE protocol to create overlapping 5- and 3- cDNA ends that can later, through judicious choice of restriction enzyme sites, be joined together through subcloning to form a full-length cDNA. It is also possible to use the sequence information gained from acquisition of the 5' and 3' cDNA ends to make new primers representing the extreme 5'- and 3'-ends of the cDNA, and to employ them to amplify a *de novo* copy of a full-length cDNA directly from the "3'-end cDNA pool." Despite the added expense of making two more primers, there are several reasons why the second approach is preferred.

First, a relatively high error rate is associated with the PCR conditions for which efficient RACE amplification takes place, and numerous clones may have to be sequenced to identify one without mutations. In contrast, two specific primers from the extreme ends of the cDNA can be used under inefficient, but low-error-rate conditions *(39)* for a minimum of cycles to amplify a new cDNA that is likely to be free of mutations. Second, convenient restriction sites are often not available, thus making the subcloning project difficult. Third, by using the second approach, the synthetic polyA tail (if present) can be removed from the 5'-end of the cDNA. Homopolymer tails appended to the 5'-ends of cDNAs have in some cases been reported to inhibit translation. Finally, if alternate promoters, splicing, and polyadenylation signal sequences are being used and result in multiple 5'- and 3'-ends, it is possible that one might join two cDNA halves that are never actually found together in vivo. Employing primers from the extreme ends of the cDNA as described confirms that the resulting amplified cDNA represents an mRNA actually present in the starting population.

9. The following are some ideas regarding potential problems with the reverse transcription and prior steps.

   a. Damaged RNA: Electrophorese RNA in 1% formaldehyde minigel, and examine integrity of the 18S and 28S ribosomal bands. Discard the RNA preparation if ribosomal bands are not sharp.

   b. Contaminants: Ensure that the RNA preparation is free of agents that inhibit reverse transcription, e.g., lithium chloride and sodium dodecyl sulfate *(40)*, regarding the optimization of reverse transcription reactions.

   c. Bad reagents: To monitor reverse transcription of the RNA, add 20 µCi of ^{32}P-dCTP to the reaction, separate newly created cDNAs using gel electrophoresis, wrap the gel in Saran wrap™, and expose it to X-ray film. Accurate estimates of cDNA size can best be determined using alkaline agarose gels, but a simple 1% agarose minigel will suffice to confirm that reverse transcription took place and that cDNAs of reasonable length were generated. Note that adding ^{32}P-dCTP to the reverse transcription reaction results in the detection of cDNAs synthesized both through the specific priming of mRNA and through RNA self-priming. When a gene-specific primer is used to prime transcription (5'-end RACE) or when total RNA is used as a template, the majority of the labeled cDNA will actually have been generated from RNA self-priming. To monitor extension of the primer used for reverse transcription, label the primer using T4 DNA kinase and ^{32}P-γATP prior to reverse transcription. Much longer exposure times will be required to detect the labeled primer-extension products than when ^{32}P-dCTP is added to the reaction.

   To monitor reverse transcription of the gene of interest, one may attempt to amplify an internal fragment of the gene containing a region derived from two or more exons, if sufficient sequence information is available.

10. The following are some ideas regarding potential tailing problems.

    a. Bad reagents: Tail 100 ng of a DNA fragment of approx 100–300 bp in length for 30 min. In addition, mock tail the same fragment (add everything but the TdT). Run both samples in a 1% agarose minigel. The mock-tailed fragment should run as a tight band. The tailed fragment should have increased in size by 20–200 bp and should appear to run as a diffuse band that trails off into higher-mol-wt products. If this not observed, replace reagents.

    b. Mock tail 25% of the cDNA pool (add everything but the TdT). Dilute to the same final concentration as the tailed cDNA pool. This serves two purposes. First, although amplification products will be observed using both tailed and untailed cDNA templates, the actual pattern of bands observed should be different. In general, discrete bands are observed using untailed templates after the first set of cycles, and a broad smear of amplified cDNA accompanied by some individual bands is typically observed using tailed templates. If the two samples appear different, this confirms that tailing took place and that the oligo(dT)-tailed Q_T primer is annealing effectively to the tailed cDNA during PCR. Second, observing specific products in the tailed amplification mixture that are not present in the untailed amplification mixture indicates that these products are being synthesized off the end of an A-tailed cDNA template, rather than by annealing of the dT-tailed primer to an A-rich sequence in or near the gene of interest.

11. Some potential amplification problems:

   a. No product: If no products are observed for the first set of amplifications after 30 cycles, add fresh *Taq* polymerase and carry out an additional 15 rounds of amplification (extra enzyme is not necessary if the entire set of 45 cycles is carried out without interruption at cycle 30). Product is always observed after a total of 45 cycles if efficient amplification is taking place. If no product is observed, carry out a PCR reaction using control templates and primers to ensure the integrity of the reagents.

   b. Smeared product from the bottom of the gel to the loading well: too many cycles or too much starting material.

   c. Nonspecific amplification, but no specific amplification: Check sequence of cDNA and primers. If all are correct, examine primers (using computer program) for secondary structure and self-annealing problems. Consider ordering new primers. Determine whether too much template is being added, or if the choice of annealing temperatures could be improved.

   Alternatively, secondary structure in the template may be blocking amplification. Consider adding formamide *(41)* or [7]aza-GTP (in a 1:3 ratio with dGTP) to the reaction to assist polymerization. [7]aza-GTP can also be added to the reverse transcription reaction.

   d. The last few base pairs of the 5'-end sequence do not match the corresponding genomic sequence: Be aware that reverse transcriptase and T7 and T3 RNA polymerase can add on a few extra template-independent nucleotides.

   e. Inappropriate templates: To determine whether the amplification products observed are being generated from cDNA or whether they derive from residual genomic DNA or contaminating plasmids, pretreat an aliquot of the RNA with RNase A.

Acknowledgments

Portions of this chapter have been adapted and reprinted by permission of the publisher from "Cloning PCR products" by Michael A. Frohman in *The Polymerase Chain Reaction*, pp. 14–37, copyright ©1994 by Birkhauser, Boston, and from "On beyond RACE (rapid amplication of cDNA ends)" by Michael A. Frohman in *PCR Methods and Applications*, copyright ©1995 by Cold Spring Harbor Laboratory Press, Cold Spring Harbor, NY.

References

1. Frohman, M. A., Dush, M. K., and Martin, G. R. (1988) Rapid production of full-length cDNAs from rare transcripts by amplification using a single gene-specific oligonucleotide primer. *Proc. Natl. Acad. Sci. USA* **85**, 8998–9002.
2. Loh, E. L., Elliott, J. F., Cwirla, S., Lanier, L. L., and Davis, M. M. (1989) Polymerase chain reaction with single sided specificity: analysis of T cell receptor delta chain. *Science* **243**, 217–220.
3. Ohara, O., Dorit, R. I., and Gilbert, W. (1989) One-sided PCR: The amplification of cDNA. *Proc. Natl. Acad. Sci. USA* **86**, 5673–5677.
4. Frohman, M. A. (1989) Creating full-length cDNAs from small fragments of genes: amplification of rare transcripts using a single gene-specific oligonucleotide primer, in *PCR Protocols and Applications: A Laboratory Manual* (Innis, M., Gelfand, D., Sninsky, J., and White, T., eds.), Academic, San Diego, CA, pp. 28–38.
5. Frohman, M. A. and Martin, G. R. (1989) Rapid amplification of cDNA ends using nested primers. *Techniques* **1**, 165–173.
6. Dumas, J. B., Edwards, M., Delort, J., and Mallet, J. (1991) Oligodeoxyribonucleotide ligation to single-stranded cDNAs: a new tool for cloning 5' ends of mRNAs and for constructing cDNA libraries by *in vitro* amplification. *Nucleic Acids Res.* **19**, 5227–5233.
7. Fritz, J. D., Greaser, M. L., and Wolff, J. A. (1991) A novel 3' extension technique using random primers in RNA-PCR. *Nucleic Acids Res.* **119**, 3747.
8. Borson, N. D., Salo, W. L., and Drewes, L. R. (1992) A lock-docking oligo(dT) primer for 5' and 3' RACE PCR. *PCR Methods Appl.* **2**, 144–148.
9. Jain, R., Gomer, R. H., and Murtagh, J. J., Jr. (1992) Increasing specificity from the PCR-RACE technique. *BioTechniques* **12**, 58–59.
10. Rashtchian, A., Buchman, G. W., Schuster, D. M., and Berninger, M. S. (1992) Uracil DNA glycosylase-mediated cloning of PCR-amplified DNA: application to genomic and cDNA cloning. *Anal. Biochem.* **206**, 91–97.
11. Schuster, D. M., Buchman, G. W., and Rastchian, A. (1992) A simple and efficient method for amplification of cDNA ends using 5' RACE. *Focus* **14**, 46–52.
12. Bertling, W. M., Beier, F., and Reichenberger, E. (1993) Determination of 5' ends of specific mRNAs by DNA ligase-dependent amplification. *PCR Methods Appl.* **3**, 95–99.
13. Frohman, M. A. (1993) Rapid amplification of cDNA for generation of full-length cDNA ends: thermal RACE. *Methods Enzymol.* **218**, 340–356.
14. Monstein, H. J., Thorup, J. U., Folkesson, R., Johnsen, A. H., and Rehfeld, J. F. (1993) cDNA deduced procionin-structure and expression in protochordates resemble that of procholecystokinin in mammals. *FEBS Lett.* **331**, 60–64.
15. Templeton, N. S., Urcelay, E., and Safer, B. (1993) Reducing artifact and increasing the yield of specific DNA target fragments during PCR-RACE or anchor PCR. *BioTechniques* **15**, 48–50.
16. Frohman, M. A. (1994) Cloning PCR products: strategies and tactics, in *Methods in Molecular Biology: PCR: The Polymerase Chain Reaction* (Mullis, K. B., Ferre, F., and Gibbs, R. A., eds.), Birkhäuser, Boston, MA, pp. 14–37.

17. Datson, N. A., Duyk, G. M., Van Ommen, J. B., and Den Dunnen, J. T. (1994) Specific isolation of 3'-terminal exons of human genes by exon trapping. *Nucleic Acids Res.* **22,** 4148–4153.
18. Ruberti, F., Cattaneo, A., and Bradbury, A. (1994) The use of the RACE method to clone hybridoma cDNA when V region primers fail. *J. Immunol. Methods* **173,** 33–39.
19. Tessier, D. C., Brousseau, R., and Vernet, T. (1986) Ligation of single-stranded oligodeoxyribonucleotides by T4 RNA ligase. *Anal. Biochem.* **158,** 171–178.
20. Mandl, C. W., Heinz, F. X., Puchhammer-Stockl, E., and Kunz, C. (1991) Sequencing the termini of capped viral RNA by 5'-3' ligation and PCR. *BioTechniques* **10,** 484–486.
21. Volloch, V., Schweizer, B., Zhang, X., and Rits, S. (1991) Identification of negative-strand complements to cytochrome oxidase subunit III RNA in Trypanosoma brucei. *Biochemistry* **88,** 10,671–10,675.
22. Brock, K. V., Deng, R., and Riblet, S. M. (1992) Nucleotide sequencing of 5' and 3' termini of bovine viral diarrhea virus by RNA ligation and PCR. *Virol. Methods* **38,** 39–46.
23. Bertrand, E., Fromont-Racine, M., Pictet, R., and Grange, T. (1993) Visualization of the interaction of a regulatory protein with RNA *in vivo*. *Proc. Natl. Acad. Sci. USA* **90,** 3496–3500.
24. Fromont-Racine, M., Bertrand, E., Pictet, R., and Grange, T. (1993) A highly sensitive method for mapping the 5' termini of mRNAs. *Nucleic Acids Res.* **21,** 1683,1684.
25. Liu, X. and Gorovsky, M. A. (1993) Mapping the 5' and 3' ends of tetrahymena-thermophila mRNAs using RNA Ligase mediated amplification of cDNA ends (RLM-RACE). *Nucleic Acids Res.* **21,** 4954–4660.
26. Sallie, R. (1993) Characterization of the extreme 5' ends of RNA molecules by RNA ligation-PCR. *PCR Methods Applic.* **3,** 54–56.
27. Skinner, T. L., Kerns, R. T., and Bender, P. K. (1994) Three different calmodulin-encoding cDNAs isolated by a modified 5'-RACE using degenerate oligo-deoxyribonucleotides. *Gene* **151,** 247–251.
28. Frohman, M. A., Dickinson, M. E., Hogan, B. L. M., and Martin, G. R. (1993) Localization of two new and related homeobox-containing genes to chromosomes 1 and 5, near the phenotypically similar mutant loci *dominant hemimelia* (*Dh*) and *hemimelic extra-toes* (*Hx*). *Mouse Genome* **91,** 323–325.
29. Crowe, J. S., Cooper, H. J., Smith, M. A., Sims, M. J., Parker, D., and Gewert, D. (1991) Improved cloning efficiency of polymerase chain reaction (PCR) products after proteinase K digestion. *Nucleic Acids Res.* **19,** 184.
30. Coleclough, C. (1987) Use of primer-restriction end adapters in cDNA cloning. *Methods Enzymol.* **154,** 64–83.
31. Don, R. H., Cox, P. T., Wainwright, B. J., Baker, K., and Mattick, J. S. (1991) Touchdown PCR to circumvent spurious priming during gene amplification. *Nucleic Acids Res.* **19,** 4008.
32. Mead, D. A., Pey, N. K., Herrnstadt, C., Marcil, R. A., and Smith, L. A. (1991) A universal method for direct cloning of PCR amplified nucleic acid. *Biotechnology* **9,** 657–663.

33. Marchuk, D., Drumm, M., Saulino, A., and Collins, F. S. (1991) Construction of T-vector, a rapid and general system for direct cloning of unmodified PCR products. *Nucleic Acids Res.* **19**, 1154.
34. Kovalic, D., Kwak, J. H., and Weisblum, B. (1991) General method for direct cloning of DNA fragments generated by the polymerase chain reaction. *Nucleic Acids Res.* **19**, 4650.
35. Holton, T. A. and Graham, M. W. (1991) A simple and efficient method for direct cloning of PCR products using ddT-tailed vectors. *Nucleic Acids Res.* **19**, 1156.
36. Stoker, A. W. (1990) Cloning of PCR products after defined cohesive termini are created with T4 DNA polymerase. *Nucleic Acids Res.* **18**, 4290.
37. Iwahana, H., Mizusawa, N., Ii, S., Yoshimoto, K., and Itakura, M. (1994) An end-trimming method to amplify adjacent cDNA fragments by PCR. *BioTechniques* **16**, 94–98.
38. Thweatt, R., Goldstein, S., and Reis, R. J. S. (1990) A universal primer mixture for sequence determination at the 3' ends of cDNAs. *Anal. Biochem.* **190**, 314.
39. Eckert, K. A. and Kunkel, T. A. (1990) High fidelity DNA synthesis by the *Thermus aquaticus* DNA polymerase. *Nucleic Acids Res.* **18**, 3739–3745.
40. Sambrook, J., Fritsch, E. F., and Maniatis, T. (1989) *Molecular Cloning: A Laboratory Manual.* Cold Spring Harbor Laboratory Press, Cold Spring Harbor, NY.
41. Sarker, G., Kapelner, S., and Sommer, S. S. (1990) Formamide can dramatically improve the specificity of PCR. *Nucleic Acids Res.* **18**, 7465.

7

Inverse PCR Approach to Cloning cDNA Ends

Sheng-He Huang

1. Introduction

Since the first report on cDNA cloning in 1972 *(1)*, this technology has been developed into a powerful and universal tool in the isolation, characterization, and analysis of both eukaryotic and prokaryotic genes. But the conventional methods of cDNA cloning require much effort to generate a library that is packaged in phage or plasmid and then survey a large number of recombinant phages or plasmids. There are three major limitations in those methods. First, substantial amount (at least 1 µg) of purified mRNA is needed as starting material to generate libraries of sufficient diversity *(2)*. Second, the intrinsic difficulty of multiple sequential enzymatic reactions required for cDNA cloning often leads to low yields and truncated clones *(3)*. Finally, screening of a library with hybridization technique is time consuming.

Polymerase chain reaction (PCR) technology can simplify and improve cDNA cloning. Using PCR with two gene-specific primers, a piece of known sequence cDNA can be specifically and efficiently amplified and isolated from very small numbers (<10^4) of cells *(4)*. However, it is often difficult to isolate full-length cDNA copies of mRNA on the basis of very limited sequence information. The unknown sequence flanking a small stretch of the known sequence of DNA cannot be amplified by the conventional PCR. Recently, anchored PCR *(5–7)* and inverse PCR (IPCR) *(8–10)* have been developed to resolve this problem. Anchored PCR techniques have a common point: DNA cloning goes from a small stretch of known DNA sequence to the flanking unknown sequence region with the aid of a gene-specific primer at one end and a universal primer at other end. Because of only one gene-specific primer in the anchored PCR, it is easier to get a high level of nonspecific amplification by PCR than with two gene-specific primers *(10,11)*. The major advantage of

From: *Methods in Molecular Biology, Vol. 69: cDNA Library Protocols*
Edited by: I. G. Cowell and C. A. Austin Humana Press Inc., Totowa, NJ

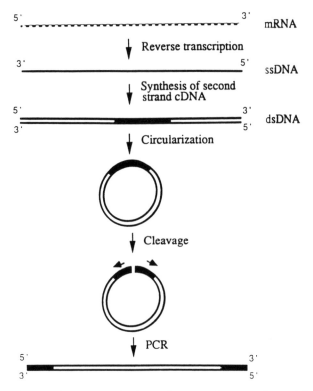

Fig. 1. Diagram of IPCR for cDNA cloning. The procedure consists of five steps: reverse transcription, synthesis of second-stranded cDNA, circularization of double-stranded cDNA, reopen the circle DNA, and amplification of reverse DNA fragment. The black and open bars represent the known and unknown sequence regions of double-stranded cDNA, respectively.

IPCR is to amplify the flanking unknown sequence by using two gene-specific primers.

At first, IPCR was successfully used in the amplification of genomic DNA segments that lie outside the boundaries of known sequence *(8,9)*. There is a new procedure that extends this technique to the cloning of unknown cDNA sequence from total RNA *(10)*. Double-stranded cDNA is synthesized from RNA and ligated end to end (Fig. 1). Circularized cDNA is nicked by selected restriction enzyme or denatured by NaOH treatment *(12,13)*. The reopened or denatured circular cDNA is then amplified by two gene-specific primers. Recently, this technique has been efficiently used in cloning full-length cDNAs *(14–16)*. The following protocol was used to amplify cDNA ends for the human stress-related protein ERp72 *(10)* (Fig. 2).

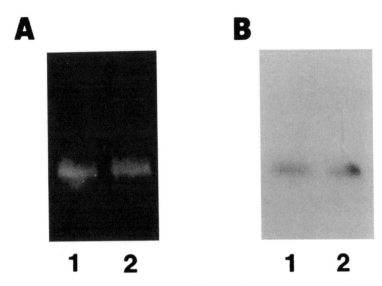

Fig. 2. Application of IPCR to amplifying the joining region (280 bp) from 5' (160 bp) and 3' (120 bp) sequences of human ERp72 cDNA. Amplified DNAs from CCRF/CEM cells sensitive (lane 1) and resistant (lane 2) to cytosine arabinoside stained by ethidium bromide (**A**) or hybridized with ^{32}P-labeled ERp72 cDNA (**B**). *See* text for the sequences of the primers and the parameters of IPCR.

2. Materials

2.1. First-Strand cDNA Synthesis

1. Total RNA prepared from human CCRF/CEM leukemic lymphoblast cells *(17,18)*.
2. dNTP Mix (10 mM of each dNTP).
3. Random primers (Boehringer Mannheim, Indianapolis, IN). Prepare in sterile water at 1 µg/µL. Store at –20°C.
4. RNasin (Promega, Madison, WI).
5. Actinomycin D (1 mg/mL): Actinomycin D is light sensitive and toxic. It should be stored in a foil-wrapped tube at –20°C.
6. Moloney Murine Leukemia Virus (MMLV) reverse transcriptase.
7. 5X First-strand buffer: 250 mM Tris-HCl, pH 8.3, 375 mM KCl, 50 mM MgCl$_2$, 50 mM dithiothreitol (DTT), and 2.5 mM spermidine. The solution is stable at –20°C for more than 6 mo.

2.2. Second-Strand Synthesis

1. 10X Second-strand buffer: 400 mM Tris-HCl, pH 7.6, 750 mM KCl, 30 mM MgCl$_2$, 100 mM (NH$_4$)$_2$SO$_4$, 30 mM DTT, and 0.5 mg/mL of bovine serum albumin (BSA). The solution is stable at –20°C for at least 6 mo.

2. 1 m*M* NAD.
3. 2 U/μL RNase H.
4. 5 U/μL *Escherichia coli* DNA polymerase I.
5. 1 U/μL *E. coli* DNA ligase.
6. Nuclease-free water.
7. T4 DNA polymerase.
8. 200 m*M* EDTA, pH 8.0.
9. GeneClean (Bio 101, La Jolla, CA).
10. TE buffer: 10 m*M* Tris-HCl, pH 7.6, 1 m*M* EDTA. Sterile filter.
11. DNA standards: Prepare 1-mL aliquots of a purified DNA sample at 1, 2.5, 5, 10, and 20 μg/mL in TE buffer. Store at –20°C for up to 6 mo.
12. TE/ethidium bromide: 2 μg/mL of ethidium bromide in TE buffer. Store at 4°C for up to 6 mo in a dark container.

2.3. Circularization and Cleavage or Denaturation

1. 5X Ligation buffer (supplied with T4 DNA ligase).
2. 1 U/μL T4 DNA ligase.
3. 4 μg/μL T4 RNA ligase.
4. 15 μ*M* Hexaminecobalt chloride.
5. Phenol:CHCl$_3$:isoamyl alcohol (25:24:1).
6. 3*M* sodium acetate, pH 7.0.
7. Absolute ethanol.
8. 70% Ethanol.

2.4. IPCR

1. 10X PCR buffer: 100 m*M* Tris-HCl, pH 8.3, 500 m*M* KCl, 15 m*M* MgCl$_2$, 0.01% (w/v) gelatin.
2. 15 m*M* MgCl$_2$.
3. Deoxyoligonucleotides were synthesized on an Applied Biosystems (Foster City, CA) 380B DNA synthesizer and purified by OPEC column from the same company. The primer pairs were selected from the 5' and 3' sequence of the cDNA coding for human ERp72 stress-related protein (5'-primer: 5'-TTC CTCCTCCTCCTCCTCTT-3'; 3'-primer: 5'-ATCTAAATGTCTAGT-3') *(10)*.
4. Light mineral oil.
5. *Taq* DNA polymerase.

3. Methods
3.1. First-Strand cDNA Synthesis (19)

Perform reverse transcription in a 25-μL reaction mixture, adding the following components: 5.0 μL 5X first-strand buffer, 2.5 μL dNTP mix, 2.5 μL random primers, 1.0 U RNasin, 1.25 μL actinomycin D, 250 U MMLV reverse transcriptase, 15–25 μg of total RNA (heat denature RNA at 65°C for 3 min prior to adding to reaction), and nuclease-free water to 25 μL final volume.

3.2. Second-Strand Synthesis (20)

1. Add components to the first-strand tube on ice in the following order: 12.5 μL 10X second strand buffer, 12.5 μL NAD (1 mM), 0.5 μL RNase H (2 μ/μL), 5.75 μL *E. coli* DNA polymerase I (5 μ/μL), 1.25 μL *E. coli* ligase (1μ/μL), and 67.5 μL nuclease-free water.
2. Incubate at 14°C for 2 h.
3. Heat the reaction mix to 70°C for 10 min, spin for a few seconds, and then put in ice.
4. Add 4 U of T4 DNA polymerase and incubate at 37°C for 10 min to blunt the ends of double-stranded cDNA.
5. Stop the reaction with 12.5 μL of 0.2M EDTA and 200 μL sterile water.
6. Concentrate and purify the sample with GeneClean. Resuspend the DNA in 100–200 μL of sterile water.
7. Estimate the DNA concentration by comparing the ethidium bromide fluorescent intensity of the sample with that of a series of DNA standards on a sheet of plastic wrap *(21)*. Dot 1–5 μL of sample onto the plastic wrap on a UV transilluminator. Also dot with 5 μL of DNA standards. Add an equal volume of TE buffer containing 2 μg/mL of ethidium bromide, and mix by repipetting up and down. Use proper UV shielding for exposed skin and eyes.

3.3. Circularization and Cleavage (see Notes 1–4)

1. Set up the circularization reaction mix containing the following components: 100 μL (100 ng DNA) of the purified sample, 25 μL of 5X ligation buffer, and 6 μL of T4 DNA ligase. Finally, add 2 μL of T4 RNA ligase or 15 μL of 15 μM hexaminecobalt chloride (*see* Note 5).
2. Incubate at 18°C for 16 h.
3. Boil the ligated circular DNA for 2–3 min in distilled water or digest with an appropriate restriction enzyme to reopen circularized DNA.
4. Purify the DNA sample with GeneClean as described in step 6 in Section 3.2. or extract with water-saturated phenol/CHCl$_3$ and then precipitate with ethanol *(20)*.

3.4. IPCR (see Note 6)

1. Add $^1/_{10}$ of the purified cDNA to 100 μL of amplification mix (22): 10 μL 10X PCR buffer, 10 μL 15 mM MgCl$_2$, 10 μL dNTP mix (2.5 mM of each), 10 μL 5'-primer (10 pmol/μL), 10 μL 3'-primer (10 pmol/μL), 10 μL cDNA, 39.5 μL nuclease-free water, and 0.5 μL *Taq* DNA polymerase (2.5 μ/μL).
2. Cap and vortex the tubes to mix. Spin briefly in a microfuge. Cover each reaction with a few drops of light mineral oil to prevent evaporation.
3. Put a drop of mineral oil into each well of the thermal cycler block that will hold a tube. Load the reaction tubes.
4. Amplify by PCR using the following cycle profile: 25 cycles: 94°C for 1 min (denaturation), 65°C for 2 min (annealing), and 72°C for 4 min (elongation).

4. Notes

1. For maximum efficiency of intramolecular ligation, a low concentration of cDNA should be used in the ligation mix. High density of cDNA may enhance the level of heterogeneous ligation, which creates nonspecific amplification.
2. Cleavage or denaturation of circularized double-stranded cDNA is important since circular double-stranded DNA tends to form supercoil and is poor template for PCR *(23)*. Circularized double-stranded DNA is only good for amplification of a short DNA fragment.
3. The following three ways can be considered to introduce nicks in circularized DNA. Boiling is a simple and common way. However, because of the unusual secondary structure of some circular double-stranded DNA, sometimes this method is not sufficient in nicking and denaturing circular double-stranded DNA. A second method is selected restriction enzyme digestion. The ideal restriction site is located in the known sequence region of cDNA. In most cases, it is difficult to make the right choice of a restriction enzyme because the restriction pattern in unidentified region of cDNA is unknown. If an appropriate enzyme is not available, EDTA-oligonucleotide-directed specific cleavage may be tried *(24,25)*. Oligonucleotide linked to EDTA-Fe at T can bind specifically to double-stranded DNA by triple-helix formation and produce double-stranded cleavage at the binding site.
4. Alkali denaturation has been successfully used to prepare plasmid DNA templates for PCR and DNA sequencing *(12,13,26)*. This method should be feasible in denaturing circularized double-stranded cDNA.
5. Inclusion of T4 RNA ligase or hexaminecobalt chloride can enhance the efficiency of blunt-end ligation of double-stranded DNA catalyzed by T4 DNA ligase *(27)*.
6. IPCR can be used to efficiently and rapidly amplify regions of unknown sequence flanking any identified segment of cDNA or genomic DNA. This technique does not need construction and screening of DNA libraries to obtain additional unidentified DNA sequence information. Some recombinant phage or plasmid may be unstable in bacteria and amplified libraries tend to lose them *(23)*. IPCR eliminates this problem.

Acknowledgments

This chapter is an updated version of a chapter previouly published in the *Methods in Molecular Biology* series. I acknowledge the contributions of John Holcenberg, Chun-Ha Wu, and Bing Cai from the earlier version. I acknowledge John Holcenberg for his invaluable comments and generous support. I especially thank Chun-Hua Wu and Bing Cai for their technical assistance and Kai-Jin Wu for her art work.

References

1. Verma, I. M., Temple, G. F., Fan, H., and Baltimore, D. (1972) In vitro synthesis of double-stranded DNA complimentary to rabbit reticulocyte 10S RNA. *Nature* **235,** 163–169.

2. Akowitz, A. and Mamuelidis, L. (1989) A novel cDNA/PCR strategy for efficient cloning of small amounts of undefined RNA. *Gene* **81,** 295–306.
3. Okayama, H., Kawaichi, M., Brownstein, M., Lee, F., Yokota, T., and Arai, K. (1987) High-efficiency cloning of full-length cDNA: construction and screening of cDNA expression libraries for mammalian cells. *Methods Enzymol.* **154,** 3–28.
4. Brenner, C. A., Tam, A. W., Nelson, P. A., Engleman, E. G., Suzuki, N., Fry, K. E., and Larrick, J. W. (1989) Message amplification phenotyping (MAPPing): a technique to simultaneously measure multiple mRNAs from small numbers of cells. *BioTechniques* **7,** 1096–1103.
5. Frohman, M. A. (1990) RACE: rapid amplification of cDNA ends, in *PCR Protocols: A Guide to Methods and Applications* (Innis, M. A., Gelfand, D. H., Sninsky, J. J., and White, T. J., eds.), Academic, San Diego, CA, pp. 28–38.
6. Shyamala, V. and Ames, G. F.-L. (1989) Genome walking by single-specific-primer polymerase chain reaction: SSP-PCR. *Gene* **84,** 1–8.
7. Huang, S.-H., Jong, A. Y., Yang, W., and Holcenberg, J. (1993) Amplification of gene ends from gene libraries by PCR with single-sided specificity. *Methods Mol. Biol.* **15,** 357–363.
8. Ochman, H., Gerber, A. S., and Hartl, D. L. (1988) Genetic applications of an inverse polymerase chain reaction. *Genetics* **120,** 621–625.
9. Triglia, T., Peterson, M. G., and Kemp, D. J. (1988) A procedure for in vitro amplification of DNA segments that lie outside the boundaries of known sequences. *Nucleic Acids Res.* **16,** 8186.
10. Huang, S.-H., Hu, Y. Y., Wu, C.-H., and Holcenberg, J. (1990) A simple method for direct cloning cDNA sequence that flanks a region of known sequence from total RNA by applying the inverse polymerase chain reaction. *Nucleic Acids Res.* **18,** 1922.
11. Delort, J., Dumas, J. B., Darmon, M. C., and Mallet, J. (1989) An efficient strategy for cloning 5' extremities of rare transcripts permits isolation of multiple 5'-untranslated regions of rat tryptophan hydroxylase mRNA. *Nucleic Acids Res.* **17,** 6439–6448.
12. Cusi, M. G., Cioe', L., and Rovera, G. (1992) PCR amplification of GC-rich templates containing palindromic sequences using initial alkali denaturation. *BioTechniques* **12,** 502–504.
13. Lau, E. C., Li, Z.-Q., and Slavkin, S. C. (1993) Preparation of denatured plasmid templates for PCR amplification. *BioTechniques* **14,** 378.
14. Green, I. R. and Sargan, D. R. (1991) Sequence of the cDNA encoding bovine tumor necrosis factor-α: problems with cloning by inverse PCR. *Gene* **109,** 203–210.
15. Zilberberg, N. and Gurevitz, M. (1993) Rapid Isolation of full length cDNA clones by "inverse PCR": purification of a scorpion cDNA family encoding α-neurotoxins. *Anal. Biochem.* **209,** 203–205.
16. Austin, C. A., Sng, J.-H., Patel, S., and Fisher, L. M. (1993) Novel HeLa topoisomerase II is the IIβ isoform: complete coding sequence and homology with other type II topoisomerases. *Biochim. Biophys. Acta* **1172,** 283–291.

17. Delidow, B. C., Lynch, J. P., Peluso, J. J., and White, B. A. (1993) Polymerase chain reaction: basic protocols. *Methods Mol. Biol.* **15,** 1–29.

18. Davis, L. G., Dibner, M. D., and Battey, J. F. (1986) *Basic Methods in Molecular Biology,* Elsevier, New York.

19. Krug, M. S. and Berger, S. L. (1987) First strand cDNA synthesis primed by oligo(dT). *Methods Enzymol.* **152,** 316–325.

20. Promega (1991) *Protocols and Applications,* 2nd ed., Madison, WI, pp. 199–238.

21. Sambrook, J., Fritch, E. F., and Maniatis, T. (1989) *Molecular Cloning,* 2nd ed., Cold Spring Harbor Laboratory, Cold Spring Harbor, NY.

22. Saiki, R. K., Gelfand, D. H., Stoffel, S., Scharf, S. J., Higuchi, R., Horn, G. T., Mullis, K. B., and Erlich, H. A. (1988) Primer-directed enzymatic amplification of DNA with a thermostable DNA polymerase. *Science* **239,** 487–491.

23. Moon, I. S. and Krause, M. O. (1991) Common RNA polymerase I, II, and III upstream elements in mouse 7SK gene locus revealed by the inverse polymerase chain reaction. *DNA Cell Biol.* **10,** 23–32.

24. Strobel, S. A. and Dervan, P. B. (1990) Site-specific cleavage of a yeast chromosome by oligonucleotide-directed triple-helix formation. *Science* **249,** 73–75.

25. Dreyer, G. B. and Dervan, P. B. (1985) Sequence-specific cleavage of single-stranded DNA: oligodeoxynucleotide-EDTA.Fe(II). *Proc. Natl. Acad. Sci. USA* **82,** 968–972.

26. Zhang, H., Scholl, R., Browse, J., and Somerville, C. (1988) Double strand DNA sequencing as a choice for DNA sequencing. *Nucleic Acids Res.* **16,** 1220.

27. Sugino, A., Goodman, H. M., Heynecker, H. L., Shine, J., Boyer, H. W., and Cozzarelli, N. R. (1977) Interaction of bacteriophage T4 RNA and DNA ligases in joining of duplex DNA at base-paired ends. *J. Biol. Chem.* **252,** 3987.

8

Cloning Gene Family Members
Using the Polymerase Chain Reaction
with Degenerate Oligonucleotide Primers

Gregory M. Preston

1. Introduction
1.1. What Are Gene Families?

As more and more genes are cloned and sequenced, it is apparent that nearly all genes are related to other genes. Similar genes are grouped into families, such as the collagen and globin gene families. There are also gene superfamilies. Gene superfamilies are composed of genes that have areas of high homology and areas of high divergence. Examples of gene super families include the oncogenes, homeotic genes, and the myosin genes. In most cases, the different members of a gene family carry out related functions. A detailed protocol for the cloning by degenerate oligonucleotide polymerase chain reaction (PCR) of members of the *Aquaporin* family of membrane water channels *(1,2)* is discussed here.

1.2. Advantages of PCR Cloning of Gene Family Members

There are several considerations that must be taken into account when determining the advantages of using PCR to identify members of a gene family over conventional cloning methods of screening a library with a related cDNA, a degenerate primer, or an antibody. It is recommended that after a clone is obtained by PCR, this template should be used to isolate the corresponding cDNA from a library, since mutations can often be introduced in PCR cloning. Alternatively, sequencing two or more PCR clones from independent reactions will also meet this objective. The following is a list of some of the advantages of cloning gene family members by PCR.

From: *Methods in Molecular Biology, Vol. 69: cDNA Library Protocols*
Edited by: I. G. Cowell and C. A. Austin Humana Press Inc., Totowa, NJ

1. Either one or two degenerate primers can be used in PCR cloning. When only one of the primers is degenerate, the other primer must be homologous to sequences in the phage or bacteriophage cloning vector *(3,4)* or to a synthetic linker sequence as with rapid amplification of cDNA ends (RACE) PCR (*see* Chapter 6). The advantage to using only one degenerate primer is that the resulting clones contain all of the genetic sequence downstream from the primer (be it 5' or 3' sequence). The disadvantage to this anchor PCR approach is that one of the primers is recognized by every gene in the starting material, resulting in single-strand amplification of all sequences. This is particularly notable when attempting to clone genes that are not abundant in the starting material. This disadvantage can often be ameliorated in part by using a nested amplification approach with two degenerate primers to amplify desired sequences preferentially.

2. It is possible to carry out a PCR reaction on first-strand cDNAs made from a small amount of RNA and, in theory, from a single cell. Several single-stranded "minilibraries" can be rapidly prepared and analyzed by PCR from a number of tissues at different stages of development, or cell cultures under different hormonal conditions. Therefore, PCR cloning can potentially provide information about the timing of expression of an extremely rare gene family member, or messenger RNA splicing variants, which may not be present in a recombinant library.

3. Finally, the time and expense required to clone a gene should be considered. Relative to conventional cloning methods, PCR cloning can be more rapid, less expensive, and in some cases, the only feasible cloning strategy. It takes at least 4 d to screen 300,000 plaques from a λgt10 library. With PCR, an entire library containing 10^8 independent recombinants (~5.4 ng DNA) can be screened in one reaction. Again, to ensure authenticity of your PCR clones, you should either use the initial PCR clone to isolate a cDNA clone from a library or sequence at least two clones from independent PCR reactions.

1.3. Degenerate Oligonucleotide Theory and Codon Usage

Because the genetic code is degenerate, primers targeted to particular amino acid sequences must also be degenerate to encode the possible permutations in that sequence. Thus, a primer to a six amino acid sequence that has 64 possible permutations can potentially recognize 64 different nucleotide sequences, one of which is to the target gene. If two such primers are used in a PCR reaction, then there are 64 × 64 or 4096 possible permutations. The target DNA will be recognized by a small fraction (^{1}/$_{64}$) of both primers, and the amplification product from that gene will increase exponentially. However, some of the other 4095 possible permutations may recognize other gene products. This disadvantage can be ameliorated by performing nested amplifications and by using "guessmer" primers. A guessmer primer is made by considering the preferential codon usage exhibited by many species and tissues (*see* Sections 3.1.). For instance, the four codons for alanine begin with GC. In the third position of this codon, G is rarely used in humans (~10.3% of the time) or rats (~8.0%), but often used in *Escherichia coli* (~35%) *(5)*. This characteristic of codon usage may be advantageously used when designing degenerate oligonucleotide primers.

1.4. Strategy for Cloning Aquaporin Gene Family Members

In a related methods chapter *(3)*, I described the cloning by degenerate primer PCR of *Aquaporin*-1 (formerly CHIP28) from a human fetal liver λgt11 cDNA library starting with the first 35 amino acids from the N-terminus of the purified protein. A full-length cDNA was subsequently isolated from an adult human bone marrow cDNA library *(4)* and, following expression in *Xenopus*, shown to encode a water-selective channel *(6)*. Today we now know that the *Aquaporin* family of molecular water channels includes genes expressed in diverse species, including bacteria, yeast, plants, insects, amphibians, and mammals *(1,2,7)*. We have recently used degenerate oligonucleotide primers designed to highly conserved amino acids between the different members of the *Aquaporin* family to clone novel *Aquaporin* gene family cDNAs from rat brain (*AQP*4) and salivary gland (*AQP*5) libraries *(8,9)*. Section 3. describes the creation of a new set of degenerate primers which we are currently using to clone, by degenerate primer PCR, *Aquaporin* homologs from a number of different tissues and species. Section 3. has been broken up into three parts.

1. The designing of the degenerate primers.
2. The PCR amplification with degenerate primers.
3. The subcloning and DNA sequencing of the specific PCR-amplified products.

2. Materials
2.1. Design of Degenerate Oligonucleotide Primers

No special materials are required here, except the amino acid sequence to which the degenerate primers are going to be designed and a codon usage table *(5)*. If the degenerate primers are to be designed according to a family of related amino acid sequences, these sequences should be aligned using a multiple-sequence alignment program. A degenerate nucleotide alphabet (Table 1) provides a single letter designation for any combination of nucleotides. Some investigators have successfully employed mixed primers containing inosine where degeneracy was maximal, assuming inosine is neutral with respect to base pairing, to amplify rare cDNAs by PCR *(10,11)*.

2.2. PCR Amplification with Degenerate Primers

For all buffers and reagents, distilled deionized water should be used. All buffers and reagents for PCR should be made up in distilled deionized 0.2-μm filtered water that has been autoclaved (PCR water) using sterile tubes and

Table 1
The Degenerate Nucleotide Alphabet

Letter	Specification	Letter	Specification
A	Adenosine	C	Cytidine
G	Guanosine	T	Thymidine
R	puRine (A or G)	Y	pYrimidine (C or T)
K	Keto (G or T)	M	aMino (A or C)
S	Strong (G or C)	W	Weak (A or T)
B	Not A (G, C, or T)	D	Not C (A, G, or T)
H	Not G (A, C, or T)	V	Not T (A, C, or G)
N	aNy (A, G, C, or T)	I	Inosine[a]

[a]Although inosine is not a true nucleotide, it is included in this degenerate nucleotide list since many researchers have employed inosine-containing oligonucleotide primers in cloning gene family members.

aerosol blocking pipet tips to prevent DNA contamination (*see* Note 1). All plastic supplies (microfuge tubes, pipet tips, etc.) should be sterilized by autoclaving or purchased sterile.

1. 10X PCR reaction buffer: 100 mM Tris-HCl, pH 8.3, at 25°C, 500 mM KCl, 15 mM MgCl$_2$, 0.1% (w/v) gelatin. Incubate at 50°C to melt the gelatin, filter-sterilize, and store at –20°C (*see* Note 2).
2. dNTP Stock solution (1.25 mM dATP, dGTP, dCTP, dTTP) made by diluting commercially available deoxynucleotides with PCR water.
3. Thermostable DNA polymerase, such as AmpliTaq DNA polymerase (Perkin-Elmer, Norwalk, CT) supplied at 5 U/μL.
4. Mineral oil.
5. A programmable thermal cycler machine, available from a number of manufacturers including Perkin-Elmer, MJ Research (Watertown, MA), and Stratagene (La Jolla, CA).
6. Degenerate oligonucleotide primers should be purified by reverse-phase high-performance liquid chromatography (HPLC) or elution from acrylamide gels, dried down, resuspended at 20 pmol/μL in PCR water, and stored at –20°C, preferably in aliquots.
7. The DNA template can be almost any DNA sample, including a single-stranded cDNA from a reverse transcription reaction, DNA from a phage library, and genomic DNA. The DNA is heat-denatured at 99°C for 10 min, and stored at 4 or –20°C.
8. Chloroform (*see* Note 3).
9. Tris-saturated phenol (*see* Note 3), prepared using ultrapure redistilled crystalline phenol as recommended by the supplier (Gibco-BRL, Gaithersburg, MD, product #5509): Use polypropylene or glass tubes for preparation and storage.

10. PC9 (*see* Note 3): Mix equal volumes of buffer-saturated phenol (pH >7.2) and chloroform, extract twice with an equal volume of 100 mM Tris (pH 9.0), separate phases by centrifugation at room temperature for 5 min at 2000g, and store at 4 to –20°C for up to 1 mo.

11. 7.5M ammonium acetate (AmAc) for precipitation of DNA: AmAc is preferred over sodium acetate because nucleotides and primers generally do not precipitate with it. Dissolve in water, filter through 0.2-μm membrane, and store at room temperature.

12. 100% Ethanol, stored at –20°C.

13. 70% Ethanol, stored at –20°C.

14. TE: 10 mM Tris-HCl, 0.2 mM EDTA, pH 8.0. Dissolve in water, filter through 0.2-μm membrane, and store at room temperature.

15. 50X TAE: 242 g Tris-base, 57.1 mL acetic acid, 18.6 g $Na_2(H_2O)_2EDTA$: Dissolve in water, adjust volume to 1 L, and filter through 0.2-μm membrane. Store at room temperature.

16. *Hae*III digested ϕX174 DNA markers: Other DNA mol-wt markers can be used depending on availability and the size of the expected PCR-amplified products.

17. 6X Gel loading buffer (GLOB): 0.25% bromophenol blue, 0.25% xylene cyanol FF, 1 mM EDTA, 30% glycerol in water. Store for up to 4 mo at 4°C.

18. Agarose gel electrophoresis apparatus and electrophoresis-grade agarose: For the optimal resolution of DNA products <500 bp in length, NuSieve GTG agarose (FMC BioProducts, Rockland, ME) is recommended.

19. Ethidium bromide (EtBr) (*see* Note 3): 10 mg/mL stock of EtBr prepared in water and stored at 4°C in a brown or foil-wrapped bottle. Use at 0.5–2.0 μg/mL in water for staining nucleic acids in agarose or acrylamide gels.

20. For the elution of specific PCR-amplified DNA products from agarose gels, several methods are available, including electroelution and electrophoresis onto DEAE-cellulose membranes *(12,13)*. Several commercially available kits will also accomplish this task. I have had some success with GeneClean II (Bio 101, La Jolla, CA) for PCR products >500 bp in length, and with QIAEX (Qiagen, Chatsworth, CA) for products from 50 bp to 5 kbp. If you do not know the approximate size of the PCR-amplified products and wish to clone all of the PCR-amplified products, the QIAquick-spin PCR purification kit is recommended (Qiagen), since this will remove all nucleotides and primers before attempting to clone. This kit is also recommended for purification of PCR products for secondary PCR amplification reactions.

2.3. Cloning and DNA Sequencing of PCR-Amplified Products

1. From Section 2.2., items 8–14 and 20.

2. pBluescript II phagemid vector (Stratagene): A number of comparable bacterial expression vector are available from several companies.

3. Restriction enzymes: *Eco*RV (for blunt-end ligation).

4. Calf intestinal alkaline phosphatase (CIP) (New England Biolabs, Beverly, MA).

5. Klenow fragment of *E. coli* DNA polymerase I (sequencing-grade preferred) and 10 m*M* dNTP solution (dilute PCR or sequencing-grade dNTPs).
6. T4 DNA ligase (1 or 5 U/µL) and 5X T4 DNA ligase buffer (Gibco-BRL).
7. Competent DH5α bacteria: Can be prepared *(12,13)* or purchased. Other bacterial strains can be substituted.
8. Ampicillin: 50 mg/mL stock in 0.2-µm filtered water, stored in aliquots at −20°C (*see* Note 4).
9. LB media: 10 g bacto-tryptone, 5 g bacto-yeast extract, and 10 g NaCl dissolved in 1 L water. Adjust pH to 7.0. Sterilize by autoclaving for 20 min on liquid cycle.
10. LB-Amp plates: Add 15 g bacto-agar to 1000 mL LB media prior to autoclaving for 20 min on the liquid cycle. Gently swirl the media after removing it from the autoclave to distribute the melted agar. **Caution:** the fluid may be superheated and may boil over when swirled. Place the media in a 50°C water bath to cool. Add 1 mL of ampicillin, swirl to distribute, and pour 25–35 mL/90-mm plate. Carefully flame the surface of the media with a Bunsen burner to remove air bubbles before the agar hardens. Store inverted overnight at room temperature, and then wrapped at 4°C for up to 6 mo.
11. Isopropylthiogalactoside (IPTG): Dissolve 1 g IPTG in 4 mL water, filter through 0.2-µm membrane, and store in aliquots at −20°C.
12. 5-bromo-4-chloro-3-indolyl-β-D-galactopyranoside (X-Gal): Dissolve 100 mg X-Gal in 5 mL dimethylformamide and store at −20°C in a foil wrapped tube (light-sensitive).
13. Plasmid DNA isolation equipment and supplies *(12,13)* or plasmid DNA isolation kits, available from many manufacturers.
14. Double-stranded DNA sequencing equipment and supplies *(12,13)*, or access to a DNA sequencing core facility.

3. Methods
3.1. Design of Degenerate Oligonucleotide Primers

1. The first step in designing a degenerate primer is to select a conserved amino acid sequence and then determine the potential nucleotide sequence (or the complement of this sequence for a downstream primer), considering all possible permutations. If the amino acid sequence is relatively long, you can potentially design two or more degenerate primers. If only one is made, make it to sequences with a high (50–65%) GC content, since these primers can be annealed under more stringent conditions (e.g., higher temperatures). Figure 1 shows an alignment of the amino acid sequences for several members of the *Aquaporin* gene family in the two most highly conserved regions. Also shown are the consensus amino acid sequence, the degenerate nucleotide sequence, and the sequence of the primers we are currently using to isolate *Aquaporin* gene family members. Interestingly, not only are these two regions highly conserved between the different members of this gene family, but they are also highly related to each other, with the conserved motif (T/S)GxxxNPAxx(F/L)G, which has been speculated to have resulted from an ancient internal duplication in a primordial bacterial

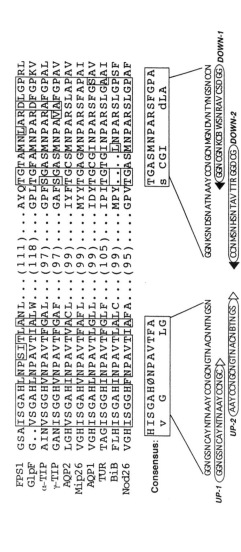

```
FPS1    GSA I SGAHL NP S I TLANL ... (111) ... AYQ TGT AMN L ARD L GPRL
GlpF    G .. VS GAHLNPAVTIALW ... (118) ... GPLTGF AMNP ARD F GPKV
α-TIP   A INVSGGHVNPAVTF GAL ... (97) .... GPF S GA SMNP ARA F GPAL
γ-TIP   GAN I SGGHVNPAVTF GAF ... (97) .... GAF S GA SMNP A V A F GPAV
AQP2    LGHVSGAHINPAVTFAFL ... (99) .... IYFTGC SMNPARS L APAV
Mip26   VGH I SGAHVNPAVTFAFL. ... (99) .... MYYTGAGMNPARSFAPAI
AQP1    VGH I SGAHLNPAVTLGLL. ... (99) .... IDYTGC I NPARSFG S AV
TUR     TAGISGGHINPAVTF GLF ... (105) ... IPITGT INPARSLGAAI
BiB     FLH I SGAHINPAVTLALC. ... (99) .... MPY . . . · L NPARSLGPSF
Nod26   VGH I SGGHF MNPAVTIAF A ... (95) .... GPV TGA SMNPARSLGPAF
```

```
Consensus:   H I SGAH Ø NP AVTF A               T GASMNPARS F GPA
                           v     G        LG              s   CGI            dLA
```

GGNGSNGCAYNTNAAYCCNGGNGTNACNNTNGSN   GGNKGSNDSNATNAAYCCNGGNMGNDVNTYNGSNCCN

UP-1 ⟨GGNGSNGCAYNTNAAYCCNGC⟩   ▼ ◀GGNCGNKCBWSNRAVCSDGG⟩ DOWN-1

UP-2 ⟨AAYCCNGCNGTNACNBTNGS⟩   ◀CCNMSNHSNTAVTTRGGDCG⟩ DOWN-2

Fig. 1. Design of degenerate primers to amplify *Aquaporin* gene family members. Top: The amino acid sequences of 10 major intrinsic protein (MIP) family proteins, including the *Saccharomyces cerevisiae* FPS1 (*23*), *E. coli* GlpF (*24*), α- and γ-tonoplast intrinsic proteins (TIP) of *Arabidopsis thaliana* (*25*), the vasopressin-responsive water channel of rat renal collecting tubules (*AQP2*) (*26*), the MIP of bovine lens fiber membranes (*27*), human *Aquaporin-1* (*4*), turgor responsive gene (TUR) 7a from *Pisum stivum* (*28*), the *Drosophila* neurogenic *big brain* protein (*29*), and the *Rhyzodium* root Nodulin-26 peribacteroid membrane protein (*30*), were aligned by the PILEUP program of progressive alignments (*31*) using a gap weight of 3.0 and a gap length of 0.1 running on a VAX computer system. The two most highly conserved regions are shown, separated by the number of intervening amino acids. The most highly conserved amino acids are enclosed. Middle: Below the aligned sequences, the consensus amino acid sequences are shown. Bottom: From part of the consensus amino acid sequences, the degenerate nucleotide sequences were determined (using the degenerate nucleotide alphabet from Table 1) followed by the sequences for the degenerate oligonucleotide primers.

103

organism, since this repeat has persisted in *Aquaporin* homologs from bacteria through plants and mammals *(1,6,14)*. These two regions are functionally related, both contributing to the formation of the water pore in *Aquaporin*-1 *(15)*.

2. The next step is to determine the number of permutations in the deduce nucleotide sequence. There are 192 permutations ([2 × 4] × 3 × 4 × 2) in the sequence 5'-YTN-ATH-GGN-GAR-3', which encodes the hypothetical amino acid sequence *Leu-Ile-Gly-Glu*. The degeneracy can be reduced by making educated guesses in the nucleotide sequence, i.e., by making a guessmer. The 3'-end of a primer should contain all possible permutations in the amino acid sequence, since *Taq* DNA polymerase will not extend a prime with a mismatch at the extending (3') end. If the above primer was to a human gene, a potential guessmer would be 5'-CTB-ATY-GGN-GAR-3', which only contains 64 permutations. This guessmer is proposed by taking into account the preferential codon usage for leucine and isoleucine in humans *(5)*.

3. The degeneracy of a primer can be reduced further by incorporating inosine residues in the place of N. The advantages of using inosine-containing primers is that they have a reduced number of permutations, and the inosine reportedly base pairs equally well with all four nucleotides, creating a single bond in all cases *(10)*. The disadvantage is that inosines reduce the annealing temperature of the primer. Inosine-containing primers were not employed in these studies.

4. It is often convenient to incorporate restriction endonuclease sites at the 5'-ends of a primer to facilitate cloning into plasmid vectors *(4,8,9)*. Different restriction sites can be added to the 5'-ends of different primers so the products can be cloned directionally. However, not all restriction enzymes can recognize cognate sites at the ends of a double-stranded DNA molecule equally well. This difficulty can often be reduced by adding a two- to four-nucleotide 5'-overhang before the beginning of the restriction enzyme site (*see also* Note 5). Some of the best restriction enzymes sites to use are *Eco*RI, *Bam*HI, and *Xba*I. Catalogs from New England Biolabs have a list of the ability of different restriction enzymes to recognize short base pair sequences. A potential pitfall of this approach would be the occurrence of the same restriction site within the amplified product as used on the end of one of the primers. Therefore, only part of the amplified product would be cloned.

5. The final consideration you should make is the identity of the 3'-most nucleotide. The nucleotide on the 3'-end of a primer should preferably be G or C, and not be N, I, or T. The reason for this is that thymidine (and supposedly inosine) can nonspecifically prime on any sequence. Guanosines and cytidine are preferred since they form three H-bonds at the end of the primer, a degree stronger than an A:T base pair.

3.2. PCR Amplification and DNA Purification

The template for these reactions can be the DNA in a phage library or the first-strand cDNA from a reverse transcription reaction on RNA. A phage library with a titer of 5×10^9 PFU/mL would contain, in a 5-μL aliquot, 2.5×10^7

PFU (~1.5 ng of DNA). Prior to PCR amplification, the DNA is heat-denatured at 99°C for 10 min.

3.2.1. PCR Reaction (see Notes 1 and 6)

In all cases, the DNA template should also be PCR-amplified with the individual degenerate primers to determine if any of the bands amplified are derived from one of the degenerate primer pools. A DNA-free control is required to assess if there is any contaminating DNA in any of the other reagents.

1. Pipet into 0.5-mL microcentrifuge tubes in the following order:
   a. 58.5 µL PCR water that has been autoclaved.
   b. 10 µL 10X PCR reaction buffer (*see* Note 2).
   c. 16 µL 1.25 m*M* dNTP stock solution.
   d. 5.0 µL Primer up-1.
   e. 5.0 µL Primer down-1.
   f. 5.0 µL Heat-denatured library or cDNA (1–100 ng).
   If several reactions are being set up concurrently, a master reaction mix can be made up, consisting of all the reagents used in all of the reactions, such as the PCR water, reaction buffer, and dNTPs.
2. Briefly vortex each sample, and spin for 10 s in a microfuge. Overlay each sample with two to three drops of mineral oil.
3. Amplify by hot-start PCR using the following cycle parameters. Pause the thermocycler in Step 4-cycle 1, and add 0.5 µL AmpliTaq DNA polymerase to each tube.
   Step 1: 95°C, 5 min (initial denaturation).
   Step 2: 94°C, 60 s (denaturation).
   Step 3: 50°C, 90 s (annealing; *see* Note 7).
   Step 4: 72°C, 60 s (extension).
   Step 5: cycle 29 times to step 2.
   Step 6: 72°C, 4 min.
   Step 7: 10°C hold.

3.2.2. DNA Isolation and Gel Electrophoresis Analysis

1. Remove the reaction tubes from the thermal cycler, and add 200 µL chloroform. Spin for 10 s in a microfuge to separate oil-chloroform layer from the aqueous layer. Carefully transfer the aqueous layer to a clean microfuge tube.
2. Remove the Ampli*Taq* DNA polymerase by extracting the aqueous phase twice with 100 µL PC9 (*see* Note 3). Spin for 2 min in a microfuge to separate the lower organic layer from the upper aqueous layer, and transfer the aqueous layer to clean microfuge tube. This step is essential before digesting the DNA with restriction enzymes for directional cloning (*see* Section 3.3.), since the polymerase can precipitate, and in the presence of nucleotides, fill in recessed 3'-termini on DNA.

3. AmAc-Ethanol precipitation: To a 100-μL DNA sample add 50 μL 7.5M AmAc (50% vol). Vortex briefly to mix. Precipitate the DNA with 350 μL 100% ethanol (2–2.5 vol). Vortex the samples for 15 s and ice for 15 min. Spin down the DNA at 12,000g for 15 min at 4°C in a microfuge. Decant the aqueous waste. Add 250 μL 70% ethanol. Vortex briefly and spin another 5 min at 4°C. Decant the ethanol, and allow the pellets to dry inverted at room temperature, or dry in a Speed-Vac for 2–10 min.
4. Resuspend in 20 μL PCR water.
5. The next step is to resolve an aliquot (2–10 μL) of the PCR fragments by gel electrophoresis. Small DNA products (<300 bp) can be resolved at high resolution on 5–10% polyacrylamide gels *(12,13)*. Moderate-sized PCR products (150–1000 bp) should be resolved on 2–4% NuSieve agarose gels (in 1X TAE buffer). Larger PCR products (>500 bp) can be resolved on 0.8–2% agarose gels (1X TAE buffer).
6. After the bromophenol blue dye has reached the end of the gel, soak the gel for 5–30 min in about 10 vol of water containing 1 μg/mL EtBr (*see* Note 3). Then view and photograph the gel under UV light. As shown in Fig. 1, there is little variability in the distance between the NPA motifs with the known members of the *Aquaporin* gene family. PCR amplification of the known *Aquaporins* cDNAs using the internal degenerate primers would generate products from 345 to 415 bp. A typical result is shown in Fig. 2.

3.2.3. Secondary PCR Amplifications and DNA Purification

Based on the results from gel electrophoresis of the PCR-amplified DNA products, a decision must be made on what to do next. The options are the following.

1. Amplify by PCR from the initial DNA sample under different conditions.
2. Amplify by PCR from a different DNA sample under the same conditions (different MgCl$_2$ concentration, annealing temperature, or primers, *see* Notes 2, 6, and 7).
3. Gel-purify a band(s) of DNA from the gel for cloning or to reamplify by PCR.
4. Purify all PCR-amplified DNA fragments for cloning or to reamplify by PCR.
5. Reamplify by PCR with the same or an internal pair of degenerate primers.

Options 1 and 2 are self-explanatory. If you want to gel-purify a particular band or group of bands from an agarose gel, a number of procedures and kits are available (*see* Section 2.2.). If you plan on immediately cloning a PCR band(s), you may want to run the rest of the initial PCR reaction on another gel to increase the recovery of DNA. It is also possible to recover specific DNA fragments from an acrylamide gel *(3,12,13)*. To purify all PCR-amplified DNA fragments from the remaining sample, a number of methods are available, including the QIAquick-spin PCR purification kit, which can be used instead of steps 1–3 in Section 3.2.2. (Qiagen). Finally, aliquots of DNA purified from a gel or from the initial PCR reaction (1–10%) can be reamplified by

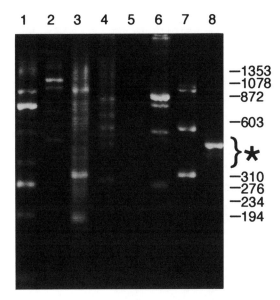

Fig. 2. Gel electrophoresis analysis of PCR-amplified DNA. DNA isolated from a human kidney cDNA library in bacteriophage λgt10 was amplified with degenerate primers UP-1 (lanes 1, 5, and 6), UP-2 (lanes 2, 7, and 8), DOWN-1 (lanes 3, 5, and 7), and DOWN-2 (lanes 4, 6, and 8). Reactions containing 5×10^6 PFU of heat-denatured phage DNA, 100 pmol of degenerate primers, and 1.5 mM MgCl$_2$ in a 100-μL vol were subjected to 40 cycles of PCR amplification under the following parameters: 94°C for 60 s, 48°C for 90 s, and 72°C for 60 s. Following chloroform extraction and ethanol precipitation, the DNA was resuspended in 20 μL of water, and 5 μL was electrophoresed into a 4% NuSieve agarose gel in 1X TAE. The gel was stained with ethidium bromide and photographed. The relative mobility of *Hae*III digested φX174 DNA markers is shown on the right. The bracket shows the size range of known members of this gene family from the primers employed.

PCR with either the same or an internal pair of degenerate oligonucleotide primers (*see* Note 1).

When attempting to identify a gene family homolog from a tissue that is known to express a homolog(s), a number of tricks can be tried to enrich the final PCR sample for new homologs. Since the degenerate oligonucleotide primers are designed from the sequence of the known gene family members, these primers will likely be biased for those homologs. *Aquaporin*-1 is abundant in the capillaries around the salivary glands and throughout the body, but absent in the salivary gland *(16)*. To identify a salivary homology of the *Aquaporin* gene family, a rat salivary gland cDNA library was used, which also contained *Aquaporin*-1 cDNAs presumably from the surrounding capillaries. The cDNA library was first amplified with an external set of degenerate primers, digested with the restriction enzyme *Pst*I (which cuts between the NPA motifs of rat *AQP*1), and reamplified with an internal pair of primers. *Pst*I was again used to digest the rat *AQP*1 DNAs, and the DNA fragments between 350 and 450 bp were cloned and sequenced *(9)*. This strategy would not work if the resulting cDNA (*AQP5*) also contained a *Pst*I site. By trying different restriction enzymes that cut DNA infrequently (6–8 bp recognition sites), a number of new homologs will preferentially be identified. Alternatively, after cloning the DNA products into bacterial expression vectors, bacterial colony lift hybridization can be used to identify colonies containing inserts for known gene family members *(3,12,13)*.

3.3. Cloning and DNA Sequencing of PCR-Amplified Products

3.3.1. Preparation of Vector for Ligation

1. For blunt-end ligations, digest 1 µg pBluescript II KS phagemid DNA (Stratagene) with 10 U *Eco*RV in a 50-µL vol. Incubate at 37°C for 2 h. For cohesive-end ligations, similarly digest the vector with the appropriate restriction enzyme(s).
2. For both blunt-end ligations and cohesive-end ligations where the vector has been digested with only one restriction enzyme, it is necessary to remove the 5'-phosphate from the vector to inhibit the vector from self-ligating. This is accomplished by treating the vector with CIP according to the manufacturer's recommendations. Note that 1 µg of a 3-kb linear DNA molecule contains 1 pmol of 5'-overhangs (*Bam*HI), blunt-ends (*Eco*RV), or 3'-overhangs (*Pst*I), depending on what enzyme it has been digested with. Afterward add EDTA to 5 m*M*, and heat-kill the enzyme at 65°C for 1 h. Adjust the volume to 50–100 µL with TE, and extract once with Tris-saturated phenol, twice with PC9, and twice with chloroform. Back-extract each organic layer with 50 µL TE, pool with the final sample. AmAc-EtOH precipitate (*see* Section 3.2.2.), and resuspend in 10 µL water.
3. If the insert is going to be directionally cloned into the vector, just extract once with 50 µL PC9, AmAc-EtOH precipitate (*see* Section 3.2.2.), and resuspend in 10 µL water.

3.3.2. Preparation of Inserts for Ligation

AmpliTaq and other thermostable DNA polymerases often fail to fill in the ends of the double-stranded DNA products completely, thus leaving recessed 3'-termini, which can be filled in with the Klenow fragment of *E. coli* DNA polymerase I. This should be done whether or not the DNA is going to be digested with restriction enzymes added to the ends of the primers for directional cloning (*see* Section 3.1.).

1. AmAc-EtOH precipitate the DNA (*see* Section 3.2.2.), and resuspend in 15 µL water.

2. Add 2 μL 10X restriction enzyme reaction buffer. Klenow DNA polymerase works well in most restriction enzyme digestion buffers (10X REact 2 or 3 from Gibco-BRL). If the DNA is going to be subsequently digested with a restriction enzyme(s), use the buffer for that enzyme.
3. Add 2 μL 10 mM dNTP solution. Then add Klenow DNA polymerase (1 U/μg DNA), and incubate at room temperature for 15 min.
4. Heat-inactivate the enzyme at 75°C for 10 min. If the DNA is going to be directly used in ligation reactions, it is not necessary to purify the DNA from the unincorporated dNTPs, since they will not inhibit T4 DNA ligase. To concentrate the DNA sample, proceed with step 6.
5. PCR products containing restriction sites on their ends should now be digested with the restriction enzymes. Incubate in the appropriate buffer, using 20 U of enzyme/μg of DNA, and incubating for 2–4 h at the proper temperature.
6. Extract the DNA once or twice with PC9 and precipitate with AmAc-EtOH as described in Section 3.2.2. Resuspend the final pellet in 5–10 μL water.

3.3.3. DNA Ligation and Bacterial Transformation

1. At this point, it is often advantageous to run a small aliquot of the different DNA fragments on a gel to assess their approximate concentrations and purity. Ideally, you want at least a 2:1 molar ratio of insert to vector in the ligation reactions. If necessary, return to the above procedures to isolate more DNA for the ligation reaction.
2. Set up the ligation reactions with the vector and insert similar to the following:
   Reaction 1: 1 μL vector (10 ng; vector control).
   Reaction 2: 1 μL vector + 1 μL insert (~10 ng insert).
   Reaction 3: 1 μL vector + 4 μL insert.
   Then add 2 μL of 5X T4 DNA ligase buffer (Gibco-BRL) and water to 9.5 μL. If the buffer is more than 4 mo old, the ATP may be depleted. Therefore, add fresh adenosine triphosphate to a final concentration of 1 mM.
3. For cohesive-end ligations, add 0.5 μL of T4 DNA ligase (1 U/μL), gently mix, spin for 5 s in a microfuge, and incubate at 15°C for 10–20 h. For blunt-end ligations, add 1 μL of T4 DNA ligase (5 U/μL), gently mix, spin for 5 s in a microfuge, and incubate at 25°C (or room temperature) for 1–12 h. Stop the reaction by heating at 75°C for 10 min, and store the samples at –20°C.
4. Set up a bacterial transformation with competent DH5α bacteria or a comparable strain of bacteria. Be sure to include a positive control (10 ng undigested vector DNA) and a negative control (water). To 1.5-mL microfuge tubes, add half of the ligation mix (5 μL) or 5 μL of control DNA or water, and 50 μL of competent bacteria (thawed slowly on ice); incubate on ice for 30 min. Heat-shock at 42°C for 2 min. Return to ice for 1 min. Add 200 μL of LB media containing 10% glycerol. Mix gently, and allow bacteria to recover and express the ampicillin resistance gene by incubating at 37°C for 1 h.
5. Prewarm LB-Amp plates at 37°C for 45 min. About 30 min before plating the bacteria on the plates, add 40 μL of X-Gal and 4 μL IPTG, and quickly spread

over the entire surface of the plate using a sterile glass spreader. Spread 20–200 μL of the transformation reactions on these plates. Allow the inoculum to absorb into the agar, and incubate the plates inverted at 37°C for 12–24 h (*see* Note 4). Afterward, placing the plates at 4°C for 2–4 h will help enhance the blue color development.

3.3.4. Plasmid DNA Minipreps and DNA Sequencing

1. Colonies that contain active β-galactosidase will appear blue, whereas those containing a disrupted *LacZ* gene will be white. Set up minicultures by inoculating individual white colonies into 2 mL of LB media containing ampicillin. After growing at 37°C overnight, isolate the plasmid DNA. Resuspend the DNA in 20–50 μL water or TE.
2. Digest 5–20 μL of the DNA with the appropriate restriction enzymes, and analyze by agarose gel electrophoresis (*see* Section 3.2.3.).
3. Perform double-stranded DNA sequencing on recombinants containing inserts in the expected size range.

4. Notes

1. All PCR reactions should be set up in sterile laminar flow hoods using pipet tips containing filters (aerosol-resistant tips) to prevent the contamination of samples, primers, nucleotides, and reaction buffers by DNA. If the PCR reaction is going to be reamplified by PCR, all possible intervening steps should also be performed in a sterile hood with the same precautions to prevent DNA contamination. These precautions should also be extended to all extractions and reactions on the nucleic acid (RNA or DNA) through the last PCR reaction. Likewise, all primers, nucleotides, and reaction buffers for PCR should be made up and aliquoted using similar precautions. All buffers for PCR should be made with great care using sterile disposable plastic or baked glass, and restricted for use with aerosol-resistant pipet tips.
2. Standard PCR reaction buffers contain 15 mM MgCl$_2$ (1.5 mM final concentration). In many cases, changes in the MgCl$_2$ concentration will have significant consequences on the amplification of specific bands. In PCR amplifying the four exons of the *AQP*1 gene, MgCl$_2$ concentrations between 0.7 and 1.0 mM gave the best results *(17,18)*; however, MgCl$_2$ concentrations between 0.5 and 5.0 mM have been reported.
3. Organic solvents and EtBr are hazardous materials. Always handle with tremendous caution, wearing gloves and eye protection. Contact your hazardous waste department for proper disposal procedures in your area.
4. Ampicillin-resistant bacteria secrete β-lactamase into the media, which rapidly inactivates the antibiotic in regions surrounding the growing bacterial colony. Thus, when bacteria are growing at a high density or for long periods (>16 h), ampicillin-sensitive satellite colonies will appear around the primary colonies (which are white in blue-white selections). This problem can be ameliorated (but not eliminated) by substitution of carbenicillin for ampicillin on agar plates.

5. When designing primers with restriction enzyme sites and 5'-overhangs, note that the 5'-overhang should not contain sequence complementary to the sequence just 3' of the restriction site, since this would facilitate the production of primer dimers. Consider the primer 5'-ggg.agatct.CCCAGCTAGCTAGCT-3', which has an *Xba*I site followed by a 5'-ggg and by a CCC-3'. These 12 nucleotides on the 5'-end are palindromic, and can therefore easily dimerize with another like primer. A better 5'-overhang would be 5'-cac.

6. When cloning a gene from a recombinant library by PCR, remember that not all genes are created equal. Genes with high G:C contents have proven more difficult to clone than most. Several researchers have made contributions in a search for factors to enhance the specificity of PCR reactions. Nonionic detergents, such as Nonident P-40, can be incorporated in rapid sample preparations for PCR analysis without significantly affecting *Taq* polymerase activity *(19)*. In some cases, such detergents are absolutely required in order to detect reproducibly a specific product *(20)*, presumably owing to inter- and intrastrand secondary structure. More recently, tetramethylammonium chloride has been shown to enhance the specificity of PCR reactions by reducing nonspecific priming events *(21)*. Commercial PCR enhancers are also available.

7. A critical parameter when attempting to clone by PCR is the selection of a primer annealing temperature. This is especially true when using degenerate primers. The primer melting temperature (T_m) is calculated by adding 2°C for A:T base pairs, 3°C for G:C base pairs; 2°C for N:N base pairs, and 1°C for I:N base pairs. Most PCR chapters suggest you calculate the T_m and set the primer annealing temperature to 5–10°C below the lowest T_m. Distantly related gene superfamily members have been cloned using this rational *(22)*. However, I have found that higher annealing temperatures are helpful in reducing nonspecific priming, which can significantly affect reactions containing degenerate primers.

Acknowledgments

I thank my colleagues, especially Peter Agre and William B. Guggino, for their support and helpful discussions. This work was supported in part by NIH grants HL33991 and HL48268 to Peter Agre.

References

1. Reizer, J., Reizer, A., and Saier, M. H., Jr. (1993) The MIP family of integral membrane channel proteins: sequence comparisons, evolutionary relationships, reconstructed pathways of evolution, and proposed functional differentiation of the two repeated halves of the proteins. *Crit. Rev. Biochem. Mol. Biol.* **28,** 235–257.

2. Knepper, M. A. (1994) The aquaporin family of molecular water channels. *Proc. Natl. Acad. Sci. USA* **91,** 6255–6258.

3. Preston, G. M. (1993) Use of degenerate oligonucleotide primers and the polymerase chain reaction to clone gene family members, in *Methods in Molecular Biology, vol. 15, PCR Protocols: Current Methods and Applications* (White, B. A., ed.), Humana, Totowa, NJ, pp. 317–337.

4. Preston, G. M. and Agre, P. (1991) Isolation of the cDNA for erythrocyte integral membrane protein of 28 kilodaltons: member of an ancient channel family. *Proc. Natl. Acad. Sci. USA* **88**, 11,110–11,114.

5. Wada, K.-N., Aota, S.-I., Tsuchiya, R., Ishibashi, F., Gojobori, T., and Ikemura, T. (1990) Codon usage tabulated from the GenBank genetic sequence data. *Nucleic Acids Res.* **18**, 2367–2411.

6. Preston, G. M., Carroll, T. P., Guggino, W. B., and Agre, P. (1992) Appearance of water channels in *Xenopus* oocytes expressing red cell CHIP28 protein. *Science* **256**, 385–387.

7. Chrispeels, M. J. and Agre, P. (1994) Aquaporins: water channel proteins of plant and animal cells. *TIBS* **19**, 421–425.

8. Jung, J. S., Bhat, B. V., Preston, G. M., Guggino, W. B., Baraban, J. M., and Agre P. (1994) Molecular characterization of an aquaporin cDNA from brain: candidate osmoreceptor and regulator of water balance. *Proc. Natl. Acad. Sci. USA* **91**, 13,052–13,056.

9. Raina, S., Preston, G. M., Guggino, W. B., and Agre, P. (1995) Molecular cloning and characterization of an aquaporin cDNA from salivary, lacrimal and respiratory tissues. *J. Biol. Chem.* **270**, 1908–1912.

10. Knoth, K., Roberds, S., Poteet, C., and Tamkun, M. (1988) Highly degenerate, inosine-containing primers specifically amplify rare cDNA using the polymerase chain reaction. *Nucleic Acids Res.* **16**, 10,932.

11. Chérif-Zahar, B., Bloy, C., Kim, C. L. V., Blanchard, D., Bailly, P., Hermand, P., Salmon, C., Cartron, J.-P., and Colin, Y. (1990) Molecular cloning and protein structure of a human blood group Rh polypeptide. *Proc. Natl. Acad. Sci. USA* **87**, 6243–6247.

12. Sambrook, J., Fritsch, E. F., and Maniatis, T. (1989) *Molecular Cloning: A Laboratory Manual*, Cold Spring Harbor Laboratory Press, Cold Spring Harbor, NY.

13. Ausubel, F. M., Brent, R., Kingston, R. E., Moore, D. D., Seidman, J. G., Smith, J. A., and Struhl, K. (eds.) (1994) *Current Protocols in Molecular Biology*, Greene Publishing/Wiley-Interscience, New York.

14. Wistow, G. J., Pisano, M. M., and Chepelinsky, A. B. (1991) Tandem sequence repeats in transmembrane channel proteins. *TIBS* **16**, 170–171.

15. Jung, J. S., Preston, G. M., Smith, B. L., Guggino, W. B., and Agre P. (1994) Molecular structure of the water channel through Aquaporin CHIP: the hourglass model. *J. Biol. Chem.* **269**, 14,648–14,654.

16. Nielsen, S., Smith, B. L., Christensen, E. I., and Agre, P. (1993) Distribution of the aquaporin CHIP in secretory and resorptive epithelia and capillary endothelia. *Proc. Natl. Acad. Sci. USA* **90**, 7275–7279.

17. Smith, B. L., Preston, G. M., Spring, F. A., Anstee, D. J., and Agre, P. (1994) Human red cell Aquaporin CHIP, I. molecular characterization of ABH and Colton blood group antigens. *J. Clin. Invest.* **94**, 1043–1049.

18. Preston, G. M., Smith, B. L., Zeidel, M. L., Moulds, J. J., and Agre P. (1994) Mutations in *aquaporin*-1 in phenotypically normal humans without functional CHIP water channels. *Science* **265**, 1585–1587.

19. Weyant, R. S., Edmonds, P., and Swaminathan, B. (1990) Effects of ionic and nonionic detergents on the *Taq* polymerase. *BioTechniques* **9**, 308–309.
20. Bookstein, R., Lai, C.-C., To, H., and Lee, W.-H. (1990) PCR-based detection of a polymorphic *Bam* HI site in intron 1 of the human retinoblastoma (RB) gene. *Nucleic Acids Res.* **18**, 1666.
21. Hung, T., Mak, K., and Fong, K. (1990) A specificity enhancer for polymerase chain reaction. *Nucleic Acids Res.* **18**, 4953.
22. Zhao, Z.-Y. and Joho, R. H. (1990) Isolation of distantly related members in a multigene family using the polymerase chain reaction technique. *Biochem. Biophys. Res. Commun.* **167**, 174–182.
23. Aelst, L. V., Hohmann, S., Zimmermann, F. K., Jans, A. W. H., and Thevelein, J. M. (1991) A yeast homologue of the bovine lens fiber MIP gene family complements the growth defect of a *Saccharomyces cerevisiae* mutant on fermentable sugars but not its defect in glucose-induced RAS-mediated cAMP signalling. *EMBO J.* **10**, 2095–2104.
24. Muramatsu, S. and Mizuno, T. (1989) Nucleotide sequence of the region encompassing the *glpKF* operon and its upstream region containing a bent DNA sequence of *Escherichia coli*. *Nucleic Acids Res.* **17**, 4378.
25. Höfte, H., Hubbard, L., Reizer, J., Ludevid, D., Herman, E. M., and Chrispeels, M. J. (1992) Vegetative and seed-specific forms of Tonoplast Intrinsic Protein in the vacuolar membrane of *Arabidopsis thaliana*. *Plant Physiol.* **99**, 561–570.
26. Fushimi, K., Uchida, S., Hara, Y., Hirata, Y., Marumo, F., and Sasaki, S. (1993) Cloning and expression of apical membrane water channel of rat kidney collecting tubule. *Nature* **361**, 549–552.
27. Gorin, M. B., Yancey, S. B., Cline, J., Revel, J.-R., and Horwitz, J. (1984) The major intrinsic protein (MIP) of the bovine lens fiber membrane: characterization and structure based on cDNA cloning. *Cell* **39**, 49–59.
28. Guerrero, F. D., Jones, J. T., and Mullet, J. E. (1990) Turgor-responsive gene transcription and RNA levels increase rapidly when pea shoots are wilted. Sequence and expression of three induced genes. *Plant Mol. Biol.* **15**, 11–26.
29. Rao, Y., Jan, L. Y., and Jan, Y. N. (1990) Similarity of the product of the *Drosophila* neurogenic gene *big brain* to transmembrane channel proteins. *Nature* **345**, 163–167.
30. Fortin, M. G., Morrison, N. A., and Verma, D. P. S. (1987) Nodulin-26, a peribacteroid membrane nodulin is expressed independently of the development of the peribacteroid compartment. *Nucleic Acids Res.* **15**, 813–824.
31. Feng, D.-F. and Doolittle, R. F. (1990) Progressive alignment and phylogenetic tree construction of protein sequences, in *Methods in Enzymology, vol. 183, Molecular Evolution: Computer Analysis of Protein and Nucleic Acid Sequences* (Doolittle, R. F., ed.), Academic, New York, pp. 375–387.

9

Subtractive Hybridization for the Isolation of Differentially Expressed Genes Using Magnetic Beads

Hans-Christian Aasheim, Ton Logtenberg, and Frank Larsen

1. Introduction

Subtractive hybridization methods provide a means to isolate genes that are specifically expressed in a cell type or tissue *(1–3)*, genes that are differentially regulated during activation or differentiation of cells *(4–7)*, or genes that are involved in pathological conditions such as cancer *(8–10)*. Such methods are suitable for the isolation of low-expression genes. The principle of these approaches is to remove the mRNA species common to different cell types or tissues by subtraction, leaving the cell type/tissue-specific mRNAs for further manipulation and analysis. Alternatively, mRNA from both the subtractor and target population can be reverse transcribed and amplified by PCR followed by subtractive hybridization and isolation of specific sequences *(9)*. In the last 20 yr, several subtractive strategies have been described *(2–5,9)*, but subtraction hybridization/cloning still remains a technically demanding, time-consuming, and labor-intensive technology, including the need for large amounts of mRNA or highly purified single-stranded DNA. Several subtractive hybridization strategies based on solid-phase hybridization on magnetic Dynabeads have previously been described *(11–14)*. In this chapter, we present one such solid-phase strategy developed by us *(12)* that is an improved version of an already published method *(12)*. Our approach takes advantage of the properties of magnetic Dynabeads allowing simple and rapid buffer changes required for optimal hybridization and enzymatic reactions. The main principles of the method presented in this chapter are outlined in Fig. 1. Purified mRNA from the subtractor cell or tissue population is isolated using magnetic Dynabeads

From: *Methods in Molecular Biology, Vol. 69: cDNA Library Protocols*
Edited by: I. G. Cowell and C. A. Austin Humana Press Inc., Totowa, NJ

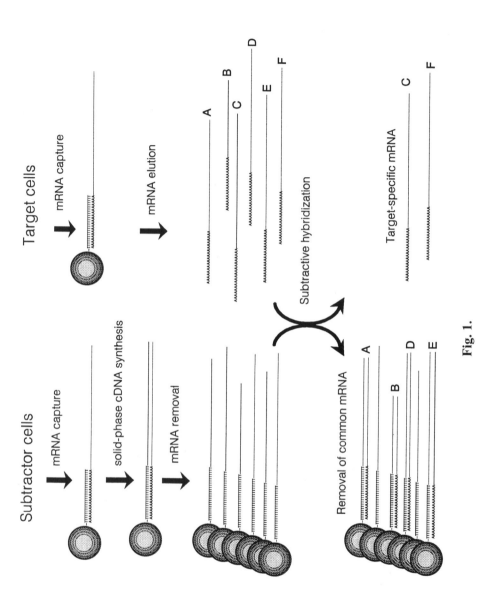

Fig. 1.

oligo(dT)25, and directly converted to immobilized first-strand cDNA (subtractor beads). mRNA from the target cell/tissue population is also isolated using Dynabeads oligo(dT)25 beads, eluted from the beads, dissolved in hybridization buffer, and mixed with subtractor beads. The target mRNA population is hybridized to the cDNA subtractor beads at 68°C for 20–24 h. Two additional hybridization rounds, using the same subtractor beads, are recommended to ensure optimal subtraction and enrichment of cell-specific sequences. After each hybridization step, the beads are regenerated by elution of the mRNA from the first-strand cDNA. After the final hybridization step, the specific mRNA left in the solution is captured with oligo(dT)25 beads and converted to a radioactive cDNA probe for the screening of cDNA libraries. Alternatively, this material can be used for the generation of a subtractive cDNA library or can be used as a source for the amplification of members of known gene families using degenerated primers directed against the conserved areas. The latter approach facilitates a directed search for differentially expressed genes. In our hands, this system provides a fast and reliable way of generating subtractive probes for the isolation of cell/tissue-specific genes. An additional advantage of this method is that the subtractor beads can readily be regenerated and used for at least three different subtractions, each involving three hybridization steps.

2. Materials

2.1. mRNA Isolation

1. Sigmacote (Sigma, St. Louis, MO, cat. no. SL-2).
2. Magnetic particle concentrator (MPC-E) (Dynal, Oslo, Norway).
3. Dynabeads oligo(dT)25 (Dynal).
4. Phosphate-buffered saline (PBS), pH 7.4: 137 mM NaCl, 2.7 mM KCl, 4.3 mM Na$_2$HPO$_4$ · 7H$_2$O, 1.4 mM KH$_2$PO$_4$.
5. TE buffer, pH 8.0: 10 mM Tris-HCl, pH 8.0, 1 mM EDTA, pH 8.0.
6. Lysis buffer: PBS with 0.5% Nonidet-P40 (Sigma).
7. 2X Lysis/binding buffer: 100 mM Tris-HCl, pH 8.0, 500 mM LiCl, 10 mM EDTA, pH 8.0, 0.2% sodium dodecyl sulfate (SDS), 5 mM dithiothretiol (DTT).
8. Wash buffer 1: 10 mM Tris-HCl, pH 7.5, 0.15M LiCl, 1 mM EDTA, 0.1% SDS.
9. Wash buffer 2: 10 mM Tris-HCl, pH 7.5, 0.15M NaCl.
10. Avian Myeloblastosis Virus (AMV) reverse transcriptase buffer: 50 mM Tris-HCl, pH 8.5 (20°C), 8 mM MgCl$_2$, 30 mM KCl, 1 mM DTT.

Fig. 1. *(previous page)* Subtractive hybridization. Schematic presentation of the method. mRNA is isolated from the subtractor cells and converted to solid-phase first-strand cDNA. mRNA is isolated from the target cells and hybridized to the subtractor beads. After hybridization, the subtractor beads with bound target mRNA are collected with a magnet, and the target-specific mRNAs are left in the supernatant. The target-specific mRNAs can be converted to radioactive cDNA probes for the screening of cDNA libraries.

2.2. First-Strand cDNA Synthesis

1. 5X AMV reverse transcriptase buffer: 250 mM Tris-HCl, pH 8.5 (20°C), 40 mM MgCl$_2$, 150 mM KCl, 5 mM DTT.
2. AMV reverse transcriptase (Boehringer Mannheim, Mannheim, Germany).
3. Thermus thermophilus (rTth) heat-stable polymerase (Perkin-Elmer, Branchburg, NJ, or Epicentre Technology, Madison, WI).
4. RNasin (Promega, Madison, WI).
5. First-strand cDNA synthesis reaction mix: 20 μL 5X first-strand buffer, 10 μL 10 mM dNTP, 100 U AMV reverse transcriptase (25 U/μL, Boehringer Mannheim) and H$_2$O to 100 μL.

2.3. Subtractive Hybridization

1. 20X SSPE: 3M NaCl, 0.15M NaH$_2$PO$_4$, 0.02M EDTA, pH 7.4.
2. Hybridization buffer: 4.5X SSPE, 0.2% SDS.
3. Loading buffer: 0.25% bromophenol blue, 30% glycerol, 40 mM EDTA.

2.4. Generation of a Subtractive Probe

2.4.1. Direct Labeling

1. cDNA synthesis kit (Promega or Amersham, Buckinghamshire, UK).
2. [α^{32}P]-dCTP (e.g., Amersham, cat. no. PB 165).

2.4.2. Amplification

1. Terminal deoxynucleotide transferase (TdT).
2. 5X TdT buffer: 0.5M potassium cacodylate, pH 7.2, 10 mM CoCl$_2$, 1 mM DTT.
3. *Taq* polymerase mixture: 10 mM Tris-HCl, pH 8.3, 50 mM KCl, 100 μg/mL bovine serum albumin, 0.05% Triton X-100, 2.5 mM MgCl$_2$, 0.2 mM dNTPs, 2.5 U *Taq* polymerase (Perkin-Elmer).
4. T-primer: 5'-dTTGCATTGACGTCGACTATCCAGGTTTTTTTTTTTTTTT-3'.
5. L-primer: 5'-dTTGCATTGACGTCGACTATCCAGG-3'.

3. Methods

In this section, we present a detailed description of the subtraction method starting with the isolation of mRNA from subtractor cells and the generation of a solid-phase bound subtractor cDNA library, followed by the subtractive hybridization and the generation of subtracted probes (*see* Notes 1, 2, and 3).

3.1. Preparation of Immobilized First-Strand Subtractor cDNA

Construction of a subtracted cDNA library requires high-quality mRNA from the tissue or cells of interest. The method described here is rapid and minimizes the risk of degradation of nucleic acid. mRNA can be isolated directly from crude lysate obviating the need first to isolate total RNA *(15)*.

The average yield of polyadenylated mRNA per gram of solid tissue is reported to vary between 200 and 400 µg (*see* Note 4). In this chapter, we present as an example the subtraction between cells of lymphoid origin.

3.1.1. Isolation of Subtractor mRNA (see Note 5)

1. Collect 50 million lymphoid cells (cell line) by centrifugation, and wash once in 10-mL cold RNase-free PBS.
2. Lyse in 0.5 mL cold PBS with 0.5% NP-40 (lysis buffer), transfer to a microcentrifuge tube, and keep on ice for 2 min.
3. Spin the lysate for 2 min at maximum speed (10,000g) in a cold microcentrifuge to pellet the nuclei and transfer the supernatant to a new microfuge tube on ice, leaving the nuclei behind. (Successful lysis leaves a viscous pellet that cannot be resuspended.)
4. Wash 5 mg (1 mL) of Dynabeads oligo(dT)25 once in 0.5 mL cold 2X lysis/binding buffer, resuspend the beads in 0.5 mL of the same buffer, and combine with the cell lysate without nuclei.
5. Mix the beads and lysate for 5 min at 4°C to achieve hybridization of polyadenylated mRNA to the oligo(dT) on the beads. The incubation can be performed on ice, including mixing the tube twice during the incubation, or alternatively, we prefer to perform the incubation on a rolling wheel at 4°C to obtain optimal mixing.
6. After binding of mRNA, wash the beads twice in 0.5–1 mL wash buffer 1 and then twice with wash buffer 2. Transfer the suspension to a new RNase-free tube between the two last washing steps.
7. Finally, wash the mRNA/Dynabead complexes once in reverse transcriptase buffer or three times if 1% detergent was used in the lysis buffer. The extensive washing is important to remove all traces of LiCl and detergent, which will inhibit subsequent enzymatic reactions. The yield of subtractor mRNA should be approx 10 µg (*see* Note 6).

3.1.2. Solid-Phase First-Strand cDNA Synthesis

The oligo(dT) sequences covalently bound the to bead surfaces are used both to capture the polyadenylated mRNA and as primers for a reverse transcriptase reaction to synthesize the first-strand cDNA. This results in a covalently immobilized first-strand cDNA library. The mRNA bound to the oligo-dT Dynabeads is directly subjected to first-strand cDNA synthesis.

1. Remove the last washing buffer and resuspend the beads from Section 3.1.1. in 100 µL of a first-strand cDNA reaction solution composed of: 20 µL 5X first-strand buffer (Boehringer Mannheim), 10 µL 10 mM dNTP, 100 U AMV reverse transcriptase (Boehringer Mannheim), RNase inhibitor (RNasin, Promega), and H$_2$O to 100 µL.
2. Place the microfuge tube in a horizontal position in an incubator with a rotating wheel at 42°C, and incubate for 1 h to ensure efficient enzymatic reactions.

3. After incubation at 42°C, add 5 μL of 10 mM MnCl$_2$ and 2.5 U rTth heat-stable polymerase to the mixture, and incubate for 10 min at 70°C (*see* Note 7).

4. After completion of the first-strand cDNA synthesis, wash the subtractor beads carefully twice in 500 μL TE (pH 8.0). The subtractor beads can then be stored in TE (pH 8.0) at 4°C for at least 1 mo (*see* Note 8).

5. Before use in hybridization procedures, resuspend the subtractor beads in 100 μL H$_2$O, and incubate at 70°C for 5 min, and then remove the H$_2$O to ensure removal of subtractor mRNA. Do not let the beads dry out, but store them in TE or hybridization buffer.

3.2. mRNA Isolation from Target Cells (see Note 9)

1. Collect approx 2 million cells (lymphoid cell lines), and wash once in ice-cold PBS.

2. Lyse cells in 100 μL PBS containing 0.5% NP-40 for 2 min on ice.

3. Remove the nuclei by centrifugation at 10,000g in a cold microcentrifuge for 2 min.

4. Prewash 250 μg (50 μL) of Dynabeads oligo(dT)25 in 200 μL 2X lysis/binding buffer, resuspend the beads in 100 μL of the same buffer, and combine with the cell lysate without nuclei. Mix the beads with the lysate by constant rolling at 4°C for 5 min as described in Section 3.1.1.

5. After binding, wash the mRNA/Dynabead complexes twice in 200 μL wash buffer 1. It is not necessary here to perform the extensive wash step described under Section 3.1.1., since the target mRNA is not subjected to any enzymatic reactions at this stage.

6. Elute the mRNA from the beads in 30 μL H$_2$O for 5 min at 70°C. The beads can be immediately collected by a magnet.

7. Transfer the mRNA-containing supernatant to a new microcentrifuge tube on ice, and prepare for the subtraction by adding 400 μL hybridization buffer. The final concentration of the hybridization buffer should be 4.5X SSPE and 0.1% SDS (*see* Note 10).

3.3. Subtractive Hybridization (see Note 11)

To prevent evaporation, we recommend to use microcentrifuge tubes with screw caps for subtractive hybridization. As described, isolated mRNA from the target population is dissolved in 400 μL hybridization buffer (4.5X SSPE, 0.1% SDS).

1. Heat both the subtractor cDNA beads in TE and the target mRNA in hybridization buffer simultaneously to 70°C for 3 min.

2. Remove the TE from the subtractor beads by capture of the beads with a magnet, and resuspend the beads in the hybridization buffer containing the target mRNA. Seal the cap of the microtube in Parafilm™ to prevent evaporation, and incubate the hybridization mixture at 68°C for 20–24 h under constant rolling (upside down) to achieve even and proper mixing of the subtractor beads with target mRNA.

3. After the first round of hybridization, transfer the tube from the 68°C incubator directly to ice, and collect the mRNA/cDNA bead complex using a magnet on ice.

4. Transfer the hybridization solution containing target-specific mRNA to a new tube, and store on ice, awaiting further rounds of hybridization.
5. Regenerate the subtractor beads by adding 25 µL H_2O followed by 5 min heating at 70°C (elution of mRNA). After removing the H_2O, resuspend the beads in TE.
6. After elution of mRNA, resuspend the subtractor beads in 100 µL TE ready for further rounds of hybridization.
7. Heat the subtractor beads and the hybridization solution separately to 70°C as in step 1, and resuspend the beads in the hybridization solution containing target mRNA.
8. Perform the hybridization as described in steps 1–6.
9. Repeat the process once more—in total, three rounds of hybridization.
10. After the last hybridization step, remove the subtractor beads from the hybridization solution by a magnet. The supernatant contains the cell-specific mRNA from the target population.
11. Isolate the specific target mRNA by adding to the hybridization solution 25 µL of Dynabeads oligo(dT)25, prewashed in hybridization buffer, followed by a 5-min incubation at 25°C (room temperature) preferably on a rotating wheel.
12. Wash the oligo(dT) beads with bound specific mRNA twice with wash buffer 1 and then twice with wash buffer 2 as described in Section 3.1.1. before resuspending the beads in first-strand cDNA buffer.
13. Keep the beads on ice awaiting further processing.

3.4. Generation of Subtractive Probes

Subtractive cDNA probes can be generated for screening cDNA libraries from the specific target mRNA. Here we present two different approaches. One approach directly generates a ^{32}P-labeled double-stranded cDNA from the target-specific mRNA using high amounts of radioactivity. The second approach employs the polymerase chain reaction *(16)* followed by radioactive labeling of the material.

3.4.1. Direct Labeling of Target-Specific mRNA (see Note 14)

1. Bind the specific mRNA obtained after the final hybridization to oligo-dT beads as described in Section 3.3.
2. Elute the mRNA from the beads by incubation in 15 µL H_2O for 3 min at 65°C. The mRNA is now ready for the cDNA synthesis reaction.
3. Convert the specific mRNA to a double-stranded radiolabeled cDNA probe according to a cDNA synthesis kit protocol (e.g., Promega or Amersham) or other cDNA synthesis protocols *(17)*, except that the final concentration of dCTP in the first-strand cDNA reaction is 0.25 mM instead of 1 mM, whereas 1 mCi of $[\alpha^{32}P]$dCTP (3000 Ci/mM; Amersham) is included in the reaction mixture. The radioisotope is shipped in EtOH and must be dried prior to use.
4. After the second-strand cDNA synthesis is finished, remove free $[\alpha^{32}P]$dCTP from the synthesis mixture by spinning through a Sephadex G-50 column together with 2 µg of sheared salmon sperm DNA (to avoid loss of material in the column owing to the low amount of cDNA).

5. The radiolabeled cDNA is now, after boiling, ready for use as a probe for the screening of existing cDNA libraries either in the form of plasmids (colony screening) or phages (plaque screening) according to standard protocols *(17,18)*.

3.4.2. Amplification of Target-Specific mRNA (see Note 15)

1. Prewash 25 μL oligo-dT Dynabeads in hybridization buffer, and add to the mRNA in hybridization buffer from Section 3.3., step 11.
2. Incubate at room temperature (25°C) for 5 min preferably on a rolling wheel.
3. After extensive washing of the mRNA/bead complexes (Section 3.1.2.), generate immobilized first-strand cDNA by adding first-strand cDNA synthesis mix: 5 μL 5X reverse transcriptase buffer, 2.5 μL 10 mM dNTP, 15 U reverse transcriptase (Boehringer Mannheim), RNase inhibitor, and H_2O to a final volume of 25 μL. Incubate at 42°C for 1 h under constant rolling.
4. After completion of the cDNA synthesis, wash the cDNA/bead complexes twice in TE.
5. Take the beads up in 20 μL H_2O, and incubate at 70°C for 5 min to elute the mRNA.
6. Collect the first-strand cDNA beads using the magnet, and wash once in TE.
7. To tail the cDNA with terminal transferase and dATP, set up a 20-μL reaction containing the immobilized first-strand cDNA with 22 U terminal transferase (Gibco), 1.5 mM dATP and terminal transferase buffer. Incubate for 15 min at 37°C, and then stop with the addition of 2 μL 0.5M EDTA.
8. Wash the beads once in TE before adding 50 μL of *Taq* polymerase mix containing 29 pmol of T-primer. Extend the primer for 15 min at 30°C, 15 min at 40°C, and 15 min at 72°C in an incubator under constant rolling (tube in horizontal position).
9. Discard the supernatant, and add 50 μL of fresh *Taq* polymerase mix containing 50 pmol of L-primer and 1 pmol of T-primer, and heat at 94°C for 2 min to release the second-strand cDNA.
10. Transfer the supernatant to a new tube, and add a drop of mineral oil (when thermal cyclers with heated lids are used, mineral oil can be omitted) *(see* Note 16).
11. Incubate the supernatant for 15 min at 30°C, 15 min at 40°C, and 15 min at 72°C to extend the T-primer.
12. After 2 min at 94°C, carry out 15 cycles of PCR amplification (94°C for 1 min, 72°C for 5 min) followed by a 10-min extension at 72°C.
13. Reamplify 5 μL of this reaction for 20 additional cycles (94°C for 1 min, 72°C for 5 min) in *Taq* polymerase mix containing 50 pmol of L-primer.
14. Spin the product through a Sephacryl-400 spin column (Pharmacia, Uppsala, Sweden) to remove nucleotides, perform a phenol/chloroform extraction, 'and precipitate with $^1/_{10}$ vol 5M NaCl and 2.5 vol ethanol.

The material is now ready for radioactive labeling either by nick translation or by random hexamer labeling using $^1/_5$ for a single-labeling reaction. The probe is boiled and used for screening cDNA libraries according to standard protocols *(17,18)*. A discussion of the applications of these techniques is given in Note 17.

4. Notes

1. To isolate genes that are differentially expressed successfully, it is important to follow the protocol in detail. It is also important that all solutions are RNase-free and that the isolation procedure of mRNA works well to obtain high-quality mRNA.
2. In general, Dynabeads with immobilized mRNA or cDNA should be handled with care to avoid shearing of the nucleic acids. Always resuspend the beads carefully by pipeting slowly up and down.
3. Microcentrifuge tubes are siliconized by coating them with Sigmacote solution (Sigma) followed by air-drying to avoid loss of material by nonspecific binding of nucleic acid to plastic.
4. To use the optimal ratio of magnetic beads and RNA, it is important to determine the mRNA content of the cells or tissues by optical density measurement (OD_{260}). The capacity of the Dynabeads oligo(dT)25 is 2 µg mRNA/mg of beads. In our present and previous work, we have worked with either cell lines of hematopoietic origin or normal lymphocytes. The yield of isolated mRNA from 50 million subtractor cells of lymphoid origin (lymphoid cell lines) is approx 10 µg mRNA. Cultured lymphoid cells are collected by centrifugation and washed once in ice-cold PBS before mRNA isolation. Adherent cell lines can be detached from plastic by using a scraping device, treatment with PBS/0.5 m*M* EDTA, or trypsinization followed by washing in PBS prior to mRNA isolation. Subpopulations of blood cells can be isolated by immunomagnetic separation using magnetic beads coated with cell-specific antibodies *(19)*. The captured cells are washed with cold PBS while attached to the beads and subsequently lysed before removal of the beads followed by mRNA isolation (*Dynal Technical Handbook*, 2nd ed.). When working with solid tissue, we recommend freezing the samples in liquid nitrogen immediately after dissection and storing them at −80°C until extraction. Up to 50 mg of animal tissue and 100 mg of plant tissue can be used per mg of Dynabeads oligo(dT)25 *(20)*.
5. The lysis of several types of hematopoietic cells can be done with a mild nonionic detergent, which will disrupt the cells, leaving the nuclei intact. The nuclei are removed by centrifugation before performing the mRNA isolation. The same lysis condition can be used for adherent cells, such as HepG2, HeLa, and COS cells. For most other cell types and all solid tissues, a more robust method of lysis is necessary. We recommend using lysis buffer with 1% SDS (2X lysis/binding buffer, 1% SDS). Solid tissues are ground to fine powder in a liquid nitrogen in a mortar before homogenization in a glass homogenizer with 2X lysis/binding buffer with 1% SDS.
6. To determine that the mRNA isolation procedure works optimally, one can elute $^1/_{10}$ of a large-scale isolation and quantify the amount by measuring the OD_{260}. One can also elute and load $^1/_{10}$ of one isolation, estimated to yield 10 µg mRNA, in a small well in a 1% agarose gel used for normal DNA separation (without formamide), and then run it shortly thereafter into the gel. The mRNA (approx 1 µg) will be clearly visualized as a smear between 0.5 and 4 kb. Add cold RNase-free loading buffer to the mRNA, and load it quickly into the well of the gel. Our

experience is that <0.2 µg mRNA can be detected on a 1% agarose gel if loaded in a small well and the gel is run for a short time. This means that it is possible to visualize both the first and second elution of hybridized target mRNA from the subtractor beads after the subtractive hybridizations. Alternatively, the sensitivity can be increased by staining the gel with Sybgreen II (Molecular Probes, Eugene, OR) instead of ethidium bromide. Both optical density measurement and visualization of mRNA on a gel are crude ways of assessing the quality of mRNA. Performing a Northern blot and hybridization of a probe directed toward the 5'-end of a moderate to abundantly expressed mRNA will assess more accurately the quality of the isolated mRNA.

7. The rTth DNA polymerase can act as a reverse transcriptase in the presence of Mn^{2+}. In our hands, the addition of rTth polymerase increases the efficiency of first-strand cDNA synthesis by 60–80% as measured by the incorporation of radioactive nucleotides (^{32}P-dCTP). However, the addition of the rTth polymerase is considered an optional step.

8. With an estimated 25% efficiency of the first-strand cDNA synthesis reaction, approx 2.5 µg of cDNA are generated from 10 µg mRNA.

9. From the target cells, about 0.3 µg of mRNA is isolated. This gives an 8- to 10-fold excess of subtractor over target mRNA. Cultured cells, blood cells, or solid tissue can be used as a target for mRNA isolation as described for the subtractor mRNA isolation (Section 3.1.1.).

10. Alternatively, larger amounts of mRNA from the target cells can be isolated and stored in ethanol at −70°C or in the hybridization buffer at the same temperature ready for use.

11. Subtraction hybridization of mRNA to immobilized cDNA is presumed to follow a second-order kinetic reaction *(21)* and is dependent on such factors as nucleic acid concentration, temperature, salt concentration, and time. It is uncertain if the presence of the Dynabeads has any direct influence on the hybridization process. During the subtraction, abundantly expressed mRNAs will be removed first. This means that if the material is not optimally subtracted, the detection of differentially expressed sequences will be difficult. We therefore chose to perform three hybridization steps on the same material to ensure optimal subtraction. The mRNA can be visualized after each round of subtraction on an agarose gel, showing a clear decrease in intensity as visualized on UV light after the third round compared with the first round of hybridization. In case no decrease is observed, we recommend proceeding to a fourth round of subtractive hybridization. The subtractive hybridization is performed in a buffer containing 0.75M NaCl and at 68°C. One can imagine that under more stringent salt conditions, the loss of highly homologous nucleic acids in subtractor and target population will diminish. We have observed that lowering the hybridization temperature to 65°C results in the loss of homologous nucleic acids. The subtractor beads need to be completely removed from the hybridization solution after the last hybridization step before oligo(dT)25 beads are added to bind the target-specific mRNA. This is important to avoid the cDNA on the subtractor beads giving you any problems

when generating a subtractive probe, in particular when the specific target mRNA is amplified by PCR before the generation of a probe. We therefore recommend performing a second collection of the subtractor beads from the hybridization solution. When working with low amounts of beads, a fast spin (30 s) at maximum speed in a microcentrifuge facilitates the concentration of beads, and facilitates the separation of beads from the supernatant. We normally use 70°C when eluting mRNA from the subtractor beads, but the temperature can be raised to 85–90°C to assure complete elution of mRNA.

12. At this point, the mRNA eluted from the subtractor beads can be visualized on a normal 1% agarose gel used for DNA separation. Add 3 μL of 6X RNase-free loading buffer to the eluted mRNA and load directly in a small well of the agarose gel. Perform a short electrophoresis, and visualize the mRNA on UV light. The mRNA should now be seen as a smear between 0.5 and 4 kb in size.

13. We add 250 ng of yeast tRNA to the hybridization solution prior to the last round of subtraction to avoid loss of the low amount of specific mRNA through nonspecific binding to the beads or to the microcentrifuge tube.

14. As mentioned in Section 3.4.1., the generation of a probe from specific target mRNA left after the third round of hybridization may be difficult owing to the limited amount of material. In those instances, it is recommended either to scale up the protocol (starting with 20 μg of mRNA from the subtractor cells and 0.6 μg of the target population) or pool material from two or three independent subtractions. The PCR-based approach for the generation of a radiolabeled probe allows a qualitative assessment of the subtraction procedure. One-tenth of the PCR-amplified material is subjected to Southern blotting *(17)* and the filter probed with an abundantly expressed gene, like actin. If no signal or greatly diminished signal is obtained with an actin probe compared to nonsubtracted material, we consider the material to be enriched and proceed to the screening of a cDNA library using this probe. All cDNA clones isolated from the cDNA library should be used as probes in Northern blots to assure their specific expression pattern.

15. We have slightly modified a protocol recently described by Lambert and Williamson *(16)* that is suitable for the amplification of mRNA from small amounts of cDNA. This method was originally described for the generation of cDNA libraries from small amounts of mRNA. The principle of this method is to generate solid-phase first-strand cDNA on Dynabeads from the enriched target mRNA. This cDNA is polyA-tailed by terminal transferase, and a second-strand cDNA is synthesized with a T-primer containing an anchor (*see* Section 2.4.2.). Subsequently, the material is amplified with the T-primer and its anchor, L-primer (*see* Section 2.4.2.).

16. The beads are stored in TE at 4°C and can be reused to generate another second-strand cDNA.

17. A directed search for differentially expressed genes may be performed using degenerate primers to highly conserved regions of members of the gene family of interest as has been described for protein tyrosine kinases *(22)*. In this case, the

solid-phase bound first-strand cDNA from the enriched target mRNA serves as a template in the PCR reaction. Although the amount of specific target mRNA may be limited, up to 50 cycles in the PCR allows the selective amplification of members of, e.g., the protein tyrosine kinase family, including novel members. The removal of most of the common sequences by subtractive hybridization and reduction of the complexity of the mRNA population left results in less problems with background when performing low stringency degenerate PCR. One can overcome the bias in this type of PCR owing to differences in expression level of the genes and/or more favorable usage of the degenerated primers for some sequences of the gene family than for others. By this approach, subtraction and enrichment can lead to the discovery of novel and differentially expressed sequences, e.g., the usage of PTK1 and PTK2 primers *(22)* to isolate protein tyrosine kinases has been suggested to have reached the extent of its usefulness, since the sequences amplified using this primer set tend to be the same using different tissues *(23)*. Using the same primer set, we have isolated several novel tyrosine kinase sequences that we are now in the process of characterizing (Aasheim et al., unpublished). In addition, immobilized cDNA gives less background problems in the PCR than cDNA generated in solution without beads. One reason for this is that the first-strand cDNA beads can be washed after cDNA synthesis to remove contaminants (e.g., free oligo[dT] primers, enzymes, buffer differences) from the reverse transcriptase reaction. These contaminants will still be present after cDNA synthesis in solution and may generate a background that appears as a smear on an agarose gel. Low background also makes it easier to clone the material in the vector of choice for further sequencing.

Acknowledgment

This work was supported by the Norwegian Council of Research.

References

1. Hedrick, S. M., Cohen, D. I., Nielsen, E. A., and Davies, M. M. (1984) Isolation of cDNA clones encoding T-cell specific membrane associated proteins. *Nature* **308,** 149–153.
2. Kwon, B. S., Kim, G. S., Ptrystowsky, M. B., Lancki, D. W., Sabath, D. S., Panand, J., and Weissman, S. M. (1987) Isolation and initial characterization of multiple species of T-lymphocyte subset cDNA clones. *Proc. Natl. Acad. Sci. USA* **84,** 2896–2900.
3. Swaroop, A., Xu, J., Agarwal, N., and Weissman, S. M. (1991) A simple and efficient cDNA library procedure: isolation of human retina-specific cDNA clones. *Nucleic Acids Res.* **19,** 1954.
4. Schneider, C., King, R. B., and Philipson, L. (1988) Genes specifically expressed at growth arrest of mammalian cells. *Cell* **54,** 787–793.
5. Yancopoulos, G. D., Oltz, E. M., Rathbun, G., Berman, J., Lansford, R. D., Rothman, P., Okada, A., Lee, G., Morrow, M., Kaplan, K., Prockop, S., and Alt, F. (1990) Isolation of coordinately regulated genes that are expressed in discrete stages of B cell development. *Proc. Natl. Acad. Sci. USA* **87,** 5759–5763.

6. Zipfel, P. F., Irving, S. G., Kelly, K., and Siebenlist, U. (1989) Complexity of the primary genetic response to mitogenic activation of human T cells. *Mol. Cell. Biol.* **9**, 1041–1048.

7. Zipfel, P. F, Balke, J., Irving, S. G., Kelly, K., and Siebenlist, U. (1990) Mitogenic activation of human T cells induces two closely related genes which share structural similarities with a family of secreted factors. *J. Immunol.* **142**, 1582–1590.

8. Duguid, J. R. and Dinauer, M. C. (1990) Library subtraction of in-vitro cDNA libraries to identify differentially expressed genes in scrapie infection. *Nucleic Acids Res.* **18**, 2789–2792.

9. Timblin, C., Battey, J., and Kuehl, W. M. (1990) Application for PCR technology to subtractive cDNA cloning: identification of genes expressed specifically in murine plasmacytoma cells. *Nucleic Acids Res.* **18**, 1587–1593.

10. Schweinfest, C. W., Henderson, K. W., Gu, J.-R., Kottaridis, S. T., Besbeas, S., Panotopoulou, E., and Papas, T. S. (1990) Subtraction hybridization cDNA libraries from colon carcinomas and hepatic cancer. *Genet. Ann. Technol. Appl.* **7**, 64–70.

11. Rodriguez, I. R. and Chader, G. J. (1992) A novel method for the isolation of tissue specific genes. *Nucleic Acids Res.* **20**, 3528.

12. Aasheim, H.-C., Deggerdal, A., Smeland, E. B., and Hornes, E. (1994) A simple subtraction method for the isolation of cell-specific genes using magnetic monodisperse polymer particles. *BioTechniques* **4**, 716–721.

13. Sharma, P., Lonneborg, A., and Stougaard, P. (1993) PCR-based construction of subtractive cDNA libraries using magnetic beads. *BioTechniques* **15**, 610–611.

14. Shraml, P., Shipmann, R., Stultz, P., and Ludwig, C. U. (1993) cDNA subtraction library construction using a magnet-assisted subtraction technique (MAST). *Trends Genet.* **9**, 70,71.

15. Jacobsen, K. S., Breivold, E., and Hornes, E. (1990) Purification of mRNA directly from crude plant tissue in 15 minutes using magnetic oligo dT microspheres. *Nucleic Acids Res.* **18**, 3669.

16. Lambert, K. N. and Williamson, V. M. (1993) cDNA library construction from small amounts of RNA using paramagnetic beads and PCR. *Nucleic Acids Res.* **21**, 775,776.

17. Sambrook, J., Fritch, E. F., and Maniatis, T. (1989) *Molecular Cloning: A Laboratory Manual.* Cold Spring Harbor Laboratory Press, Cold Spring Harbor, NY.

18. Grunstein, M. and Hogness, D. (1975) Colony hybridization: A method for the isolation of cloned DNA's that contain a specific gene. *Proc. Natl. Acad. Sci. USA* **72**, 3961–3965.

19. Funderud, S., Nustad, K., Lea, T., Vartdal, F., Gaudernack, G., Stenstad, P., and Ugelstad, J. (1987) Fractionation of lymphocytes by immunomagnetic beads, in *Lymphocytes: A Practical Approach* (Klaus, G. G. B., ed.), IRL, Oxford, UK, pp. 55–65.

20. Jacobsen, K. S., Haugen, M., Saeboe-Larsen, S., Hollung, K., Espelund, M., and Hornes, E. (1994) Direct mRNA isolation using magnetic oligo(dT) beads: a protocol for all types of cell cultures, animal and plant tissues, in *Advances in Biomagnetic Separation* (Uhlen, M., Hornes, E., Olsvik, O. E., eds.), Eaton Publishing, pp. 61–72.

21. Milner, J. J., Cecchini, E., and Dominy, P. J. (1995) A kinetic model for subtractive hybridization. *Nucleic Acids Res.* **23,** 176–187.

22. Wilks, A. F. (1989) Two putative protein-tyrosin kinases identified by application of the polymerase chain reaction. *Proc. Natl. Acad. Sci. USA* **86,** 1603–1607.

23. Larsom-Blomberg, L. and Dzierzak, E. (1994) Isolation of tyrosine kinase related genes expressed in the early hematopoetic system. *FEBS Lett.* **348,** 119–125.

10

Preparation of Competent Cells for High-Efficiency Plasmid Transformation of *Escherichia coli*

Yasuhiko Nakata, Xiaoren Tang, and Kazunari K. Yokoyama

1. Introduction

Transformation of *Escherichia coli* was first described by Mandel and Higa *(1)*, who reported that *E. coli* cells, after treatment with calcium chloride, can take up bacteriophage λ DNA and produce viable phage particles. The conditions for the transfer of exogenous DNA into *E. coli* have been examined in detail in studies of bacteriophage transfection, genetic transformation, and plasmid transformation. Modifications that improve the efficiency of transformation include prolonged exposure of cells to $CaCl_2$ *(2)*, substitution of Ca^{2+} ions by other cations, such as Rb^+ *(3)*, Mn^{2+}, and K^+, and addition of other compounds, such as dimethyl sulfoxide (DMSO), dithiothreitol, and cobalt hexaminechloride *(4)*. The improvements in the frequency of transformation have been made solely as a consequence of empirical experimentation.

In the basic protocol, the preparation of competent cells for high-efficiency transformation appears to require:

1. Harvesting of bacterial cells at the logarithmic phase of growth;
2. Incubation of cells on ice throughout the procedure; and
3. Prolonged exposure of $CaCl_2$.

With respect to cell growth, it is accepted that cells should be harvested at the logarithmic phase of growth and their density should not exceed 1×10^8 viable cells/mL ($OD_{600} = 0.3$–0.4). However, it is unknown why this density of *E. coli* cells (1×10^8 viable cells/mL) is so suitable for inducing a transient state of "competence." To examine this issue, we prepared competent cells from cultures with densities from 5×10^7 to 4×10^8 cells/mL ($OD_{600} = 0.3$–0.98) and transformed them with supercoiled pBR322 plasmid DNA by the standard

From: *Methods in Molecular Biology, Vol. 69: cDNA Library Protocols*
Edited by: I. G. Cowell and C. A. Austin Humana Press Inc., Totowa, NJ

"calcium chloride" method with slight modifications for an analysis of the competence of the cells *(5)*. We found that a graph of the efficiency of transformation against cell density yield a biphasic curve, with maxima at two optical densities (OD_{600}), namely, 0.3–0.4 and 0.94–0.95. The cells harvested at an optical density of 0.94–0.95 had twice the transformation efficiency of cells harvested at the logarithmic phase of growth. This new and unexpected observation allows us to prepare competent cells that are able to take up DNA at higher efficiency than when the usual protocol is followed and to harvest four to eight times more competent cells from a given culture.

2. Materials
2.1. Basic Protocol Using Calcium Chloride

1. Single colony of *E. coli* cells.
2. Lucia-Bertani (LB) medium: 10 g tryptone, 5 g yeast extract, 10 g sodium chloride/L. Sterilize by autoclaving.
3. SOB medium: 20 g tryptone, 5 g yeast extract, 0.5 g sodium chloride/L. Sterilize by autoclaving, and add 5 mL of $2M$ $MgCl_2$.
4. $0.1M$ $CaCl_2$.
5. LB plates prepared with ampicillin: 10 g tryptone, 5 g yeast extract, 10 g sodium chloride, 15 g bacteriological agar/L. Sterilize by autoclaving, and add ampicillin at 50 mg/mL.
6. Plasmid DNA.
7. 50-mL Polypropylene tubes (Falcon 2070, Becton Dickinson).
8. 15-mL Polypropylene tubes (Falcon 2069, Becton Dickinson).

2.2. Additional Materials for the One-Step Preparation and Transformation of Competent Cells

1. 2X Transformation and storage solution (TSS): dilute a sterile 40% (w/v) solution of polyethylene glycol (PEG) 3350 to 20% PEG with sterile LB medium that contains 100 mM $MgCl_2$. Add DMSO to 10% (v/v), and adjust to pH 6.5.
2. Ice-cold LB medium (*see* Section 2.1.) that contains 20 mM glucose; identical LB medium with addition of 20 mM glucose.

2.3. High-Efficiency Transformation by Calcium Chloride

1. Single colony of *E. coli* cells.
2. LB medium (*see* Section 2.1.).
3. SOC medium: identical to SOB medium (*see* Section 2.1.) with the addition of 20 mM glucose.
4. 80 mM and $0.1M$ $CaCl_2$.
5. 50 mM $MgCl_2$.
6. 20 mM $MgSO_4$.
7. 15% Glycerol.

8. 10 mM Piperazine-N,N'-bis(2-ethane sulfonic acid) (PIPES), pH 7.0.
9. LB plates containing ampicillin (*see* Section 2.1.).
10. Plasmid DNA.
11. 50-mL Polypropyrene tubes (Falcon 2070, Becton Dickinson).
12. 15-mL Polypropylene tubes (Falcon 2096, Becton Dickinson).

3. Method

3.1. Basic Protocol Using Calcium Chloride

The basic procedure of transformation of *E. coli* cells was originally described by Mandel and Higa *(1)* and Cohen et al. *(6)*. This procedure is frequently used to prepare batches of competent cells that yield 5×10^6–5×10^7 transformed colonies/µg of supercoiled plasmid DNA. This efficiency of transformation is high enough to allow all routine cloning in plasmids to be performed with ease. The procedure works well with most strains of *E. coli*; it is rapid, simple, and reproducible. Competent cells generated by this protocol can be preserved at $-70°C$, although there can be some reduction during storage in the efficiency of subsequent transformation.

3.1.1. Preparation of Competent Cells

1. Pick a single colony (2–3 mm in diameter) from a plate of cells that have been freshly grown for 16–20 h at 37°C, and transfer it to 50 mL of LB medium in a sterile 300-mL flask. Incubate the culture for 16–20 h at 37°C with moderate shaking (50–60 cycles/min on a rotary shaker).
2. Inoculate 1 mL of the culture into 100 mL of LB medium in a sterile 500-mL flask. Grow cells at 37°C for about 3 h with vigorous shaking (300 cycles/min on a rotary shaker). For efficient transformation, it is essential that the number of viable cells do not exceed 10^8 cells/mL. To monitor the growth of the cells, determine OD_{600} every 20–30 min (*see* Note 1).
3. Transfer the cells aseptically to two 50-mL prechilled, sterile polypropylene tubes. Leave the tubes on ice (0°C) for 10 min (*see* Note 2).
4. Centrifuge cells at 2400g for 10 min at 4°C in a CX-250 or its equivalent (Tomy, Tokyo, Japan).
5. Pour off the supernatant, resuspend each pellet of cells in 10 mL of ice-cold 0.1M CaCl$_2$, and store on ice (*see* Note 3).
6. Recover the cells by centrifugation at 2400g for 10 min at 4°C as indicated in step 4. Discard the supernatant, and resuspend each pellet in 10 mL of an ice-cold solution of 0.1M CaCl$_2$. Keep the resuspended cells on ice for 30 min.
7. Centrifuge cells at 2400g as in step 4 for 10 min at 4°C. Discard the supernatant, and resuspend each pellet in 2 mL of the ice-cold solution of 0.1M CaCl$_2$ (*see* Note 4).
8. Dispense cells into prechilled, sterile polypropylene tubes (aliquots of ~250 mL are convenient). Freeze immediately at $-70°C$.

Dagert and Ehrlich *(3)* showed that cells can be stored at 4°C in a solution of 0.1M CaCl$_2$ for 24–48 h. The efficiency of transformation increases four- to six-fold during the first 12–24 h of storage and then decreases to the original level.

3.1.2. Check the Competence of Cells

1. Use 10 ng of pBR322 to transform the competent cells in 200 μL of the suspension of cells by the following procedure described. Plate appropriate aliquots (1, 10, and 25 mL) of the transformed culture on LB/plates with ampicillin, and incubate at 37°C overnight.
2. Count the number of transformed colonies per microgram of DNA (*see* Note 5).

3.1.3. Transformation of Competent Cells

1. Place 10 ng of DNA in a volume of 10 μL (no more than 50 ng in a volume of 10 μL or less) in several sterile 15-mL, round-bottomed test tubes and place on ice.
2. Rapidly thaw competent cells by warming in your hand and dispense aliquots of 200 μL of the suspension of cells immediately into the test tubes that contain DNA. Mix the contents of the tubes by swirling gently. Place the tubes on ice for 30 min (*see* Note 6).
3. Subject the cells to heat shock by placing tubes in a water bath of 42°C for 2 min. Do not shake the tubes (*see* Note 7).
4. Rapidly transfer the tubes to an ice bath. Allow the cells to chill for 1–2 min.
5. Add 1 mL of LB medium (or SOC medium) into each tube. Place each tube on a rotary shaker operated at 75g for 1 h at 37°C to allow the bacteria to recover and to express the antibiotic resistance marker encoded by the plasmid.
6. Plate aliquots of the transformed culture on LB plus ampicillin plates or other appropriate antibiotic-containing plates.
7. Leave the plates at room temperature until the liquid has been adsorbed.
8. Invert the plates and incubate at 37°C. Colonies should appear in 12–16 h (*see* Notes 8 and 9).

3.2. Alternative Protocol for the One-Step Preparation and Transformation of Competent Cells

Chung et al. *(7)* reported a procedure that is considerably easier and simpler than the basic protocol. It eliminates the need for centrifugation, washing, heat shock, and long periods of incubation. Moreover, competent cells prepared by this simple procedure can be frozen directly at −70°C for long-term storage. A variety of strains can be rendered competent by this procedure.

1. Dilute a fresh overnight culture of bacteria 100-fold in LB medium and incubate at 37°C until the cells reach an OD$_{600}$ of 0.3–0.4.
2. Add a volume of ice-cold 2X TSS equal to that of the cell suspension, and mix gently on ice.

3. Place 100 μL of the suspension of competent cells and 1–5 μL of a solution of DNA (0.1–100 ng) in an ice-cold polypropylene tube, and incubate for 5–60 min at 4°C (*see* Note 10).

4. Add 0.9 mL of LB medium that contains 20 m*M* glucose and incubate for 30–60 min at 37°C with gentle shaking to allow expression of the antibiotic resistance gene (*see* Note 11).

3.3. High-Efficiency Transformation by Calcium Chloride

3.3.1. Preparation of Competent Cells

1. Pick a single colony (2–3 mm in diameter) from a plate of cells that have been freshly grown for 16–20 h at 37°C, transfer it to 50 mL of LB medium in a sterile 300-mL flask, and incubate the culture for 16–20 h with moderate shaking.

2. Inoculate 1 mL of the suspension of cells into 100 mL of LB medium in a sterile 500-mL flask. Incubate cells at 37°C with vigorous shaking (300 cycles/min on a rotary shaker). To monitor the growth of the culture, measure OD_{600} every 20 min.

3. Collect samples at different times during the growth cycle, between optical densities at 600 nm of 0.85 and 0.98. An optical density of 0.94–0.95 is usually best for transformation (*see* Fig. 1).

4. Transfer the cells to four 50-mL prechilled, sterile polypropylene tube. Leave the tubes on ice (0°C) for 5–10 min (*see* Note 12).

5. Centrifuge cells at 2400*g* for 10 min at 4°C.

6. Discard the supernatant, and resuspend each pellet in 2.5 mL of an ice-cold solution of 80 m*M* $CaCl_2$ and 50 m*M* $MgCl_2$, and incubate for 10 min (*see* Notes 13 and 14).

7. Repeat steps 5 and 6 twice.

8. Finally, resuspend the cell pellet in ice-cold 0.1*M* $CaCl_2$ at 5×10^9 cells/mL, and then mix the suspension with an equal volume of 50% glycerol to yield a suspension of competent cells.

9. Transfer competent cells into prechilled, sterile polypropylene tubes (aliquots of ~250 mL are convenient). Freeze immediately at –70°C. The cells retain competence under these condition, although the efficiency of transformation may drop slightly during prolonged storage.

3.3.2. Transformation of Competent Cells

1. Place 80 μL of the suspension of competent cells and 50 pg of pBR322 plasmid DNA in a 15-mL tube, and store on ice for 30 min (*see* Notes 15–17).

2. Subject the cells to heat shock by placing tubes in a water bath of 42°C for 40 s, and then transfer the mixture rapidly to an ice bath for 1 min (*see* Note 18).

3. Add 400 μL of liquid SOC medium to each tube. Place each tube on a rotary shaker at 75*g* for 30 min at 37°C to allow the bacteria to recover and to express the antibiotic resistance marker encoded by the plasmid.

4. Transfer an aliquot of each sample to agar-SOB medium that contains 20 m*M* $MgSO_4$ and the appropriate antibiotic(s), spread the mixture over the surface of an agar plate, and incubate at 37°C overnight (*see* Notes 19–21).

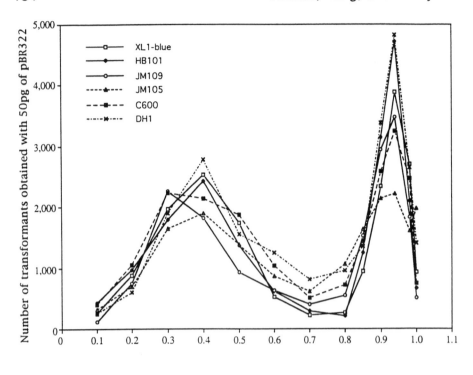

Optical density (OD600 nm)

Fig. 1. Graphs showing the numbers of transformants per 50 pg of supercoiled pBR322 DNA when competent cells of six different strains of *E. coli* were prepared at different times during the course of their culture. The cell density of cultures was monitored in terms of OD_{600} or by counting cells. An optical density of 600 nm of 0.94–0.95 during the culture of *E. coli* cells seems to correspond to the best time for harvesting potentially competent cells of most strains of *E. coli*.

4. Notes

1. This procedure requires that cells are growing rapidly (i.e., it requires cells at early or midlog arithmic phase). Accordingly, it is very important that the growing cells have sufficient air. Overgrowth of the culture decreases the efficiency of transformation. Because the relationship between OD_{600} and the number of viable cells per milliliter varies substantially from strain to strain, it is essential to calibrate the spectrophotometer by measuring OD_{600} of a particular strain of *E. coli* at different times during its growth cycle and determining the number of viable cells at each of these times by plating dilutions of the culture on LB agar plates in the absence of antibiotics.
2. Cells should be kept cold for all subsequent steps.
3. Resuspension should be performed very gently, and cells should be kept on ice.

4. It is important to resuspend this final pellet well. The suspension may be left on ice for several days. For many strains of *E. coli* (one exception is MC1061), competence increases during storage on ice, and it reaches a maximum at 12–24 h. MC1061 cells should be frozen immediately.

5. Transformation efficiencies of 10^7–10^8 and 10^6–10^7 can be obtained for *E. coli* MC1061 and DH1, respectively. The competence of cells decreases very slowly during storage for several months.

6. Be sure to include the following controls in the experiment:

   a. Competent bacteria that are exposed to a known amount of a standard preparation of supercoiled plasmid DNA.

   b. Competent bacteria that are not exposed to any at all.

7. Alternatively, you can incubate cells at 37°C for 5 min for heat shock.

8. When selecting for resistance to ampicillin, transformed cells should be plated at low density ($<10^4$ colonies/90-mm plate), and the plates should not be incubated for more than 20 h at 37°C. β-Lactamase, secreted into the medium by ampicillin-resistant transformants, rapidly inactivates the antibiotic in regions that surround the colonies. Thus, plating cells at high density or incubating them for long periods results in the appearance of ampicillin-sensitive satellite colonies.

9. At least 30 min of cell growth in nonselective medium (outgrowth) and, often, heat shock are necessary if plasmids with the pBR322 tetracycline resistance promoter and gene are to express sufficient protein to allow the cells to form colonies with an efficiency of 1.0 on tetracycline-containing plates. Cells expressing the common plasmid-encoded gene for ampicillin resistance (β-lactamase) may not require such prolonged outgrowth for subsequent formation of colonies on ampicillin-containing plates.

10. The transformation frequency, as measured by monitoring transformation with 1 μg of DNA, is relatively constant for amounts of DNA below 10 ng. However, the frequency decreases at higher concentrations of DNA. The duration of incubation at 4°C is not important.

11. It is unnecessary to expose the transformation mixture to heat shock. The expected frequency of transformation ranges between 10^7 and 10^8 colonies/ng DNA.

12. Cells should be kept cold during all subsequent steps.

13. The pH of the transformation buffer does not affect the frequency in the range of 5.6–7.2. The ability of other cations to replace Ca^{2+} ions has been studied. Salts, such as LiCl, KCl, RbCl, and $HACoCl_3$, had little effect on the frequency of transformation. Both $MnCl_2$ and $MgCl_2$ could replace $CaCl_2$, but many divalent cations other than those of the alkaline earth metals were not effective substitutes (e.g., Zn^{2+}, Co^{2+}, Mo^{2+}).

14. The presence of Mg^{2+} ions at 10–20 m*M* in all media considerably enhances the efficiency of transformation. The optimum concentration of DMSO was 7% for the cold shock and storage of the competent cells. Freezing in liquid N_2 (cold shock) enhanced the transformation efficiency four- to fivefold. Competent cells frozen in the presence of 7% DMSO were stable for more than 2 mo.

15. Be sure to include the following controls in the experiment:
    a. Competent bacteria that are exposed to a known amount of a standard preparation of supercoiled plasmid DNA.
    b. Competent bacteria that are not exposed to plasmid DNA.
16. Only 3–10% of cells are competent to incorporate plasmid DNA. Transformation frequencies decrease with increasing plasmid size, as reported by Hanahan *(4)*. Plasmids up to 10 kb in length are able to transform competent cells with up to a twofold difference in frequency. The transformation frequency is almost equal for plasmids of 4.4–8 kb in size on a molar basis.
17. The dose–response curve is nearly linear over a range of cell to plasmid ratios of 1:2000 to 1:1. Thus, a single plasmid is sufficient to transform a cell. The curve for the *rec⁻* strain DH1 becomes nonlinear at ratios above 2 plasmids/cell, and saturation occurs above ~300 plasmids/viable cell. An increase in the plasmid-to-cell ratios from 1:1 to 200:1 only increases the number of transformed cells several-fold. Thus, exposing cells to large numbers of plasmids relative to the number of cells does not significantly improve the efficiency of transformation.
18. The number of transformants obtained usually increases linearly with increasing numbers of plasmid molecules up to a point, which is reached at about 10 ng of DNA. Beyond this point, the number of transformants does not increase linearly with increasing numbers of plasmid DNA molecules. The volume of the solution of DNA (up to 20 µL) has little effect on the transformation frequency. In our efforts to optimize heat-shock conditions, cells were subjected to a heat pulse at 40, 42, and 44°C for 0–90 s. The highest efficiency was obtained with a heat pulse at 42°C for 40 s.
19. If more than 200 µL of the suspension of transformed competent cells are to be plated on a 90-mm plate, the cells should be concentrated by centrifugation and gently resuspended in an appropriate volume of LB or SOC medium. The entire transformation mixture can be spread on a single plate (or plated in top agar) if resistance to tetracycline is used as the selectable marker. However, in the case of selection for resistance to ampicillin, only a portion of the culture should be spread on a single plate. The number of ampicillin-resistant colonies does not increase in direct proportion to the number of cells applied to the plate, perhaps because of growth-inhibiting substances released from cells that are killed by the antibiotic.
20. We have found that the efficiency of transformation is greatest with cells from cultures at two optical densities at 600 nm (OD_{600}), namely, 0.3–0.4 (5×10^7–1×10^8 cells/mL) and 0.94–0.95 (4×10^8 cells/mL). Moreover, cells harvested at an optical density of 0.94 to 0.95 usually can be transformed with twice the efficiency of cells harvested at an optical density of 0.3–0.4 (7×10^7 vs 4×10^7).
21. The applicability of this method to various strains of *E. coli* has been assessed. A biphasic pattern of transformation efficiency was observed in each case. Thus, there are no restrictions with regard to the strain of *E. coli*.

We have described a protocol for preparing competent *E. coli* cells that can be transfected at extremely high efficiency by plasmids. It is possible to prepare competent cells at a level four to eight times higher than those obtained with the basic protocol from any given culture.

References

1. Mandel, M. and Higa, A. (1970) Calcium-dependent bacteriophage DNA infection. *J. Mol. Biol.* **53,** 159–162.
2. Dagert, M. and Ehrlich, S. D. (1974) Prolonged incubation in calcium chloride improves competence of *Escherichia coli* cells. *Gene* **6,** 23–28.
3. Kushner, S.-R. (1978) An improved method for transformation of *Escherichia coli* with *col* E1-derived plasmids, in *Genetic Engineering* (Boyer, H. W. and Nicosia, S., eds.), Elsevier/North Holland, Amsterdam, pp. 17–23.
4. Hanahan, D. (1983) Studies on transformation of *Escherichia coli* with plasmids. *J. Mol. Biol.* **166,** 557–580.
5. Tang, X., Nakata, Y., Li, H.-O., Zhang, M., Gao, H., Fujita, A., Sakatsume, O., Ohta, T., and Yokoyama, K. (1994) The optimization of preparation of competent cells for transformation of *E. coli*. *Nucleic Acids Res.* **22,** 2857,2858.
6. Cohen, S. N., Chang, A. C. Y., and Hsu, L. (1972) Nonchromosomal antibiotic resistance in bacteria: genetic transformation of *Escherichia coli* by R-factor DNA. *Proc. Natl. Acad. Sci. USA* **69,** 2110–2114.
7. Chung, C. T., Niemela, S. L., and Millen, R. H. (1989) One-step preparation of competent *Escherichia coli*: transformation and storage of bacterial cells in the same solution. *Proc. Natl. Acad. Sci. USA* **86,** 2172–2175.

11

The Tetramethylammonium Chloride (TMAC) Method for Screening cDNA Libraries with Highly Degenerate Oligonucleotide Probes Obtained by Reverse Translation of Amino Acid Sequences

Bent Honoré and Peder Madsen

1. Introduction

If an unknown protein is purified and available in relatively small amounts, it is possible to determine the sequences of short internal peptides *(1)*. In order to determine the whole sequence of the protein by cDNA cloning, one of the peptides of perhaps five to seven amino acids may be reverse translated into nucleotide sequence resulting in a 15–21 base long deoxyribonucleotide. Because of codon degeneracy, the number of possible oligonucleotides may be more than several hundred, which must be present in order to ensure that the correct sequence is represented. Only one of these oligonucleotides corresponds to the correct sequence. Traditionally long stretches of DNA are hybridized in buffered saline solution where physicochemical parameters affecting the annealing are well known *(2,3)*. However, for shorter DNA sequences, the melting temperature of each oligonucleotide depends on the G + C content, since G:C base pairs possessing three hydrogen bonds interact more strongly than A:T base pairs with two hydrogen bonds. The different oligonucleotides in a mixture will thus possess different melting temperatures. This means that in buffered saline solution, one usually chooses a melting temperature that is so low that the oligonucleotide with the lowest G + C content can hybridize. However, in doing so, it is possible that oligonucleotides with a higher G + C content may form stable hybrids with mismatches resulting in the cloning of artifact cDNAs. Even though this procedure has been used success-

From: *Methods in Molecular Biology, Vol. 69: cDNA Library Protocols*
Edited by: I. G. Cowell and C. A. Austin Humana Press Inc., Totowa, NJ

fully *(4–6)*, it is more convenient to use a different buffer type that contains tetramethylammonium chloride (TMAC), since it has been reported that this salt selectively binds to and stabilizes A:T base pairs so that their melting temperature becomes similar to that of G:C base pairs *(7–9)*.

The successful cloning of a large number of cDNAs shows that the technique works for 15- to 20-mer oligonucleotides with a degeneracy up to 512 and a G + C content between 27 and 61% in the cloned cDNA *(10)*. Although sequence-dependent stability of very short DNA sequences in TMAC have been reported recently *(11)*, such features might be averaged out in the longer oligonucleotides used here. Additionally, we have found that the TMAC technique works with oligonucleotides containing deoxyinosine as a neutral or slightly destabilizing base at highly ambiguous positions.

2. Materials

2.1. Plating of cDNA Library and Replica Lifts

1. Luria-Bertani (LB) medium: 10 g/L peptone, 5 g/L yeast extract, 5 g/L NaCl, and LB-MM medium: LB medium supplemented with 2 g/L maltose and 2 g/L $MgCl_2$ for bacterial growth.
2. LB agar: LB medium with 15 g/L of agar and LB-MM top agar or agarose: LB-MM medium with 7 g/L of agar or agarose for plating on screening plates.
3. Saline magnesium (SM) buffer for diluting λ phages: 100 mM NaCl, 10 mM $MgCl_2$, 50 mM Tris-HCl, pH 7.5.
4. 500 cm^2 Screening plates (cat. no. 240 835, Nunc, Roskilde, Denmark): 245 × 245 mm.
5. Nylon filters (Hybond-N, Amersham, Amersham, UK): 220 × 220 mm and MESH sheets to interpose between the filters in hybridization tubes.
6. Denaturation solution: 0.5M NaOH, 1.5M NaCl. Store at room temperature.
7. Renaturation solution: 0.5M Tris-HCl, 1.5M NaCl, pH 7.5. Store at room temperature.
8. 20X SSC: 3.0M NaCl, 0.3M sodium citrate. Store at room temperature.

2.2. Labeling and Purification of Oligonucleotide

1. TE buffer: 10 mM Tris-HCl, pH 8.0, 0.1 mM EDTA, pH 8.0. Store at 4°C.
2. T4 polynucleotide kinase (Amersham), and 10X kinase buffer: 500 mM Tris-HCl, pH 7.5, 100 mM $MgCl_2$, 50 mM DTT, 1 mM spermidine HCl, 1 mM EDTA, pH 8.0. Store aliquots at –20°C.
3. [γ-^{32}P]Adenosine-5'-triphosphate (ATP) (ICN Radiochemicals, Irvine, CA), 7000 Ci/mmol, 167 μCi/μL, 24 pmol ATP/μL.
4. Formamide/dye mix: 10 mM EDTA, 0.5 g/L bromphenol blue, 0.5 g/L xylene cyanol in deionized formamide. Store at –20°C.
5. 10X TBE buffer: 1M Tris, 1M borate, 0.01M EDTA, pH 8.0. Store at room temperature.
6. 20% Polyacrylamide/7M urea: mix 25 mL of 40% polyacrylamide (380 g/L acrylamide, 20 g/L bis-acrylamide), 21.25 g urea, 5 mL 10X TBE buffer, and add water to 50 mL.

7. TEMED (20–25 µL) and 10% ammonium persulfate (APS) (200–250 µL).
8. Kodak X-ray films, Saran Wrap™, NaBH$_4$ (10%), plastic bags.

2.3. Hybridization of Oligonucleotide

1. TMAC (Fluka [Buchs, Switzerland] or Merck [Darmstadt, Germany]): 5M in water (*see* Note 1). Store at room temperature. Note that this chemical is hazardous with a strong odor and should only be handled in a fume hood.
2. Yeast RNA (Boehringer Mannheim, Mannheim, Germany): 100 mg/mL in water (*see* Note 2). Store at –20°C.
3. 100X Denhardt's solution: 20 g/L bovine serum albumin (BSA), 20 g/L Ficoll 400 (Pharmacia, Uppsala, Sweden), 20 g/L polyvinylpyrollidone. Store at –20°C.
4. Prehybridization buffer: 6X SSC, 5X Denhardt's solution, 20 mM sodium phosphate, pH 7.0, and 1.5 mg/mL yeast RNA. Store at 4°C.
5. Hybridization buffer: 3M TMAC, 5X Denhardt's solution, 50 mM sodium phosphate, pH 7.0, 1 mM EDTA, 0.6% sodium dodecyl sulfate (SDS) and 0.25 mg/mL yeast RNA. Store at room temperature.
6. Washing buffer 1: 3M TMAC, 50 mM Tris-HCl, pH 8.0, 0.2% SDS.
7. Washing buffer 2: 2X SSC, 0.1% SDS.
8. 2X YT-medium: 16 g/L peptone, 10 g/L yeast extract, 10 g/L NaCl.
9. Petri dishes (90 mm in diameter, Nunc) with LB agar supplemented, after autoclaving and cooling to 45°C, with 100 µg/mL of ampicillin for growth of bacteria containing pBluescript plasmids with ampicillin resistance.

3. Methods
3.1. Plating of cDNA Library and Replica Lifts

The library used should contain the cDNA encoding the protein to be cloned. The first choice is a cDNA library made with RNA purified from the same cell line or tissue as has been used to sequence the protein, although other cDNA libraries may be used (*see* Note 3).

1. Plate the library (e.g., λZAP) at a density of up to 1.5 × 10^5 PFU/screening plate (245 × 245 mm with 0.25 L of LB agar) using an overnight culture of *Escherichia coli* XL-1-Blue or LE392 and 30 mL of LB-MM top agar or agarose, which is allowed to set for 30 min. The plates are then incubated at 37°C overnight.
2. On the next day, place the plates for 1 h at 4°C (*see* Note 4). Plaque transfer to filters, denaturation, and renaturation are performed largely as described in the blotting and hybridization protocols for Hybond™ membranes from Amersham. In detail, one nylon filter (first replica filter) is marked with a few lines with India ink defining a unique pattern in a band from the edges of the filter and about an inch to the center, and then placed on the plate for 1 min. The pattern of lines are then marked on the reverse side of the plate with India ink. The filter is carefully lifted off the plate, placed with the plaque side up at filter paper, and a second filter (second replica filter) is placed on the plate. Now the line pattern from the plate is marked on the filter with India ink, and the filter is left on the

plate for 3 min or more. The filters are then placed, still with the plaque side up, on top of filter paper, lightly soaked with denaturation solution and left for 7 min. The nylon filters are then transferred to filter paper lightly soaked with renaturation solution for 3 min. This step is repeated, and finally the filters are washed in 2X SSC, air-dried, and crosslinked with UV light from a transilluminator for 1.5 min on each side. For long-term storage of agar plates, *see* Note 5.

3.2. Labeling and Purification of Oligonucleotide

The amino acid sequence of the peptide is reverse translated to the fully degenerated nucleotide sequence, and a suitable oligonucleotide is selected. The principles for the selection of the best possible oligonucleotide are described in Note 6.

1. Label 20–100 pmol of oligonucleotide in a volume of 10 µL by mixing the oligonucleotide with 1 µL of 10X kinase buffer, 3–5 µL of [γ-^{32}P]ATP (500–835 µCi, 72–120 pmol), TE buffer to 10 µL, and finally 1 µL of T4 polynucleotide kinase (10 U). The mix is incubated at 37°C for 1 h (*see* Note 7).
2. The labeled oligonucleotide may be purified from the unlabeled on a polyacrylamide/7M urea gel. Add 12 µL of formamide/dye mix to the labeling mix, then heat to 95°C for 3 min, and load on a 20% polyacrylamide/7M urea gel, which has been prerun with 1X TBE buffer at 18 W for 20 min prior to loading.
3. Run the gel at a maximum 18 W for a few hours until the blue color disappears. Dismount the gel from the glass plates and cover with Saran Wrap™. A few pieces of filter paper are taped to the gel and lightly soaked with 10% NaBH$_4$.
4. Expose an X-ray film for 5–10 s and develop. One strong band that represents the labeled oligonucleotide should be present. This band can then be cut out from the gel with a scalpel by superimposing the X-ray film, which contains marks from the filter paper soaked with NaBH$_4$.
5. Place the gel piece in an Eppendorf tube with 1 mL hybridization buffer and shake. In order to extract the labeled oligonucleotide from the gel, change, save, and pool the buffer several times. The efficiency of the extraction can be checked with a monitor.

3.3. Hybridization of Oligonucleotide

1. Interpose the nylon filters between MESH sheets and incubate in prehybridization buffer for 1–2 h. Replace the prehybridization buffer with fresh buffer and incubate for a further 1–2 h (*see* Note 8).
2. Wash the filters thoroughly four times for about 15 min in hybridization buffer and incubate overnight or longer in hybridization buffer together with the labeled oligonucleotide at the calculated hybridization temperature, which depends on the length of the oligonucleotide (*see* Note 9).
3. Pour off the hybridization buffer with labeled probe and save at –20°C (*see* Note 10), then wash the filters in washing buffer 1 at the hybridization temperature with four exchanges each for 15 min and then follow by 1–2 h of incubation.

4. Wash the filters two times for 2 min in washing buffer 2 at room temperature and seal in plastic bags. The marks on the filters are drawn up with radioactive ink, and autoradiography is performed at −70°C with intensifying screens for 24 h to a few weeks.
5. After development of the films, they are aligned and stronger spots appearing on both replica films (usually strongest on the first replica) likely represent correct hybridizing clones. Cut out about 1 cm^2 or less of the agar/agarose and put in Eppendorf tubes for each positive clone. Add 300 µL of SM buffer, crush the agar/agarose with a glass rod, and shake the tube for 15 min. Dilute 1 µL from this solution with 1500 µL of SM buffer, and plate 1, 10, and 100 µL together with 300 µL of bacteria and 3 mL of LB-MM top agar on smaller Petri dishes. Put nylon filters on these as previously described. It is usually not necessary to use replica filters, but it may be advantageous for the beginner.
6. Well-isolated positive plaques may be used for in vivo excision of pBluescript plasmids with the cDNA insert from the λZAP phages. Cut out the plaque from the plate with a scalpel and place in an Eppendorf tube. Add 300 µL of SM buffer, crush the agar/agarose with a glass rod, and shake the tube for 15 min.
7. Perform self-excising of the pBluescript plasmid from the phage DNA as described by Stratagene by mixing 100 µL of the λZAP phages, 200 µL of XL-1 Blue bacteria, and 1 µL of the ExAssist helper phage in a 50-mL tube, incubated at 37°C for 15 min. Add 2X YT medium and grow the mixture in a shaking incubator at 300 rpm for 2–2.5 h. Heat the contents to 70°C for 20 min, centrifuge, and save the supernatant containing the pBluescript plasmid, packaged as filamentous phage particles, at 4°C. Mix 1 µL of the phage supernatant with 200 µL of SOLR bacteria incubated at 37°C for 15 min and finally plate on Petri dishes with agar containing 100 µg/mL of ampicillin. After overnight growth, the colonies may be transferred to Hybond-N filters in a similar way to that described for plaque transfer (Section 3.1., step 2). After the colony transfer, the Petri dishes should be grown again at 37°C for some hours until the colonies are again visible by the naked eye. Plasmid preparations may then be prepared by growing positive colonies in larger scales by standard procedures.

4. Notes

1. The TMAC is hygroscopic, so the water is added directly to the bottles to dissolve the powder. The solution is then filtered to remove undissolved material, and water is finally added to bring the concentration to 5*M* by taking the weight as indicated on the bottles. We do not find it necessary to determine the actual concentration by measuring the refractive index of the solution as used in some protocols.
2. The yeast RNA blocking solution is made by dissolving 10 g RNA in 50 mL 0.3*M* sodium acetate (pH 6.0). This solution is extracted twice with 0.8 vol of phenol and twice with 0.8 vol of chloroform (note that the lower phase is the RNA-containing phase). Precipitation is performed by adding 2 vol of ethanol. The solution is centrifuged, the pellet dried, redissolved in 50 mL 0.3*M* sodium

acetate (pH 6.0), and 2 vol of ethanol are added for reprecipitation. After centrifugation, the pellet is washed several times with ethanol/water (80/20, v/v), dried, and dissolved in water at a concentration of 100 mg/mL as measured by the UV light absorbance at 260 nm.

3. The first choice is to use a cDNA library made with mRNA purified from the tissue or cell line from which the amino acid sequencing results have been obtained. However, the same protein may be expressed in other cell lines or tissues as well. The presence of a protein in a cell line or a tissue may be analyzed by two-dimensional gel electrophoresis, separating the proteins in the first dimension according to pI and in the second dimension according to molecular mass *(12)*. Two proteins from two different sources that migrate to the same position in the gel may indicate that they represent the same protein or are closely related proteins. This can be further confirmed by reaction with a specific antibody against the protein (immunoblotting) or cutting of the Coomassie brilliant blue-stained spot from the dried 2D gel for amino acid microsequencing *(1)*.

4. Bacterial plates, especially those made with top agar, are placed at 4°C in order to solidify the agar further. Without this step, a part of the agar with plaques may stick to the nylon filter with the result that the plaques in that area are lost.

5. The agar plates may be stored for a few days at 4°C. For long-term storage (years), the plates can be put directly at −70°C. The filters can then be stored at −20°C and reused for screening with other oligonucleotides later or immediately if the probe is first stripped off the filters, e.g., by adding water that is boiled, slightly cooled, and brought to 0.1% SDS (blotting and hybridization protocols for Hybond™ membranes, Amersham). The filters should never dry out.

6. The oligonucleotide used to screen the library should in principle be as long as possible and as little degenerated as possible. We have successfully used oligonucleotides with lengths from 15–20 bases and up to 23 for inosine-containing oligonucleotides *(10)*. In order to make the oligonucleotide as little degenerated as possible, it may be advantageous to terminate the oligonucleotide at codon position 2, since the third codon position usually is ambiguous. If possible, avoid oligonucleotides reverse translated from peptides with Leu, Arg, or Ser, since the codons for these amino acids are degenerated at two or all three positions. We have successfully used oligonucleotides with a degeneracy up to 512 *(10)*. If all oligonucleotides are equally represented in this mixture, the concentration of each different oligonucleotide will be <0.2% of the total concentration. If one or a few nucleotides are unintentionally preferred at the ambiguous positions during the synthesis, the mixture may become "biased" at some of the ambiguous positions, so that the correct oligonucleotide may be underrepresented in the mixture. This means that the concentration of the correct oligonucleotide may be even lower than 0.2% of total. It is important to keep in mind that the quality of the oligonucleotide preparation is crucial for a successful result. We have found that different batches of oligonucleotides that should recognize the same cDNA sequence may in fact hybridize to the plaques with very different strengths and some of them very poorly indeed. Thus, if one batch of oligonucleotides works

unsuccessfully, it may be worth trying one or two other batches for screening, perhaps from another supplier.

7. The equilibrium constant for the labeling reaction:

$$[\gamma\text{-}^{32}\text{P}]\text{-ATP} + 5'\text{-oligonucleotide} \leftrightarrow \text{ADP} + [^{32}\text{P}]\text{oligonucleotide} \tag{1}$$

is about unity *(13)*. The specific activity of the labeled oligonucleotide can be increased by increasing the concentration of $[\gamma\text{-}^{32}\text{P}]$-ATP, whereas the amount of transferred $[^{32}\text{P}]$ to the oligonucleotide can be increased by increasing the oligonucleotide concentration. The unlabeled oligonucleotide will hybridize as strong as the labeled one, and it is therefore necessary to purify the labeled oligonucleotide, e.g., on a denaturing polyacrylamide gel where the labeled oligonucleotide that contains a 5'-phosphate will move faster than the unlabeled oligonucleotide.

8. The prehybridization buffer is used to block the filters with RNA in order to get a low background for the following autoradiography, which may be performed for several days or weeks. The blockage does not work sufficiently in TMAC and is therefore performed in saline sodium citrate.

9. The hybridization temperature of the oligonucleotide is determined as follows. First the irreversible melting temperature is calculated as described by Jacobs et al. *(9)*:

$$T_i = -682 \times (L^{-1}) + 97°C \tag{2}$$

where L is the number of nucleotides in the oligonucleotide and T_i is the irreversible melting temperature. The hybridization temperature, T_h, is usually about 10°C below the T_i. Thus, for a 15-mer oligonucleotide, the T_i is calculated to 52°C and T_h is accordingly 42°C. If the oligonucleotide used contains inosines, these should be regarded as neutral or slightly destabilizing, e.g., a 20-mer oligonucleotide with three inosines is regarded as a 17-mer or shorter when calculating T_h.

10. The oligonucleotide probe may be reused for the purification of plaques if performed within 1 or 2 wk. The half-life of the probe is 14 d.

References

1. Vandekerckhove, J. and Rasmussen, H. H. (1994) Internal amino acid sequencing of proteins recovered from 1D or 2D-gels, in *Cell Biology: A Laboratory Handbook*, vol. 3 (Celis, J. E., ed.), Academic, San Diego, CA, pp. 359–368.
2. Marmur, J. and Doty, P. (1962) Determination of the base composition of deoxyribonucleic acid from its thermal denaturation temperature. *J. Mol. Biol.* **96**, 109–118.
3. Schildkraut, C. and Lifson, S. (1965) Dependence of the melting temperature of DNA on salt concentration. *Biopolymers* **3**, 195–208.
4. Singer Sam, J., Simmer, R. L., Keith, D. H., Shively, L., Teplitz, M., Itakura, K., Gartler, S. M., and Riggs, A. D. (1983) Isolation of a cDNA clone for human X-linked 3-phosphoglycerate kinase by use of a mixture of synthetic oligodeoxyribonucleotides as a detection probe. *Proc. Natl. Acad. Sci. USA* **80**, 802–806.

5. Lin, F. K., Suggs, S., Lin, C. H., Browne, J. K., Smalling, R., Egrie, J. C., Chen, K. K., Fox, G. M., Martin, F., Stabinsky, Z., Badrawi, S. M., Lai, P.-H., and Goldwasser, E. (1985) Cloning and expression of the human erythropoietin gene. *Proc. Natl. Acad. Sci. USA* **82,** 7580–7584.

6. Honoré, B., Rasmussen, H. H., Celis, A., Leffers, H., Madsen, P., and Celis, J. E. (1994) The molecular chaperones HSP28, GRP78, endoplasmin, and calnexin exhibit strikingly different levels in quiescent keratinocytes as compared to their proliferating normal and transformed counterparts: cDNA cloning and expression of calnexin. *Electrophoresis* **15,** 482–490.

7. Melchior, W. B., Jr. and Von Hippel, P. H. (1973) Alteration of the relative stability of dA-dT and dG-dC base pairs in DNA. *Proc. Natl. Acad. Sci. USA* **70,** 298–302.

8. Wood, W. I., Gitschier, J., Lasky, L. A., and Lawn, R. M. (1985) Base composition-independent hybridization in tetramethylammonium chloride: a method for oligonucleotide screening of highly complex gene libraries. *Proc. Natl. Acad. Sci. USA* **82,** 1585–1588.

9. Jacobs, K. A., Rudersdorf, R., Neill, S. D., Dougherty, J. P., Brown, E. L., and Fritsch, E. F. (1988) The thermal stability of oligonucleotide duplexes is sequence independent in tetraalkylammonium salt solutions: application to identifying recombinant DNA clones. *Nucleic Acids Res.* **16,** 4637–4650.

10. Honoré, B., Madsen, P., and Leffers, H. (1993) The tetramethylammonium chloride method for screening of cDNA libraries using highly degenerate oligonucleotides obtained by backtranslation of amino-acid sequences. *J. Biochem. Biophys. Methods* **27,** 39–48.

11. Riccelli, P. V. and Benight, A. S. (1993) Tetramethylammonium does not universally neutralize sequence dependent DNA stability. *Nucleic Acids Res.* **21,** 3785–3788.

12. O'Farrell, P. H. (1975) High-resolution two dimensional gel electrophoresis of proteins. *J. Biol. Chem.* **250,** 4007–4021.

13. Sambrook, J., Fritsch, E. F., and Maniatis, T. (1989) *Molecular Cloning: A Laboratory Manual.* Cold Spring Harbor Laboratory Press, Cold Spring Harbor, NY.

12

Screening cDNA Libraries by Hybridization with Double-Stranded DNA Probes and Oligonucleotides

Caroline A. Austin

1. Introduction

Probably the most commonly used method to screen a cDNA library is hybridization to a labeled DNA probe. This probe may be a single-stranded oligonucleotide or a double-stranded cDNA or polymerase chain reaction (PCR) product. The DNA may be either radioactively or nonradioactively labeled. The sequence of an oligonucleotide probe may be derived from a number of sources, for example, degenerate probes may be obtained by back-translating a peptide sequence of an unknown protein, or may be a short conserved region of sequence within a cDNA from another member of a multigene family or from a cognate cDNA from another species (*see* Note 1). Double-stranded DNA probes may be a partial cDNA obtained by screening another library, or a PCR product, or a cDNA from another member of a gene family or from another species.

This chapter concentrates on the methods for labeling DNA probes and hybridization to filter bound library DNA. A protocol for plating and taking replica filter lifts from plasmid libraries is included; handling λ-phage libraries and making replica lifts are described in Chapters 11 and 13–15.

2. Materials
2.1. Plasmid Library Plating and Colony Lifts

1. Titered plasmid cDNA library.
2. Luria-Bertani (LB) medium: 10 g/L bacto-tryptone, 5 g/L yeast extract, 10 g/L NaCl. Autoclave to sterilize.

From: *Methods in Molecular Biology, Vol. 69: cDNA Library Protocols*
Edited by: I. G. Cowell and C. A. Austin Humana Press Inc., Totowa, NJ

3. LB agar: LB medium with 15 g/L of agar. Autoclave to sterilize. If antibiotics, such as ampicillin, are required, allow agar to cool to 55°C and then add antibiotic to appropriate concentration from stock solution, e.g., add ampicillin to 50–100 µg/mL from a 50 mg/mL stock prior to pouring plates.
4. Plates: 15-cm diameter, triple-vent Petri dishes.
5. Nylon filters, such as Hybond-N⁺ (Amersham, Amersham, UK) or Protan BA 85 (Schleicher and Schuell, Dassel, Germany).
6. Whatman 3MM paper.
7. Three trays or large dishes.
8. Denaturation buffer: $0.5M$ NaOH, $1.5M$ NaCl.
9. Renaturation buffer: $1M$ Tris-HCl, pH 7.0, $1.5M$ NaCl.
10. 20X Standard saline citrate (SSC): $3M$ NaCl, $0.3M$ sodium citrate.
11. 3X SSC.
12. 3X SSC, 0.1% sodium dodecyl sulfate (SDS).

2.2. Screening by Hybridization with Labeled Oligonucleotide

2.2.1. Oligonucleotide Labeling

1. Chosen oligonucleotide (*see* Note 1) with free 3'-OH.
2. T4 polynucleotide kinase.
3. 10X T4 kinase buffer: $1M$ Tris-HCl, pH 7.6, 100 mM MgCl$_2$, 200 mM 2-mercaptoethanol, store at –20°C.
4. [γ^{32}P]-Adenosine triphosphate (ATP) 3000 Ci/mmol (Amersham).

2.2.2. Hybridization with a Labeled Oligonucleotide Probe

1. Wash solution: 3X SSC, 0.1% SDS.
2. Wetting solution: 3X SSC.
3. 100X Denhardt's solution: 20 g/L bovine serum albumin (BSA), 20 g/L Ficoll 400, 20 g/L polyvinylpyrollidone. Store at –20°C.
4. Prehybridization buffer 6X SSC, 1X Denhardt's, 0.5% SDS, 100 µg/mL denatured salmon sperm DNA, 0.05% sodium pyrophosphate.
5. Hybridization buffer 6X SSC, 1X Denhardt's, 0.5% SDS, 100 µg/mL denatured salmon sperm DNA, 20 µg/mL tRNA, 0.05% sodium pyrophosphate.
6. Wash buffer 6X SSC, 0.05% sodium pyrophosphate (SSC/PP).

2.3. Screening by Hybridization with a Double-Stranded DNA Probe

2.3.1. Labeling

Double-stranded probes can be labeled by random primer labeling or by nick translation. Kits for both are available from several companies, including Amersham and Stratagene.

2.3.2. Hybridization of a Double-Stranded DNA Probe

1. 3X SSC.
2. 100X Denhardt's solution: 20 g/L BSA, 20 g/L Ficoll 400, 20 g/L polyvinylpyrrolidone. Store at –20°C.

3. Prehybridization buffer containing 5X SSC, 10X Denhardt's, 7% SDS, 100 μg/mL denatured salmon sperm DNA, 20 mM sodium phosphate.
4. Hybridization buffer 5X SSC, 10X Denhardt's, 7% SDS, 100 μg/mL denatured salmon sperm DNA, 20 mM sodium phosphate, 8% dextran sulfate.
5. 2X SSC/0.1% SDS.

3. Methods

λ-Phage libraries are by far the most common type of cDNA library for "standard" screening methods, and they present a number of advantages, such as ease of constructing large libraries and of plating large numbers of recombinants. However, for some purposes, plasmid libraries are appropriate; in addition, a number of recent specialized screening systems, such as those described by Cowell (Chapter 16) and Gudkov and Roninson (Chapter 18), employ plasmid libraries, and it may sometimes be of use to screen such libraries using DNA hybridization techniques. Methods for titering, plating, and making replica lifts of bacteriophage λ libraries are given in Chapters 11 and 13–15. Methods are given for making colony lifts of plasmid libraries in Section 3.1. The protocols for probing with labeled oligonucleotides or DNA fragments described in Sections 3.2. and 3.3., respectively, are appropriate for either type of library.

In order to ensure that the library used will contain the desired cDNA, it should have been prepared from mRNA extracted from a cell type or tissue in which the message or encoded protein is known to be expressed. Libraries can be made "in house," possibly using one of the commercial kits that are available. Libraries may be obtained from other workers or a resource center (Chapter 23), or purchased, since many libraries are now available commercially.

3.1. Plating of cDNA Library and Colony Lifts

3.1.1. Transformation and Plating Plasmid Libraries

Plasmid libraries require very high efficiency transformation either by electroporation *(1)* or using highly competent cells made, for example, by the method of Hanahan *(1–3)*. Highly competent and electrocompetent cells can also be obtained commercially from suppliers, such as Stratagene or Promega. For library screening, the library is plated directly onto filters placed on plates containing the appropriate antibiotic *(1)*; for amplification the library is plated without filters.

1. Plate the library (*see* Note 2) by pipeting 0.2 mL of LB containing up to 2×10^5 CFU onto a nylon filter laid on a 15-cm LB agar plate containing ampicillin at 50–100 g/mL (or other selective antibiotic as appropriate for the plasmid type in

which the library is constructed). Spread the liquid evenly over the surface of the filter using a bent glass rod sterilized by dipping in ethanol, flaming with a Bunsen burner, and allowing to cool for 30 s.
2. Leave the plates lid side up on the bench to allow the surplus liquid to soak in or evaporate (about 1 h).
3. Incubate the plates lid side down overnight, by which time distinct colonies should have appeared.

3.1.2. Replica Plating (3–5)

1. Remove the master filter from each plate using blunt-ended forceps, and place colony side up on a sheet of dry filter paper such as Whatman 3MM. For each master filter, wet a new nylon filter by placing on a fresh LB plate. Using two blunt-ended forceps, remove the new filter from each plate and place on top of the master filter. Place a piece of Whatman 3MM paper on top of the pair of filters and then a glass plate. Apply pressure by hand or with a heavy book for a few seconds.
2. Prior to separating the replica and the master, make orientation marks by piercing both the master and replica with a needle. Make the needle hole in an asymmetric pattern to allow later orientation of the replica with the master.
3. Separate the master and replica filters by peeling apart with blunt-ended forceps, and place the replica colony side up onto a new LB antibiotic plate.
4. Make a second replica from the master in the same way without further growth of the master. Remember to make the orientation holes prior to pulling apart.
5. Place the second replica onto a new LB antibiotic plate.
6. Return the master filter to its LB plate and store at 4°C.
7. Place both replicas at 37°C until colonies are easily visible (0.5–1 mm diameter). This usually takes 3–5 h.

3.1.3. Lysis of Bacteria on Filters

1. Place Whatman 3MM paper in three trays, the first saturated with $0.5M$ NaOH, $1.5M$ NaCl, the second with $1M$ Tris-HCl, pH 7.0, $1.5M$ NaCl, and the third with 3X SSC.
2. Place the filters colony side up on the first tray ($0.5M$ NaOH, $1.5M$ NaCl) and leave for 10–15 min. The NaOH/NaCl soaks through the filter and lyses the bacterial colonies. Remove the filters from tray 1, removing excess liquid from the underside of the filter on the tray edge.
3. Transfer the filters to the second tray containing Whatman 3MM soaked in $1M$ Tris-HCl, $1.5M$ NaCl, and leave for 2–5 min. Then transfer to the third tray containing 3X SSC for 2–5 min.
4. Transfer to a piece of 3MM paper, and wipe the colony side with a tissue saturated in 3X SSC, 0.1% SDS, to remove bacterial debris (Note 3).
5. Remove the filters to a fresh sheet of 3MM paper to air-dry.
6. To fix the DNA to the filter, either bake for 1 h at 80°C or UV crosslink in a suitable UV crosslinker, e.g., UV Stratalinker 2400 (Stratagene).

3.2. Screening by Hybridization with Labeled Oligonucleotide

3.2.1. Oligonucleotide Labeling

1. Add 200 ng of oligonucleotide to a microfuge tube containing 2 μL 10X T4 kinase buffer, 100 μCi [γ^{32}P]-ATP, 10 U T4 polynucleotide kinase, and water up to 20 μL. Incubate at 37°C for 1 h.
2. Add the whole mixture to the hybridization buffer, or purify labeled oligonucleotide from free ATP (*see* Note 4).

3.2.2. Hybridization with a Labeled Oligonucleotide Probe (6)

3.2.2.1. PREWASHING FILTERS

1. Wash filters in 3X SSC, 0.1% SDS at 65°C for 16–20 h. Large volumes and several changes of buffer aid removal of bacterial debris (Note 3).
2. After this wash, filters may be prehybridized immediately or stored dry at 4°C.

3.2.2.2. PREHYBRIDIZATION

1. If necessary, wet the filters in 3X SSC.
2. Place the filters in a suitable container (Note 5) in prehybridization buffer, allowing 100 μL/cm^2 of filter.
3. Incubate at 37°C for at least 2 h.

3.2.2.3. HYBRIDIZATION

1. Remove prehybridization buffer and add hybridization buffer, allowing 50 μL/cm^2 of filter.
2. Add the labeled oligonucleotide probe to the filters at approx 4 ng/mL.
3. Incubate for 24 h at an appropriate temperature (Note 6).

3.2.2.4. WASHING FILTERS

1. Carefully remove the hybridization solution, pour into a thick glass bottle (e.g., Duran), and store at –20°C. This can be reused for second-round screens if these are performed less than a week later.
2. Filters are washed in 6X SSC, 0.05% SSC/PP. Place filters in a sandwich box or similar container that is large enough to allow movement of the filters, and wash with large volumes of SSC/PP at appropriate temperatures (Note 7).
3. Blot the filters to remove excess liquid, but do not allow to dry. Wrap in Saran Wrap™ or cling film, and expose to film at –70°C with intensifying screens. Use radioactive or fluorescent markers to aid alignment of the developed autorad with the filters. Strong signals should appear overnight, but exposures of up to a week may be required *(7)* (*see* Note 6).

3.2.3. Picking Positives and Colony Purification

1. Use the radioactive or fluorescent markers to align the developed autorads with the probed filters, identify each filter on the autorad and mark the needle holes in the filters on the autorad, to enable the film to be aligned with the master filters.

2. Identify positive spots that appear in duplicate, i.e., on both filter lifts. As a guide, genuine positives will be colony/plaque shape and size and will give duplicate positive signals, although one filter may give a stronger signal than the other.
3. For plasmid libraries, pick positive colonies or an area containing a positive. Use a sterile toothpick to transfer the colonies to a series of sterile microfuge tubes containing 200 μL each of LB. Disperse the colonies by vortexing, and replate on 9-cm LB agar plates containing antibiotic, spreading the cell suspension on nylon filters as before. It may save time to replate each putative positive at two or more densities aiming to obtain a plate with 50–200 colonies to increase the chances of being able to pick a well-isolated positive colony at the next stage. *See* Chapter 14 for details of picking positive plaques and plaque purification with λ-phage libraries.
4. Incubate the plates overnight at 37°C, and make replica filters as described in Section 3.1.2.
5. Continue until well-isolated single colonies can be picked.

3.3. Screening by Hybridization with a Double-Stranded DNA Probe

3.3.1. Labeling

Labeling of double-stranded probes can be accomplished by random primer labeling or by nick translation using commercial kits according to the manufacturer's instructions.

3.3.2. Hybridization of a Double-Stranded DNA Probe

3.3.2.1. PREHYBRIDIZATION

1. Wet the filters in 3X SSC.
2. Place the filters in a suitable container (Note 5) in prehybridization buffer, allowing 100 μL/cm^2 of filter.
3. Incubate at 65°C for at least 1 h.

3.3.2.2. HYBRIDIZATION

1. Prepare hybridization buffer allowing 50 μL/cm^2 of filter. Heat the mixture to 90°C, or place in a boiling water bath for 10 min and allow to cool to 65°C.
2. Remove prehybridization buffer, and add hybridization buffer.
3. Heat the labeled probe in a boiling water bath for 5 min, and add to the hybridization solution at approx 1 ng/mL.
4. Incubate overnight at 65°C preferably with shaking or other movement to allow the hybridization solution to move over the filters (Note 5).

3.3.2.3. WASHING FILTERS

1. Carefully remove the hybridization solution, pour into a thick glass bottle (e.g., Duran), and store at –20°C. This can be reused for second screens if these are performed less than 1 wk later.

2. Wash filters in preheated 2X SSC, 0.1% SDS at 65°C two or three times for 20 min each, and for high stringency, wash further in 0.1X SSC, 0.1% SDS at 65°C for another 20 min.
3. Blot filters dry, and wrap in Saran Wrap or cling film autoradiograph at −70°C as described in Section 3.2.2.4.
4. Pick positive colonies or area, and rescreen as above until colony-purified positives are obtained (*see* Chapter 14 for details of picking positive plaques and plaque purification with λ-phage libraries).

4. Notes

1. Probably the most common use of oligonucleotides in screening cDNA libraries is to isolate a cDNA for which some partial peptide sequence data are available or to screen for additional members of a gene family where specific amino acid motifs are particularly well conserved. In both cases, the oligo sequence is derived from a back-translated amino acid sequence. The main obstacle to this approach is the problem of codon degeneracy. This is discussed in Chapter 11 by Honoré and Madsen, but degeneracy can be minimized by selection where possible of less degenerately coded regions of amino acid sequence. In addition, codon usage tables can be consulted to "guess" the correct base at particular points. The paucity of the sequence CG in vertebrate genomes may also allow the complexity to be reduced. Oligonucleotides can be synthesized to contain deoxyinosine in place of T or G to reduce probe complexity further.
2. The number of colonies to aim for depends on the complexity of the library and the species of origin. As a rule of thumb, for a vertebrate cDNA library that has not been normalized in any way and that contains 10^6 or more independent recombinants, around 10^6 colonies should be plated to isolate a moderate to low abundance cDNA. For some purposes, such as screening a library containing cDNA generated with a specific primer, far fewer colonies need to be screened.
3. Bacterial debris on the filters can cause background problems, which can lead to false-positive signals. Wiping the filters with very wet tissue in one direction and prewashing of the filters decreases this background.
4. Alternatively, labeled oligonucleotide may be separated from free ATP by such methods as spun-column chromatography. Plug a 1-mL plastic syringe with a small disk cut from a Whatman GF/C filter, and fill with Sephadex G-50M in TE. The syringe is then placed in a 15-mL Falcon tube and spun at 1000g for 4 min. Add TE (100 µL) to the top of the compressed Sephadex and spin as before. Cut the lid from a 1.5-mL microfuge tube, and place at the bottom of the Falcon tube to collect the flowthrough from the column. After adding 80 µL of TE to the probe, mix carefully, pipet it onto the spun column, and spin again as above. Collect the flowthrough, and discard the column as appropriate for radioactive waste. The specific activity of the probe should be in excess of 10^8 cpm/µg.
5. Several types of containers can be used. Screw-cap bottles for hybridization ovens (e.g., Hybaid or Appligene) are routinely used in many molecular biology laboratories (*see* Chapter 11). However if no such oven is available, a water bath at the

required temperature may be used with the filters either in a plastic sandwich box or sealed in plastic bags.

6. The hybridization temperature for an oligonucleotide probe is crucially important. However, the optimum may have to be determined by trial and error. A rough rule of thumb for oligonucleotide probes is 20°C for a 14-mer, 37°C for a 17-mer, and 42°C for a 23-mer *(6)*, although this is affected by the G/C content of the probe. Alternatively a temperature 3–5°C below the T_m may be used, where an approximation of the T_m is calculated by $T_m = 2(A + T) + 4(G + C)$ *(7)*. However, for a degenerate oligonucleotide, it may not be possible to calculate the T_m in this way, since the base composition of the hybridizing species is unknown.

7. Washing temperatures also vary depending on the length of the oligonucleotide probe and base content, and may have to be determined empirically: for a 17-mer wash at 37°C for 1 h and at 47°C for a further 10 min, for a 14-mer wash at 30°C for 1 h and at 37°C for a further 10 min *(6)*. These suggested temperatures assume that there are no mismatches between the labeled oligonucleotide and target sequence. As an alternative approach, wash four times for 5 min each at room temperature in 6X SSC, 0.1% SDS, autoradiograph the filters, and then rewash the filters in a series of steps of increasing temperature, autoradiographing after each wash. The later washes should only be for 5 min. With an efficiently labeled probe, an exposure of 4–6 h should be sufficient at −70°C with intensifying screens for the initial exposure.

References

1. Sambrook, J., Frisch, E. F., and Maniatis, T. (1989) *Molecular Cloning: A Laboratory Manual.* Cold Spring Harbor Laboratory, Cold Spring Harbor, NY.
2. Hanahan, D. (1983) Studies on transformation of *Escherichia coli* with plasmids. *J. Mol. Biol.* **166,** 557–580.
3. Hanahan, D., Jessee, J., and Bloom, F. R. (1991) Plasmid transformation in *Escherichia coli* and other bacteria. *Methods. Enzymol.* **204,** 63–113.
4. Hanahan, D. and Meselson, M. (1980) Plasmid screening at high colony density. *Gene* **10,** 63–67.
5. Grosveld, F. G., Dahl, H.-H. M., de Boer, E., and Flavell, R. A. (1981) Isolation of β-globin related genes from a human cosmid library. *Gene* **13,** 227–237.
6. Woods, D. (1984) Oligonucleotide screening of cDNA libraries. *Focus* **6(3),** 1–2.
7. Brown, T. A. (1991) *Molecular Biology LABFAX.* Blackwell Scientific Publications, Oxford, UK.

13

Immunological Screening
of λ-Phage cDNA Expression Libraries

Helen C. Hurst

1. Introduction

When trying to obtain a cDNA clone to a novel protein, the only handle one may have is an antibody that recognizes the protein of interest. Consequently, the obvious approach is to screen a λ-phage expression library using the antibody. In general, polyclonal sera give better results, but a mixture of monoclonal antisera could be used instead. Fortunately, a number of convenient commercial vectors are available that can be used to generate an expression library where the cloned inserts are induced to produce protein within bacteria, usually as a fusion protein with β-galactosidase. Alternatively, it may be possible to buy a library of a cell line or tissue known to express the protein of interest in reasonable abundance. The library should have a complexity of at least 10^6, since in theory only one in six inserts will clone in the orientation and frame required to produce protein, although modern vectors often allow directional cloning. Thus, one in three of the clones to the protein of interest may be recognized by the antibody. Attention should also be given to the average size of the inserts, particularly for a large protein, since even polyclonal sera to the full protein may only recognize one epitope, which could lie at the N-terminus. Random-primed libraries tend to contain more N-terminal sequences and fewer extraneous sequences such as long 3'-untranslated regions than oligo-dT-primed libraries, but the former are rarely directional. A suitable compromise may be to screen a mixed library that has been generated using both priming methods, and many of these are available commercially.

Having obtained the library, the essence of the technique is to infect a bacterial lawn much as for DNA probe screening. Initially, expression of the cloned inserts is suppressed in case the protein products are toxic. Once phage lysis is

From: *Methods in Molecular Biology, Vol. 69: cDNA Library Protocols*
Edited by: I. G. Cowell and C. A. Austin Humana Press Inc., Totowa, NJ

detected, the plates are overlaid with nitrocellulose filters and switched to conditions that will allow fusion protein production onto the filter. Subsequently, the filter can be probed with the antibody as for a Western blot to detect phage plaques producing the cognate protein. False positives can be excluded by making replica filters. Repeated rounds of screening will allow the purification of a cDNA clone to the protein of interest.

2. Materials
2.1. Library Plating and Replica Lifts

1. Phage expression library of known titer: The protocol described here is suitable for λgt11 *(1)* or λZAP (Stratagene, La Jolla, CA) libraries.
2. Appropriate bacterial host, e.g., Y1090 (amp^R) for λgt11; XL1-blue (tet^R) for λZAP.
3. Bacterial culture media and well-dried plates: L-broth with 10 mM MgSO$_4$, 0.2% maltose for growing plating cells; L-broth agar or NZY agar plates; top agarose: 0.7% agarose in L-broth with 10 mM MgSO$_4$, 0.2% maltose.
4. SM phage dilution buffer: 100 mM NaCl, 10 mM MgSO$_4$, 50 mM Tris-HCl, pH 7.5, 0.01% gelatin. Autoclave.
5. Isopropyl β-D-thiogalactopyranoside (IPTG). 1M stock in water. Store at –20°C.
6. Nitrocellulose filters; 132- and 82-mm circles.
7. TBST: 20 mM Tris-HCl, pH 7.5, 150 mM NaCl, 0.05% Tween-20.
8. Block: TBST plus 1% bovine serum albumin (BSA).

2.2. Antibody Screening

1. High-titer polyclonal primary antibody or two or more monoclonal antisera.
2. Bacterial lysate.
3. Antibody detection system: usually requiring incubation with an enzyme-conjugated secondary antibody followed by either a color reaction (e.g., Stratagene's Picoblue immunodetection kit) or a photochemical system (e.g., Amersham's [Amersham, UK] ECL kit). Commercially available screening kits often contain positive controls (phage clones and appropriate primary antibodies). If the antibody to the protein of interest is in short supply, it is a good idea to run through the technique using these controls before experimenting with a precious antiserum!

3. Methods

Before using the antibody to screen a library, it is important to check its specificity on Western blots and to determine the optimum dilution factor and incubation conditions (*see* Note 1). It is also important to check that there is minimal crossreaction with proteins in an *Escherichia coli* lysate to avoid too high a background. If the antibody does show some recognition of bacterial proteins this can usually be pre-absorbed out (*see* Note 2), and mild crossreactivity will be progressively reduced if the antibody dilution is kept and reused in subsequent screens. This is particularly desirable if the serum is precious—a mouse polyclonal serum, for example.

3.1. Library Plating and Replica Lifts

1. Grow a fresh overnight culture of host bacteria in L-broth with maltose. Spin cells down and resuspend in a half volume of SM. Keep at 4°C until required.
2. Plate out 10^6 plaques at 50,000/15-cm Petri dish using the plating cells (600 µL/plate) from above and top agarose (Note 3). Incubate plates upside down at 42°C for 3–4 h until the plaques are just visible as pinpricks on the bacterial lawn.
3. Soak one set of 132-mm nitrocellulose circles in 10 mM IPTG, and pat dry. Taking one plate at a time from the 42°C incubator, quickly lay a numbered damp (but not wet) filter on each plate trying to avoid air bubbles, and place in a 37°C incubator for 2–4 h (Note 4). The temperature of the agar should not drop below 37°C, or protein production will be resuppressed.
4. Prepare a further set of IPTG-soaked filters and 500 mL each of TBST and of Block. Again working with one plate at a time, stab through the first filter, and agar with a syringe needle four or five times to orient the two. Peel off this filter and place in a large lunch box or plastic container containing the TBST. Place a second numbered filter on the agar as before, and return to the 37°C incubator for a further 1–3 h (*see* Note 4). Before removing these filters, place each plate on a light box, and use a permanent marker pen to mark dots over the needle holes visible in the agar. After removing the filters, it is wise to use a needle to pierce through the inked dots!
5. On removal, each set of filters should be rinsed briefly in TBST—this will reduce background—and then transferred individually to the Block. Swirl the container as each filter is added to prevent them from sticking together. Filters can be left in Block at 4°C overnight (and probably up to 2–3 d as long as they do not dry out). The plates should be wrapped in cling film and stored at 4°C until the screening is complete.

3.2. Antibody Screening

1. Make an appropriate dilution of the primary antibody in 50 mL of fresh Block. Incubate five to six filters at a time for 1 h at room temperature (*see* Note 1) on a slow shaker. An ideal container is a 150-mm high-sided Petri dish. Again, remember to swirl the dish as a filter is added to ensure there is always a layer of fluid between each. If large supplies of antisera are available, set up sufficient parallel incubations to probe all the filters; otherwise, probe in sequential batches. Afterward, the antibody dilution can be stored at –20°C and reused for subsequent screens, although it may need "pepping up" with some fresh antibody occasionally.
2. Remove the filters to a large container of TBST. These can be left until all the filters have been incubated with antibody. Wash all the filters three times for 5 min each wash in a large excess of TBST, ensuring that they do not stick together during shaking.

3. Dilute the detection system's antibody–enzyme conjugate into Block, and incubate as recommended (usually 1 h at room temperature).
4. Wash the filters as in step 2.
5. Give the filters a final rinse in Tris-buffered saline to remove Tween-20.
6. Develop the filters using either color detection or chemiluminesence. In either case, a small number of spots should appear (*see* Note 3). Check that spots are duplicates by orienting the stab holes—for detection systems using color development, this can most easily be done by tracing the orientation holes and the positive spots onto pieces of clear plastic. For photochemical detection, it is important to include fluorescent markers on the mounting used to autoradiograph the filters to enable you to orient the film to the filters.
7. Pick all duplicate positive signals (*see* Note 1), and replate out a dilution on 9-cm agar plates as described in Chapter 14. By repeating the processes, you will ultimately be able to pick a pure plaque for further authentication.

4. Notes

1. The library screen is essentially a Western blot, so it is important to do a series of these to determine the optimum incubation conditions (usually 1 h at room temperature) and the appropriate dilution factor for the primary antibody. Excess antibody will only give an unworkable background. It is also important to assess whether the washing conditions described will be sufficient to detect only the protein of interest in a blot using total cell extract. If a number of other proteins are also detected, then the antibody is probably unsuitable to use for screening since too many unrelated clones will be purified. However, more stringent washing may improve the situation, and if so, these conditions should also be used in the screen. Sometimes a purified protein used to raise an antibody contains a low-level contaminant that proves to be very antigenic, giving a bright signal on the Western in addition to the protein of interest. In this situation, it is possible to proceed, but when picking plaques after the first-round screen, it will be important to select both bright and faint spots to ensure that the interesting clones are not overlooked. Another issue is that if the protein of interest is thought to be covalently modified (phosphate, carbohydrate), then it is wise to check that the antisera will still recognize the native protein, since obviously these modifications will not be reproduced in the bacterially made protein during the screen.
2. Polyclonal antisera are often reactive with bacterial and phage proteins. This can be tested by dotting some phage-infected bacterial lysate (often provided in screening kits) on the corner of a Western blot prior to blocking or running some on a spare lane in the initial sodium dodecyl sulfate-polyacrylamide gel electrophoresis gel. If a strong signal is observed, it will be necessary to preabsorb out the reacting antibodies. This can be done by incubating three to four strips of nitrocellulose (5 × 10 cm) in bacterial lysate and then blocking them for 1 h at room temperature before rinsing three times in TBST. Dilute a reasonable amount of the primary antibody 1:5 in TBST, and incubate this with one of the filters for 15 min at room temperature. Remove the filter, and discard and replace with

another. Repeat until all of the prepared filters have been used. Store the final antibody solution, and recheck on a Western for the appropriate dilution and also that the crossreactivity has been reduced sufficiently.

3. Top agarose is much easier to use than top agar, which frequently sticks to the nitrocellulose and tears when the filters are removed.

4. Optimal incubation times are difficult to determine, since so many variables will influence the amount of protein and the ease of detection of the signal (density of plaques on the plate, specific activity of antibody for protein, protein size, etc.). I readily detected clones to a 70-kDa protein using a 3-h incubation for the first lifts and a 2-h incubation for the second lifts. As the plaques grow and the bacteria in the middle lyse, less and less protein is made in the center of the plaque. Consequently, after development, the final signals often do not look like uniform spots, but more like "donuts," that is, a bright ring around a clear center.

Reference

1. Huynh, T. V., Young, R. A., and Davis, R. W. (1988) Constructing and screening cDNA libraries in λgt10 and λgt11, in *DNA Cloning: A Practical Approach*, vol. 1 (Glover, D. M., ed.), IRL, Oxford, UK, pp. 49–78.

14

Cloning Sequence-Specific DNA-Binding Factors from cDNA Expression Libraries Using Oligonucleotide Binding Site Probes

Ian G. Cowell

1. Introduction

The method described in this chapter has been used in the molecular cloning of transcription factors and other factors with DNA-binding activity toward specific double-stranded DNA sequences. The protocol is based on the method of Singh et al. *(1)* and shares feature with the immunological approach to screening cDNA expression libraries (*see* Chapter 13). The principle is to probe a cDNA expression library (usually a λ-phage expression library) with a labeled double-stranded DNA probe containing the sequence recognized by the factor in question. Recombinants expressing a protein capable of binding the probe sequence in the presence of nonspecific competitor DNA are thus identified and can be isolated.

The most commonly used vectors for the construction of cDNA expression libraries are the bacteriophage vectors λgt11 (Huynh et al. *[2]*) and λ ZAP (Stratagene, La Jolla, CA). In both of these vectors, cDNA inserts are cloned into the coding region of the *LacZ* gene. In the case of λgt11, the cloning site is just upstream of the *LacZ* translational stop signal and in λ ZAP, which employs *LacZ* α complementation for blue/white selection of recombinants, cDNAs are cloned into the *LacZ* α peptide-coding sequence. Thus, in both vectors, the cloned cDNAs are potentially expressed as β-galactosidase fusion proteins (*see* Note 1). Problems of underrepresentation of cDNAs encoding polypeptides that are toxic or that reduce cell growth are minimized by the use of host cells expressing the *Lac* operon repressor to prevent *LacZ*-directed expression of the fusion protein until plaque formation is visibly under way.

From: *Methods in Molecular Biology, Vol. 69: cDNA Library Protocols*
Edited by: I. G. Cowell and C. A. Austin Humana Press Inc., Totowa, NJ

Fusion protein production is then induced for a few hours with isopropyl β-D-thiogalactopyranoside (IPTG), which inactivates the *Lac* repressor. Using λ-phage expression libraries, the use of DNA probes has been particularly successful in the cloning of members of the leucine zipper family of transcription factors *(1,3–6)*, and has also been used for other classes of DNA-binding protein *(7,8)*. The following protocols where written for λgt11 libraries *(2)*, but these methods can be employed with λ ZAP libraries (Stratagene) with minor modifications (*see* Note 1).

There are several notes and cautions that should be considered before embarking on library screening. First, if posttranslational modification, such as phosphorylation or specific proteolytic cleavage, is required for efficient DNA-binding activity, then the factor may not bind DNA with sufficiently high affinity when expressed in *Escherichia coli*. Careful selection of binding and washing conditions may circumvent this problem. Poor solubility or incorrect protein folding of a bacterially expressed factor may preclude efficient DNA binding. However, DNA-binding activity can, at least in some cases, be regenerated by denaturation with guanidinium chloride and gradual renaturation (*see* Section 3.5.). Finally, a problem with binding site screening arises if the factor of interest binds DNA efficiently only as a heterodimeric or heteromeric complex, since only a single polypeptide sequence can be expressed from any one clone. Similarly, this method is not appropriate for the cloning of components of a multisubunit transcription factor complex if they do not stably bind DNA themselves in the absence of the other components. South-western blotting can be used to determine whether DNA-binding activity resides in a single polypeptide (*see* Note 2). South-western blotting is a good preliminary experiment to carry out before attempting to screen libraries. A positive South-western result not only indicates that DNA-binding activity resides in a single polypeptide, but also proves that DNA binding activity is maintained when the protein is bound to a nitrocellulose membrane.

The method discussed herein is split into several parts: library titering, library plating and making replica lifts, probe labeling and screening, and finally plaque purification. In addition, an optional step that can improve the signal from some factors is included in Section 3.5.

2. Materials
2.1. Titering the Library

1. Expression cDNA library in λgt11 or λ ZAP (*see* Note 1).
2. Fresh plate of *E. coli* Y1090 maintained on Luria-Bertoni (LB) medium agar containing ampicillin at 50 µg/mL (for λ ZAP libraries, use *E. coli* XL1-Blue [*see* Note 1] maintained on LB agar containing tetracycline at 12.5 µg/mL).

3. LB medium containing 10 mM MgSO$_4$: 10 g tryptone, 5 g yeast extract, 10 g sodium chloride/L. Sterilize by autoclaving, and add MgSO$_4$ to 10 mM from a filter-sterilized 1M stock.

4. LB agar: 10 g tryptone, 5 g yeast extract, 10 g sodium chloride, 15 g bacteriological agar/L. Sterilize by autoclaving, and add MgSO$_4$ to 10 mM from a filter-sterilized 1M stock.

5. Well-dried LB agar 9-cm diameter plates.

6. Autoclaved top agarose: 0.7 g agarose in 100 mL LB medium. Melt in microwave, and maintain at 45°C for 20 min prior to use.

7. Maltose solution 10% (w/v) in water. Autoclave.

8. SM buffer to dilute phage: 100 mM NaCl, 10 mM MgSO$_4$, 50 mM Tris-HCl, pH 7.5, 0.01% gelatin. Autoclave.

9. Freshly streaked LB agar plate of *E. coli* Y1090 (or a fresh plate of *E. coli* XL1-Blue [Stratagene] if using λ ZAP; XL1-Blue cells should be cultured on LB agar containing tetracycline at 12.5 μg/mL).

10. Falcon 2063 tubes (Becton Dickinson).

2.2. Library Plating and Replica Lifts

1. Titered λgt11 or λ ZAP expression library (*see* Sections 2.1. and 3.1. and Note 1).

2. 10–20 Well-dried LB agar 15- and 9-cm diameter plates.

3. Autoclaved top agarose: 0.7 g agarose in 100 mL LB medium. Melt in microwave and maintain at 45°C for 20 min prior to use.

4. Maltose solution 10% (w/v): Sterilize by autoclaving.

5. SM buffer to dilute phage: 100 mM NaCl, 10 mM MgSO$_4$, 50 mM Tris-HCl, pH 7.5, 0.01% gelatin. Autoclave.

6. Fresh plating cells (*see* Section 3.1.1. and Note 1).

7. Nitrocellulose filters (Schleicher and Schuell, Dassel, Germany): 132- and 82-mm circles.

8. IPTG: 1M stock in water. Store at –20°C.

9. Wash buffer: 50 mM NaCl, 10 mM Tris-HCl, pH 7.5, 1 mM EDTA, and 0.05% lauryl dimethylamide oxide (LDAO; Calbiochem, La Jolla, CA), 1 mM dithiothreitol (DTT) (added just before use).

10. South-western block: 2.5% dried milk powder, 50 mM HEPES, pH 8.0, 10% (v/v) glycerol, 50 mM NaCl, 0.05% LDAO, 1 mM DTT (added just before use).

11. Falcon 2059 tubes (Becton Dickinson).

2.3. Probe Labeling

2.3.1. Oligonucleotide Annealing

1. T4 polynucleotide kinase buffer (10X): 500 mM Tris-HCl, pH 7.5, 100 mM MgCl$_2$, 5 mM DTT, 1 mM spermidine. Make up freshly.

2. Complementary single-stranded binding site oligonucleotides in TE or water at approx 1 mg/mL. Store at –20°C (*see* Note 4).

3. TE: 10 mM Tris-HCl, pH 8.0, 1 mM EDTA.

2.3.2. Probe Labeling and Ligation

1. T4 polynucleotide kinase and 10X polynucleotide kinase (PNK) buffer: 500 mM Tris-HCl, pH 7.5, 100 mM MgCl$_2$, 5 mM DTT, 1 mM spermidine. Make up freshly.
2. [γ-^{32}P]-Adenosine triphosphate (ATP) at 3000–5000 Ci/mmol.
3. Sephadex G-50 M (Pharmacia, Uppsala, Sweden) autoclaved in water or TE.
4. T4 DNA ligase and 10X ligase buffer: 500 mM Tris-HCl, pH 7.5, 100 mM MgCl$_2$, 10 mM DTT, 10 mM ATP, 10 mM spermidine.

2.4. Screening and Selection of Positive Plaques

1. Poly(dA)/Poly(dT) (Pharmacia) stock at 1 mg/mL made up in water.
2. TNE-50: 50 mM NaCl, 10 mM Tris-HCl, pH 7.5, 1 mM EDTA, 1 mM DTT. Add DTT just before use and store buffer at 4°C.
3. South-western block (*see* Section 2.2.).

2.5. Guanidinium Denaturation and Renaturation

1. TNE-50 (*see* Section 2.4.).
2. Guanidinium chloride (6M) in TNE-50.
3. South-western block (*see* Section 2.2.).

3. Methods
3.1. Titering the Library

Titering of a library phage stock is critical. It allows the calculation of the number of phage particles or plaque forming units (PFU) in a given volume of stock. A known density of PFU can then be plated out for screening.

3.1.1. Preparation of Plating Cells

1. Make fresh plating cells by inoculating one colony into 20 mL of LB medium containing 0.2% maltose. For all λ work, use LB medium containing 10 mM MgSO$_4$.
2. Shake the culture vigorously at 37°C until stationary phase is reached (overnight incubation, for example).
3. Pellet the cells, and resuspend in half their original volume using SM buffer. Plating cells may be stored on ice until required (up to 24 h).

3.1.2. Plating

1. Make several 10-fold serial dilutions of the library phage stock in 100-μL aliquots of SM in Falcon 2063 tubes.
2. Add 300 μL of fresh plating cells to each tube, gently mix, and place the tubes in a water bath at 37°C for 15 min for infection to occur.
3. Add 3 mL molten top agar at 45°C to each tube, and pour onto a series of prewarmed 90-cm LB agar plates.
4. When the top agar has set, transfer the plates to a 42°C incubator (37°C for λ ZAP) and incubate overnight.

5. Count the number of plaques formed on one of the plates to calculate the titer of the original phage stock.

3.2. Library Plating and Replica Lifts

Once the titer has been ascertained, the library can be plated on 15-cm plates, and protein expression induced for transfer to nitrocellulose membranes. Once again the protocol assumes the use of a λgt11 library and *E. coli* Y1090 plating cells *(2)*, but only minor modification is required for λ ZAP libraries *(see Note 1)*.

1. Having determined the titer of the library, plate out at least 10^6 PFU at a density of up to 10^5 PFU/15-cm plate (Note 3). For each plate, dilute phage into 200 μL of SM buffer, and add 600 μL of fresh plating cells *(see Section 3.1.1.)* using tubes, such as Falcon 2059 tubes. Leave to infect for 15 min at 37°C, and then add 7.5 mL molten top agarose and pour onto prewarmed LB-agar plates.
2. Leave the plates at room temperature for the top agar to set, and transfer to a 42°C incubator until plaques are just visible as pinpricks on the bacterial lawn. This takes 3–4 h. (For λ ZAP, incubate at 37°C for 6–8 h or overnight.)
3. Soak one nitrocellulose filter for each plate in 10 m*M* IPTG and dab off the excess liquid. Working quickly, take one plate at a time, place a numbered filter, still damp with IPTG solution, on the surface of the top agar and then transfer the plate to a 37°C incubator. Incubate for 1–2 h. It is best to handle one plate at a time as the temperature of the agar should not drop below 37°C for λgt11 libraries.
4. Mark the position of each filter relative to the plate by piercing the filter and agar with three to four orientation holes using a syringe needle dipped in India ink. Using blunt-ended forceps, carefully remove the filter and transfer to a plastic box (approx 20 × 20 × 5–10-cm deep) containing 500 mL of wash buffer.
5. Place a second IPTG-soaked filter on each plate in turn, and incubate at 37°C for a further 2–3 h, again marking the filters for alignment before removing from the plates. For the second lifts, a permanent marker pen can be used to make marks on the filters over the syringe needle holes that should be visible in the agar. Use a light box to make this easier if necessary.
6. Remove the second set of filters to wash buffer. All filters should be washed for 5–10 min to remove loose pieces of top agarose and so reduce background. The filters are now ready for blocking *(see Section 3.4.)*.
7. Once the second set of filters have been removed, wrap the plates in Saran Wrap™ or equivalent, and store them at 4°C until positive plaques have been identified.

3.3. Probe Labeling

The selection of the probe is another important feature and is discussed in more detail in Note 4. The author has used a specific factor binding sequence in a 20- to 25-mer double-stranded oligonucleotide with *Hin*dIII sticky ends to allow efficient labeling and concatenation. The probe sequence is first annealed from single-stranded oligonucleotides, and the double-stranded oligonucleotide is then labeled with [γ^{32}P]-ATP using T4 PNK and concatenated by end to end ligation with T4 DNA ligase.

3.3.1. Oligonucleotide Annealing

1. Take 40 μg each of the upper and lower strand oligonucleotides made up in water, and place in a 1.5-mL microfuge tube. Add 10 μL of 10X T4 PNK buffer (500 mM, Tris-HCl, pH 8.0, 100 mM MgCl$_2$, 5 mM DTT, 1 mM spermidine). Make the volume up to 100 μL with deionized water.
2. Incubate at 90°C for 5 min, 65°C for 10 min, 37°C for 15 min, and then at room temperature for 30 min.
3. Ethanol-precipitate, and take the double-stranded oligonucleotide up in TE at 100 ng/μL.
4. Store the double-stranded oligonucleotide at –20°C until required.

3.3.2. Probe Labeling and Ligation

1. In a 1.5-mL microfuge tube mix: 4 μL double-stranded oligonucleotide (0.1 μg/μL), 2 μL 10X polynucleotide kinase buffer, 5 μL (50 μCi) [γ^{32}P]-ATP (5000 Ci/mmol), 2 μL (20 U) T4 polynucleotide kinase, 7 μL H$_2$O.
2. Incubate at 37°C for 1 h.
3. Add 80 μL of TE to the labeling reaction, and separate the labeled oligonucleotide from the excess [γ-^{32}P]-ATP by spun-column chromatography. (The author uses a 1-mL syringe plugged with polyallomer wool and filled with Sephadex G-50M. Spin at 1000–2000g in a bench-top centrifuge with the syringe supported in a 15-mL Falcon tube.)
4. The labeling reaction should yield 1–2 × 10^8 cpm in total. Check the incorporation of radioisotope by counting 1 μL of the spun-column eluate by liquid scintillation counting.
5. To the labeled oligonucleotide (approx 100 μL) add: 11 μL 10X T4 DNA ligase buffer, 2 μL 10 mM ATP, 2 μL (20 U) T4 DNA ligase.
6. Ligate overnight at 15–18°C.

3.4. Screening and Selection of Positive Plaques

Before embarking on library screening, it is advisable to determine the optimum conditions for binding in terms of specific probe sequence, nonspecific competitor DNA, and binding buffer composition. These points are considered in Notes 4–7.

1. For 10–20 filters, place 500 mL of South-western block in a tray or large sandwich box. Immerse the filters from Section 3.2. in the blocking solution one at a time, ensuring that both surfaces of each filter come into contact with the solution.
2. Block overnight at 4°C with gentle shaking.

3. Make up the probe mixture by adding the concatenated probe at $5 \times 10^5 - 1 \times 10^6$ cpm/mL to 100 mL of TNE-50 together with nonspecific competitor DNA (Poly[dA]/Poly[dT]) at 10 µg/mL (*see* Note 6).
4. Probing is most conveniently carried out in 15-cm Petri dishes containing 50 mL of probe mix each. Place the probe solution into each of two 150-mm plastic Petri dishes, and lay the filters one at a time in the probe mixture, protein side up, placing only five to six filters in each dish.
5. Incubate for 1 h at room temperature (*see* Note 7), using a slowly moving orbital platform shaker to keep the filters floating freely. Filters can be processed in batches, reusing the probe solution and keeping the filters awaiting probing in blocking solution.
6. Remove the probed filters to a large tray or lunch box, and wash three to four times for 5–10 min each time in 200–300 mL of TNE-50 at room temperature, making sure that the filters remain free in the solution and do not stick together. Washing can be performed on an orbital platform shaker with fairly vigorous movement.
7. Blot the filters dry, and expose to X-ray film, including fluorescent or radioactive markers to facilitate orientation of the filters to the developed film (*see* Notes 8 and 9).
8. The probe mixture may be retained and kept at $-20°C$ for up to 1 wk, and may be used for second-round screens after adding fresh DTT immediately prior to use.

3.5. Guanidinium Chloride (GuHCl) Denaturation/Renaturation

Some DNA-binding proteins may be inactive as bacterially produced fusion proteins because of insolubility or incorrect protein folding. DNA-binding activity can sometimes be recovered by denaturation in GuHCl followed by slow renaturation. This procedure is preferably applied immediately after making the filter lifts, or if initial screening by the standard method has been unfruitful, the filter-bound proteins may be denatured as follows and then reprobed.

1. Immerse the filters one at a time in $6M$ GuHCl using a small tray or sandwich box containing 40 mL of solution/15-cm filter.
2. Agitate on a slowly moving orbital shaker for 10 min at room temperature.
3. Remove the GuHCl solution, discard one-half of it, and make up the volume with TNE-50. Mix and add back to the filters. Shake for a further 10 min.
4. Repeat this dilution of the GuHCl every 10 min seven more times, and finally discard the solution and replace with TNE-50. Wash for a further 10 min.
5. Block the filters overnight in South-western block.
6. Probe as described in Section 3.4.

3.6. Plaque Purification

Once positives are detected on duplicate filters (*see* Note 9), corresponding plugs of agar should be picked from the plate for further rounds of screening and for phage purification.

1. Using the orientation holes in the agar, align the plate with the developed autoradiograph, and use the wide end of a Pasteur pipet to pierce the agar.
2. Transfer the resulting agar plug into 1 mL of SM buffer containing a drop of chloroform in a glass bijou bottle.
3. Shake the bottles gently to speed phage elution (2 h to overnight).
4. Plate out the titered, eluted phage, but this time at a density of 500–1000 plaques on a 9-cm plate. Duplicate lifts are not necessary for second and further screens.
5. Screen the second-round filters, and pick positive areas of agar using the narrow end of a Pasteur pipet, again into 1 mL of SM plus a drop of chloroform.
6. Replate the phage eluted from these plugs at lower density on 9-cm plates for third-round screens.
7. Pick single-well isolated plaques if possible.
8. Replate at about 50 plaques/plate. All plaques should be positive on screening. If not, continue with another round of plaque purification.

4. Notes

1. The protocols given here are written assuming that the reader will be using a λgt11 cDNA library *(2)*. However, the method can be used with λ ZAP libraries with only minor modification, specifically, the use of *E. coli* XL1-Blue rather than Y1090 for plating cells and the growth of the phage at 37°C. NZY agar is also recommended for λ ZAP growth, but the author has found LB agar supplemented with 10 mM $MgSO_4$ to be satisfactory. Suitable cDNA libraries can be constructed by the investigator, can be obtained commercially, or may be available from other laboratories. The library should originate from a cell line or tissue known to express the DNA-binding activity of interest at a reasonable level, and for vertebrate organisms, should have a complexity of at least 10^6 independent recombinants and preferably in excess of 5×10^6. Remember that unless a directional cloning method has been employed during the library's construction, only one in six inserts will be in the correct orientation and frame to generate a β-galactosidase fusion protein. However, the author has found that in practice, proteins are often translated from an internal AUG codon. Although it is not necessary for clones to be full-length, sufficient coding information to generate a complete DNA-binding domain and possibly a dimerization domain is essential. Generally speaking, mixed oligo-dT primed and random primed libraries probably give the best results in this kind of exercise. If it is known that the DNA-binding activity resides in a particularly large polypeptide, then random priming will increase the probability of encountering clones encoding the DNA-binding activity of interest. Notably, however, smaller proteins may also be difficult to clone from oligo dT-primed libraries if they possess particularly long 3'-untranslated regions.
2. South-western blotting is a variation of the traditional Western blotting technique. Briefly, following sodium dodecyl sulfate-polyacrylamide gel electrophoresis and transfer onto a nitrocellulose membrane, electrophoresed DNA-binding proteins are detected with a labeled DNA probe (rather than by immunological means). The probe can be prepared in exactly the same way as described in Sec-

tion 3.3., and the filter treated identically to the filter lifts in Sections 3.4. and 3.5. The method therefore embodies the technique used to screen the library. The sample material could be whole-cell extract, nuclear extract, or partially purified DNA-binding protein.

3. The author has found a density of 5×10^4 PFU/150-mm plate to be optimum.
4. The nature of the binding site probe is critical for successful cDNA library screening. In general, the binding site is chosen on the basis of gel retardation or other experiments, and should be bound specifically by the factor of interest. Probes containing multiple binding sites give good signals and may take the form of a single, long oligonucleotide containing several sites for the factor of interest (*see* Hai et al. *[4]* for example), a restriction fragment containing multiple cloned binding site oligonucleotides or as described in this chapter, a concatenated binding site-containing oligonucleotide *(6)*. For the construction of concatenated probes, the specific binding sequence for the factor of interest should be contained in a 20- to 25-mer double-stranded oligonucleotide with free unphosphorylated 5'-ends to facilitate labeling with [γ-^{32}P]-ATP and compatible "sticky ends" to allow ligation. A consideration when designing a probe is to avoid generating binding sites for other factors at the ligated junctions of the a concatenated probe.
5. The author found TNE-50 to be a suitable buffer for probe binding and filter washing. It is a low-salt (50 mM NaCl) buffer selected to encourage the binding of bacterially synthesized proteins, which may bind DNA relatively poorly owing to such factors as lack of posttranslational modification. Binding activity in TNE-50 should preferably be tested in advance for the factor of interest by either gel retardation or South-western blot analysis.
6. Gel retardation analysis can also be used to determine the best nonspecific competitor DNA to use. Poly (dA)/poly (dT) was suitable, but poly (dI)/poly (dC) or herring sperm DNA may be more suitable for some factors.
7. The described hybridization conditions are given as a guide. Longer incubation times possibly at a lower temperature may improve the signal from a poorly binding factor. Similarly, the washing conditions described are the ones previously found to be the most effective. Longer washes at room temperature may result in loss of signal, but longer washes at lower temperatures may reduce background without significant loss of bound probe.
8. Using Kodak X-omat AR film and an intensifying screen, an overnight exposure at −70°C should be sufficient to detect clones giving strong signals. However, if the background allows, a second, longer exposure (1 wk) is also recommended in order to detect clones giving weaker signals.
9. Genuine positive plaques will give a signal from both of the duplicate filters, although the signal may be stronger on one duplicate than on the other.

References

1. Singh, H., LeBowitz, J. H., Baldwin, A. S., and Sharp, P. A. (1988) Molecular cloning of an enhancer binding protein: isolation by screening of an expression library with a recognition site DNA. *Cell* **52**, 415–423.

2. Huynh, T. V., Young, R. W., and Davis, R. W. (1988) Construction and screening of cDNA libraries in lambda gt11 and lambda gt10, in *Cloning: A Practical Approach*, vol. 1 (Glover, D. M., ed.), IRL, Oxford, UK, pp. 49–78.
3. Maekawa, T., Sakura, H., Kanei Ishii, C., Sudo, T., Yoshimura, T., Fujisawa, J., Yoshida, M., and Ishii, S. (1989) Leucine zipper structure of the protein CRE-BP1 binding to the cyclic AMP response element in brain. *EMBO J.* **8,** 2023–2028.
4. Hai, T., Liu, F., Coukos, W. J., and Green, M. R. (1989) Transcription factor ATF cDNA clones: an extensive family of leucine zipper proteins able to selectively form DNA-binding heterodimers. *Genes Dev.* **3,** 2083–2090.
5. Poli, V., Mancini, F. P., and Cortese, R. (1990) IL-6DBP, a nuclear protein involved in interleukin-6 signal transduction, defines a new family of leucine zipper proteins related to C/EBP. *Cell* **63,** 643–653.
6. Cowell, I. G., Skinner, A., and Hurst, H. C. (1992) Transcriptional repression by a novel member of the bzip family of transcription factors. *Mol. Cell Biol.* **12,** 3070–3077.
7. Kageyama, R. and Pastan, I. (1989) Molecular cloning and characterisation of a human DNA binding factor that represses transcription. *Cell* **59,** 815–825.
8. Williams, T. M., Moolten, D., Burlein, J., Romano, J., Bhaerman, R., Godillot, A., Mellon, M., Rauscher, F. J., III, and Kant, J. A. (1991) Identification of a zinc finger protein that inhibits IL-2 expression. *Science* **254,** 1791–1793.

15

Protein Interaction Cloning by Far-Western Screening of λ-Phage cDNA Expression Libraries

Shinichi Takayama and John C. Reed

1. Introduction

One of the most powerful ways to gain novel insights into a particular protein's cellular functions is to identify other proteins with which it specifically interacts. Traditionally, this problem has been approached by use of protein biochemistry, involving purification of interacting proteins and determination of at least a portion of their amino acid sequences by microsequencing techniques, so that degenerate oligonucleotides can be prepared and cDNAs obtained by conventional hybridization screening of cDNA libraries. This biochemical approach, however, is an enormously laborious task, and in the vast majority of cases, the yields of protein are simply inadequate to make it feasible. As an alternative, cDNAs encoding interacting proteins can be obtained by direct screening of λ-phage cDNA expression libraries using recombinant proteins and methods similar to those developed for antibody-based screening. This expression cloning approach is based on the "far-Western" method (also known as ligand blotting or West-Western) in which proteins immobilized on nitrocellulose or other suitable membranes can be detected by their ability to bind to soluble recombinant proteins of interest *(1–6)*. These protein–protein interactions are conceptually similar to traditional antigen–antibody interactions involved in Western blotting, but typically occur with lower affinity.

The far-Western approach to interaction cloning also represents an alternative to the now popular yeast two-hybrid method *(see* Chapter 16) *(7)*. Though the two-hybrid approach generally possesses superior sensitivity compared to the far-Western method in terms of detecting low-affinity protein–protein interactions, "bait" proteins that have acidic domains can often transactivate reporter genes when fused with a DNA-binding domain, without

From: *Methods in Molecular Biology, Vol. 69: cDNA Library Protocols*
Edited by: I. G. Cowell and C. A. Austin Humana Press Inc., Totowa, NJ

requirement for interaction with a transactivation domain-containing fusion partner, thus creating "background" reporter gene activation that in many cases, essentially precludes use of the two-hybrid approach. A second potential advantage of the far-Western approach is that one can sometimes optimize the conditions for screening cDNA expression libraries with protein probes by first using cell lysates in a standard sodium dodecyl sulfate-polyacrylamide gel electrophoresis (SDS-PAGE)/immunoblotting format. Probing such blots with the recombinant protein of interest provides critical information about the expected molecular weights of potential interacting proteins and the best conditions for detecting them. Finally, unlike expression of mammalian proteins in yeast, where proper processing in terms of posttranslational modifications may not reliably occur, protein probes for use in far-Western screening of libraries can be conveniently prepared in insect cells using the baculovirus system and even in mammalian cells, thus increasing the chances that the protein probe assumes the appropriate conformation and undergoes proteolytic processing, phosphorylation, or other modifications that may be essential for its interactions with target proteins. However, both the far-Western and two-hybrid approaches suffer from the probable lack of posttranslational processing of candidate interacting proteins when cDNA libraries are expressed in bacteria and yeast, and thus are incapable of identifying all potential physiologically important partner proteins.

Nevertheless, the far-Western screening approach has been employed successfully for the cloning of cDNAs encoding many interesting proteins, including Rb-binding proteins, such as RBAP-1, RBP-3, and RIZ *(8–10)*, Myc partners, including Max and Mad *(11,12)*, the Fos/Jun partner protein FIP *(13)*, and the Bcl-2-binding proteins BAG-1 and Bad *(14,15)*. In this chapter, we provide a description of methods that have proven successful in our hands for interaction cloning by the far-Western approach.

Before beginning library screening, it is very useful to attempt detection of interacting proteins by far-Western blotting methods using cell lysates that have been subjected to SDS-PAGE and transferred to nitrocellulose filters. This can provide critical information regarding which types of cells and tissues may express candidate interacting proteins at high levels and thus permits rational choices about the most appropriate cDNA libraries for screening. In addition, far-Western blotting experiments performed using cell lysates can be used to optimize the conditions of preblocking, probe binding, washing, and detection of interacting proteins in solid phase (i.e., immobilized on nitrocellulose) with soluble protein probes. One particularly relevant issue is whether attempts should be made to renature target proteins immobilized on nitrocellulose using guanidine-HCl renaturation procedures *(16,17)*. A protocol for far-Western blotting is given in Section 3.4.

During these preliminary investigations, experiments can also be performed using different methods for preparing and detecting protein probes. Among the possible approaches are:

1. Direct detection of protein probes bound to interacting proteins on nitrocellulose using antibodies directed against the protein probe *(9,14,18,19)*;
2. Expression of protein probes as fusions with an appended epitope tag and detection using antibodies targeted against that heterologous tag *(10,20)*; and
3. Use of radiolabeled protein probes without detecting antibodies *(8,13,15,21,22)*.

For the first two options, it is often unnecessary to purify the protein probe. Indeed, in our experience, crude cell lysates can be employed (e.g., from Sf9 insect cells infected with the recombinant baculovirsuses) with perfectly acceptable results, provided the abundance of the protein probe is sufficient (~1–5% of total cellular protein). With the radiolabeling approach, however, the protein must first be purified and then radiolabeled either directly using ^{125}I (targets tyrosine residues) *(11)* or by expressing the protein probe as a fusion protein that contains an extrinsic phosphorylation site sequence for a commercially available kinase, such as heart-muscle protein kinase (HMK) *(8,13,21,22)*. An alternative to radiolabeling is to biotinylate purified protein probes and then detect them using streptavidin-conjugated to alkaline phosphatase (23). It may also be possible to express protein probes as fusion proteins with an enzymatic protein, such as alkaline phosphatase, for direct colorimetric probing *(24)*.

Some potential disadvantages of using a secondary antibody detection strategy should be considered. For example, more washing and incubation steps will be required, and the affinity of some protein–protein interactions may be too weak to withstand this. A potential solution here is to fix the protein–protein interaction in place on nitrocellulose filters using formaldehyde or other crosslinking agents prior to attempting antibody-based detection. Second, the colorimetric detection agents often employed may also disrupt protein–protein interactions. For example, alkaline phosphatase-based detection with 5-bromo-4-chloro-3-indolyl phosphate/nitroblue tetrazolium (BCIP/NBT) is typically performed at relatively high pH, and the H_2O_2 generated during horseradish peroxidase (HRP)-based detection methods can also result in protein–protein dissociation. Nevertheless, because we have generally encountered higher amounts of nonspecific background when using radiolabeled bait proteins, resulting in more false-positives during library screening, antibody-based detection methods remain our preferred method.

Steric interference is another issue that should be addressed during preliminary far-Western blotting experiments. In cases where antibodies targeted directly against the protein probe are employed, for example, it is important that the antibody not bind to a site on the protein probe that may be important for interactions with biologically relevant proteins. For this reason, it can be very helpful to have some knowledge (from deletional mutagenesis studies) concerning the domains in the bait protein that are required for its bioactivity. Similarly, the choice of whether an epitope or His$_6$-tag is placed at the N- or C-terminus of the protein could be steered by knowledge of which end of the protein may better tolerate modification in terms of retaining bioactivity. We have been able to detect some protein–protein interactions, for example, when a glutathione S-transferase (GST) moiety is added to the N-terminus of the bait protein, but not when the tag is situated at the C-terminus. If a C-terminal tag is required, a potential option is to fuse the bait protein with the Fc portion of immunoglobulin heavy chain, thus leaving the N-terminus of the bait protein free to interact with other proteins, and providing a convenient way to purify and detect the recombinant protein via its C-terminal Fc domain using protein A or anti-Ig reagents.

With the epitope tag approach, we often express our recombinant proteins as GST fusions in either *Escherichia coli* or insect cells, thus permitting affinity purification using glutathione-Sepharose. The GST moiety can therefore serve the dual purpose of an epitope tag, using anti-GST monoclonal antibodies (MAbs) for detection. One might also imagine constructing probes with two epitope tags, thus adding an additional level of specificity to the library screening procedure, although we have yet to try this ourselves.

Additional issues for optimizing conditions for library screening include attention to salt concentration (NaCl or KCl at 0–0.5M), addition of 10% glycerol, detergents (Tween-20), and divalent cations. All of these can be addressed prior to library screening using far-Western blot assays.

2. Materials

2.1. Library Plating and Replica Lifts

2.1.1. Titering the Phage Library

1. cDNA Expression library in λEX-lox or λgt11 (*see* Section 3.1. and Note 1).
2. Host *E. coli* strain (*E. coli* BL21 or Y1090r, respectively; *see* Note 2).
3. Luria-Bertani (LB) medium: 10 g tryptone, 5 g yeast extract, 10 g sodium chloride/L. Sterilize by autoclaving.
4. LB agar: 10 g tryptone, 5 g yeast extract, 10 g sodium chloride, 15 g bacteriological agar/L. Sterilize by autoclaving, and add MgSO$_4$ to 10 mM from a filter-sterilized 1M stock.
5. 1M MgSO$_4$: filter-sterilize.
6. Top agarose (per 100 mL): 1 g bacto-tryptone, 0.5 g yeast extract, 0.5 g NaCl, 0.7 g agarose.
7. Maltose solution 10% (w/v) in water: autoclave.
8. SM buffer to dilute phage: 100 mM NaCl, 10 mM MgSO$_4$, 50 mM Tris-HCl, pH 7.5, 0.01% gelatin. Autoclave.

2.1.2. Plating the Library and Preparation of Plaque Lifts

1. 2X YT plates (per liter): 16 g bacto-tryptone, 10 g yeast extract, 5 g NaCl, 5 g agar. Adjust pH to 7.5 with 10N NaOH and sterilize by autoclaving. Library screening will require 20 150-mm plates.
2. Top agarose: *see* Section 2.1.1.
3. SM buffer: *see* Section 2.1.1.
4. Titered λ-phage library: *see* Section 2.1.1.
5. *E. coli* strain: *see* Section 2.1.1.
6. Nitrocellulose filters: 137-mm diameter; 0.45 μm; Schleicher & Schuell (Keene, NH) (BA85).
7. 1M Isopropylthio-β-D-galactoside (IPTG): Prepare in water and store in aliquots at −20°C.

2.2. Screening by Far-Western Method Using Antibody-Based Detection

1. TBST: 50 mM Tris-HCl, pH 7.5, 150 mM NaCl, 0.05% Tween-20.
2. Hyb-75: 20 mM HEPES, pH 7.7, 75 mM KCl, 0.1 mM EDTA, 2.5 mM MgCl$_2$, 1 mM DTT, 0.05% NP-40.
3. Hyb-75 with 5 or 1% skim milk: Hyb-75 plus 5 g or 1 g/100 mL of nonfat dried milk (Difco, Detroit, MI, 0032-17-3).
4. HBB (HEPES-binding buffer): 25 mM HEPES, pH 7.7, 25 mM NaCl, 5 mM MgCl$_2$, 0.5 mM DTT.
5. HBB containing 6M guanidine hydrochloride.
6. NP40 lysis buffer: 1% NP-40, 50 mM Tris-HCl, pH 8.0, 100 mM NaCl, 2 mM EDTA.
7. Alkaline phosphatase buffer: 100 mM NaCl, 5 mM MgCl$_2$, 100 mM Tris-HCl, pH 9.5.
8. NBT/BCIP solution: Mix 10 mL of alkaline phosphatase buffer (item 7) with 66 μL NBT stock solution (item 9) and 33 μL of BCIP stock solution (item 10). Prepare immediately prior to use.
9. NBT (Sigma, St. Louis, MO, N-5514): 10-mg tablet: 5 tablets/mL of 70% dimethylformamide.
10. BCIP (Sigma B-0274): 25-mg tablet: 2 tablets/mL of 100% dimethlformamide.
11. Antimouse alkaline phosphatase-conjugated antibody (Promega, Madison, WI, S3721) dilution 1:7000 in Hyb-75 (final conc. 0.15 μg/mL).
12. Anti-GST MAb (7E5 antibody developed in our laboratory or Santa Cruz Biotechnology, Santa Cruz, CA, sc-138).

2.3. Library Screening Using Radiolabeled Probes

2.3.1. Preparation of ³²P-Labeled Probe

1. LB medium: *see* Section 2.1.1.
2. Ampicillin 50 mg/mL in water, filter-sterilize.
3. 1M IPTG.
4. 1X HMK buffer: 20 mM Tris-HCl, pH 7.5, 100 mM NaCl, 12 mM MgCl$_2$.
5. NETN buffer: 20 mM Tris-HCl, pH 8.0, 100 mM NaCl, 1 mM EDTA, 0.5% NP-40.

6. Catalytic subunit of bovine heart muscle protein kinase: Sigma P-2645 (250 U/vial). Resuspend in vial in 25 μL of 40 mM DTT. Store at 4°C for up to 1 wk.
7. [γ^{32}P]-adenosine triphosphate (ATP) (~6000 Ci/mmol).
8. Stop solution: 10 mM Na-phosphate, pH 8.0, 10 mM Na-pyrophosphate, 10 mM EDTA, 1 mg/mL bovine serum albumin.
9. 50 mM Tris-HCl, pH 8.0, containing 5 mM reduced glutathione. Prepare freshly.

2.3.2. Library Screening

1. *See* Section 2.2., items 1–5.
2. ^{32}P-labeled probe.

2.4. Far-Western Blotting for Optimizing Detection Procedures Prior to Library Screening

See Section 2.2.

3. Methods

The protocols described here are based on our successful interaction cloning of cDNAs using recombinant baculovirus Bcl-2 protein as a probe, followed by detection with anti-Bcl-2 MAb and an alkaline-phosphatase colorimetric (BCIP/NBT) assay for detection *(14)*, and on experiments where a GST-BAG-1 fusion protein was used, combined with anti-GST antibody for detection (Takayama et al., in preparation).

3.1. Library Plating and Replica Lifts

Expression cDNA libraries in either λgt11 or λEX-lox vectors can be used. We prefer the λEX-lox system (Novagen) because of the higher levels of recombinant protein production (~10× higher than λgt11; *see* Note 1) that are attainable by use of the T7 promoter, and thus, this method is described first.

3.1.1. Titering Phage Library

1. Grow a single colony of appropriate BL21 host strain (*see* Note 2) in LB medium supplemented with 0.2% maltose and 10 mM MgSO$_4$ at 37°C to an OD$_{600}$ = 1.0.
2. Store the host cells at 4°C until needed (<24 h).
3. Prepare three dilutions of phage in SM buffer so that (based on previous titering) approx 10, 100, and 1000 PFU are present in 0.1 mL.
4. Mix 200 μL of cells prepared in step 1 with 100 μL diluted phage.
5. Incubate at 37°C for 20 min (without shaking).
6. Add 3 mL of 47°C top agarose, and mix well (by inverting capped tube end-over-end).
7. Quickly pour the cell/phage mixture onto a prewarmed LB agar plate (100-mm diameter), holding the tube close to the agar surface as you pour to avoid bubbles.
8. Allow top agarose to harden at room temperature for ~30 min.
9. Place plates upside down at 37°C overnight, and calculate the phage titer (i.e., number of PFU/mL).

3.1.2. Plating the λEX-lox Library and Preparation of Plaque Lifts

For library screening, prepare 20 plates with 2×10^4–5×10^4 plaques (*see* Note 3)/150-mm diameter Petri dish for a total of approx 10^6 phage as described:

1. For each plate, mix 300 μL of phage diluted in SM (2–5×10^4 phages) with 600 μL plating cells prepared as in Section 3.1.1., and incubate at 37°C for 20 min to allow phage to adsorb to the bacteria.
2. Mix phage-adsorbed host bacteria with 7 mL molten top agarose (47°C), and quickly pour onto prewarmed 2X YT plates.
3. Leave the plates at room temperature until agarose sets (typically ~30 min). Invert plates and incubate at 37°C until plaques have reached a diameter of ~0.5–1 mm (usually takes ~6 h). Unlike λgt11, the plaques formed with lEX-lox at this point are often unclear, but proceed anyway.
4. Working rapidly to minimize cooling of the plates, remove plates from incubator, and using gloved hands, carefully apply nitrocellulose circles (137-mm diameter) that have been soaked in 10 mM IPTG and then dried. Flex the filters to create a convex surface, and apply this initially to the center of the plate. Slowly lay down the sides, allowing the filter to wet as it touches the agarose surface and avoiding bubbles. If filter is askew, do not reposition it. Just remove and start over with a fresh IPTG-soaked filter. Handle filters by edges at all times. Filters should be labeled with a pencil prior to use, and plastic plates should be correspondingly labeled with a marking pen on the bottom (not tops).
5. Incubate plates with filters upside down at 37°C for 4 h.
6. Allow plates to cool to room temperature briefly, and then incubate upside down at 4°C for 30 min to 1 h. (This minimizes sticking of agarose to nitrocellulose membranes when the filters are removed.)
7. Punch at least three holes (asymmetrically) in edges of filters using a 22-gage needle dipped in India ink to allow later orientation of the filter with the plate.
8. Carefully remove nitrocellulose membranes using flame-sterilized forceps, and put into ~30 mL TBST solution in a 150-mm Petri dish for washing off agarose and bacterial debris (1 filter/dish).
9. Nitrocellulose membranes are now ready to use for far-Western detection of phage that are producing proteins capable of binding your probe protein of interest. Store phage plates at 4°C covered in plastic wrap or in their original plastic sleeves to prevent desiccation.

3.1.3. Plating a λgt11 Library and Preparation of Plaque Lifts

3.1.3.1. PREPARATION OF HOST CELLS: *E. COLI* STRAIN Y1090R-

1. Streak Y1090r- cells on LB agar plates containing 50 μg/mL ampicillin.
2. Inoculate 2 mL of LB/ampicillin (50 μg/mL) medium with a single well-isolated colony and grow at 37°C with aeration overnight.
3. Pipet 1 mL of overnight culture into 100 mL of LB/amp medium supplemented with 0.2% maltose and 10 mM MgSO$_4$.
4. Culture at 37°C in a 0.5-L Erlenmyer flask with vigorous shaking until OD$_{600}$ = 1.0.

3.1.3.2. PLATING AND PLAQUE LIFTS

1. Dilute 2–5 × 10⁴ PFU of λgt11 phage into 300 μL of SM buffer (titered as described in Section 3.1.1.) (*see* Note 3).
2. Mix with 600 μL of host cells.
3. Incubate at 37°C for 15 min to allow phage adsorption to cells.
4. Add 7 mL of top agarose (47°C).
5. Mix and pour onto prewarmed LB/amp agar plates (150-mm diameter).
6. Incubate plates at 42°C for 3.5 h.
7. Working quickly to minimize cooling of plates, place a dry nitrocellulose filter saturated in 10 m*M* IPTG onto each plate.
8. Incubate plates at 37°C for 4 h.
9. Let plates cool to room temperature briefly, and then place at 4°C for 30 min.
10. Make three asymmetric marks on nitrocellulose membranes with a needle dipped India ink.
11. Remove nitrocellulose membranes, and rinse the filters in TBST as described in Section 3.1.2. Filters are now ready for far-Western screening. Store phage plates at 4°C covered in plastic wrap or in their original plastic sleeves to prevent desiccation.

3.2. Screening by Far-Western Method Using Antibody-Based Detection

After washing nitrocellulose membranes with TBST, proteins immobilized on filters can be subject to denaturation-renaturation using 6*M* guanidine-HCl (*see* Note 4).

1. Place 1 filter/dish in a series of Petri dishes each containing 20 mL of HBB buffer.
2. Replace solution with 20 mL of HBB containing 6*M* guanidine hydrochloride, and incubate at 4°C with rocking for 10 min.
3. Change 6*M* guanidinium solution, and incubate for an additional 15 min.
4. Remove 10 mL of solution, and add 10 mL fresh HBB solution. Continue incubation at 4°C for 15 min.
5. Repeat step 4 another four times, removing 10 mL and replacing with 10 mL of fresh HBB.
6. Decant off the solution, and immerse filters in fresh HBB solution.
7. Incubate at 4°C for 5 min and then decant off HBB.
8. Add 10–20 mL of Hyb-75 solution with 5% skim milk (for preblocking) to the Petri dishes each containing one filter. Incubate for 30 min at 4°C with gentle rocking (Note 5).
9. Remove preblocking solution from the Petri dishes, and replace with Hyb-75 containing 1% skim milk and recombinant protein probe (*see* Notes 6 and 7). Incubate for 2 h to overnight at 4°C with protein side up.
10. Wash filters twice for 1 min each in 20 mL/filter of Hyb-75 with 1% skim milk.
11. Incubate with MAb against probe protein in 1% skim milk/hyb-75 solution without DTT for 2 h at 4°C (*see* Note 8).
12. Wash two times for 1 min each with 20 mL/filter of Hyb-75 with 1% skim milk, but without DTT.

13. Incubate with 10–20 mL/filter of 0.15 µg/mL of alkaline phosphatase-conjugated antimouse Ig antibody in 1% skim milk/hyb-75 solution without DTT for 2 h at 4°C.

14. Wash two times for 1 min each with 20 mL/filter of Hyb-75 with 1% skim milk, but without DTT.

15. Add 20 mL of alkaline phosphatase buffer to each Petri dish containing a filter (This will bring pH to ~9.8.)

16. Incubate each filter separately in 10–20 mL of NBT/BCIP solution for ~30 min in the Petri dishes at room temperature. Monitor the color development during the reaction, and stop when filters become a diffuse, light blue color. If color fails to develop, place the dishes at 37°C, and allow reactions to proceed for a longer time.

17. Stop reaction by removing filters from NBT/BCIP solution and immersing in 1X PBS with 20 mM EDTA, which chelates Mg^{2+} and ceases the reaction.

18. Warm phage plates, and align needle holes with color-developed filters using a light box to illuminate plates.

19. Using the back side of a sterile Pasteur pipet, core the agarose over positive plaques, and dispense into 1 mL of SM buffer. Let phage diffuse out at room temperature for 20 min or overnight at 4°C. In addition to round signals, sometimes the shape of true positives is that of ring or half ring. These ring-plaque signals, however, are generally not seen at high phage densities.

20. For secondary screens, titer phage on BL21 or Y1090r. For λEX-lox, assume phage titer of ~10^5–10^6 PFU/mL. For λgt11, assume 10^6–10^7 PFU/mL. Then repeat screening with protein probe using ~1000 PFU/100-mm plate. Pick plaques with front end of a Pasteur pipet, and dispense into 1 mL SM buffer.

21. For tertiary screens, plate at <500 PFU/100-mm plate.

22. After phage is plaque-purified, plate again at ~200/100-mm plate and lift. Cut filters into four quarters and probe to test specificity of results as follows:
    a. Protein + primary Ab + secondary Ab;
    b. Probe + secondary Ab, without primary Ab;
    c. Control protein or cell lysate + primary Ab + secondary Ab; and
    d. Control protein or lysate + secondary antibody without primary Ab.
    If additional antibodies to other epitopes on the target protein are available, these can also be employed as a further check on the specificity of the reactions.

23. If using lEX-lox, convert phage to plasmid (Note 9).

3.3. Library Screening Using Radiolabeled Probes

3.3.1. Preparation of ^{32}P-Labeled Probe

The fusion protein should be produced with an cAMP-protein kinase phosphorylation site by subcloning into pGEX-2TK, Pharmacia (Piscataway, NJ), or other suitable vectors.

1. Inoculate a single colony of bacteria containing a pGEX-2TK-based recombinant plasmid encoding the desired bait protein as a GST fusion into 2 mL of LB containing 50 µg/mL ampicillin and culture overnight at 37°C with aeration.

2. Transfer 2 mL of overnight culture to 500 mL of prewarmed LB/amp medium, and grow with vigorous shaking at 37°C.
3. After 4–6 h when the OD_{600} = 0.6, cool cultures to 30°C and add IPTG to a final concentration of 0.1 mM.
4. Continue growing at 30°C for 3 h to overnight with vigorous shaking.
5. Pellet cells by centrifugation at ~4500g (spin at 4500g using Beckman JA-14 or Sorval GSA rotor) for 5 min at 4°C, and resuspend the pellet in 10 pellet volumes of NETN.
6. Sonicate for 1 min using probe-type sonicator at medium power on ice until solution clarifies. Additional sonication for 1-min period may be required once or twice more, but allow the solution to cool before resuming. (We use a 3-mm probe and an apparatus from Heat Systems, model XL2020, at a setting of 5 [max. 10 range].) The solution can still be viscous, since genomic DNA will be removed by centrifugation in step 7.
7. Spin at 10,000g for 20 min at 4°C.
8. Transfer supernatant to fresh tube and 1 mL of glutathione-Sepharose (1:1 suspension in NETN with 0.5% skim milk).
9. Rock at 4°C for 30 min.
10. Wash 3× in 10 mL of NETN.
11. Wash 1× in 10 mL of HMK buffer without DTT.
12. Resuspend beads in 0.75 mL of reaction buffer consisting of:
    a. 75 µL of 10X HMK buffer with DTT;
    b. 600 µL of dH$_2$O;
    c. 50 µL of [γ^{32}P]-ATP (6000 Ci/mmol); and
    d. 25 µL of HMK enzyme (resuspended in 40 mM DTT as described).
13. Incubate at 4°C for 30 min, flicking tube occasionally.
14. Add 1 mL of HMK stop buffer.
15. Spin down beads briefly at 1000g, and aspirate off supernatant with 23-gage needle and syringe. Discard radioactive supernatant appropriately.
16. Wash Sepharose beads 5× with 10 mL of NETN, aspirating off supernatant with syringe and needle as in step 16.
17. Elute labeled fusion protein in 2 mL of 50 mM Tris-HCl, pH 8.0, containing 5 mM glutathione.
18. Count a 2–5 µL aliquot using β-counter.

3.3.2. Library Screening

Essentially the same procedures are used as in Section 3.2., with the following exceptions:

1. Consider adding 50M excess of unlabeled GST non fusion protein to preblocking and binding buffers to avoid false positives caused by GST-binding proteins.
2. Incubate filters with ^{32}P-labeled GST fusion proteins at 1–2.5 × 10^4 cpm/mL in Hyb-75 with 1% skim milk.
3. Incubate filters with probe overnight at 4°C with gentle rocking.

4. Wash three times for 10 min each with Hyb-75 containing 1% skim milk.
5. Instead of incubation with primary and secondary antibodies, wrap filters moist in plastic wrap and expose to X-ray film with intensifying screens at −80°C.

3.4. Far-Western Blotting for Optimizing Detection Procedures Prior to Library Screening

1. Run total cell lysates in SDS-PAGE (20–50 µg/lane), and transfer to nitrocellulose in Tris-glycine buffer by standard methods.
2. Wash filter with TBST.
3. Perform renaturation-denaturation with guanidine-HCl as in Section 3.2., steps 1–7 (10 mL/150-cm^2 blot).
4. Incubate for 30 min in 10 mL of Hyb-75 containing 5% skim milk to preblock.
5. Transfer blot to 10 mL of Hyb-75 containing 1% skim milk and probe.
6. Incubate at 4°C for 2 h to overnight with gentle rocking.
7. Wash twice for 1 min each with 10 mL Hyb-75 containing 1% skim milk without DTT.
8. Incubate with primary antibody in 10 mL of Hyb-75 with 1% skim milk, but without DTT for 1–2 h at 4°C.
9. Wash as in step 7.
10. Incubate with alkaline phosphatase-conjugated secondary antibody in 10 mL of Hyb-75 with 1% skim milk, but without DTT for 0.5–1 h at 4°C.
11. Wash as in step 7.
12. Immerse filter into alkaline phosphatase solution.
13. Decant off solution, and add 10 mL of activated alkaline phosphatase substrate (NBT/BCIP solution).
14. Incubate at room temperature for 30 min with gentle rocking.
15. Decant off NBT/BCIP, and stop with 10 mL of PBS containing 20 mM EDTA.

4. Notes

1. There are three major λ-phage-based expression vectors available for library screening: λgt11, λEX-lox, and λZAP. The level of recombinant protein production with these vectors is generally λEX-lox > λgt11 > λZAP. For this reason, we prefer λEX-lox and do not recommend λZAP.
2. For initial attempts at library screening, we generally recommend using BL21-pLysE (Novagen) without IPTG induction. This gives high levels of expression, but the recombinant proteins produced can sometimes be toxic. If no positive clones are obtained with BL21-pLysE, then try BL21 strain with IPTG induction. The BL21-pLysS strain is generally used only for production of large quantities of probe proteins, but not for library screening.
3. Signals are generally better when screening at lower density.
4. Note that not all proteins can be renatured by this procedure, and some proteins might be further denatured. Thus, it is very important that preliminary experiments be performed to determine which approach (i.e., with or without guanidinium-HCl denaturation) is preferable by far-Western blot analysis of cellular lysates.

5. Additional reagents can be helpful during preblocking, application of protein probe, and antibody detection to avoid false positives. These include addition of 3.8 μ*M* poly-L-lysine during preblocking and during application of the protein probe, and inclusion of 1% serum (goat, mouse, or human).

6. The concentration of protein probe should be ≥0.1 μg/mL. We typically employ 1 μg/mL of recombinant probe proteins for crude lysates and 10 μg/mL for purified proteins. To estimate the amount of protein in crude lysates, run lysates derived from cells containing control ("empty" vector) and from the same cells expressing the protein of interest side-by-side in SDS-PAGE gels, fix, and stain with Coomassie blue, and dry down. Localize the band of interest based on its molecular weight and absence in control cells. Estimate percentage of total cellular protein represented by the probe band. Generally, a rough estimate is sufficient, but if more precise estimates are required, the gels can be scanned and the area under the tracings calculated (probe band/all bands), or proteins can be prepared from ^{35}S-L-methionine-labeled cells, and autoradiography can be performed followed by densitometry or β-scanning. For production of Bcl-2 (an integral membrane protein), we lysed baculovirus-infected Sf9 after 48 h of infection in a solution containing 1% NP-40 detergent. Approximately 1–2% of the total cellular protein was recombinant Bcl-2 protein. We therefore used 100 μg of crude lysate/mL in Hyb-75/1% skim milk solution. For GST fusion proteins, we typically employ 1–10 μg of purified protein/1 mL Hyb-75 solution, followed by detection with anti-GST MAb.

7. Insect cells infected by recombinant baculovirus vectors can produce native proteins that have correct folding, proper posttranslational modification, and associated cofactors that may be physiologically important. The expression level in insect cell, however, is variable and some proteins are very difficult to produce efficiently. Purification of baculovirus-produced recombinant proteins is also often difficult, but proteins can be produced with His-tags using commercially available vectors (PharMingen, San Diego, CA). Improved methods for preparation of recombinant viruses have also greatly streamlined this process recently, such as use of a color selection method using a lacZ gene (BaculoGold™ system by PharMingen).

8. When using antibody against the recombinant protein probe for detection, if possible, screen a library derived from a species that will not react with the antibody probe. For example, to identify Bcl-2-binding proteins, we screened a mouse cDNA library using the human Bcl-2 protein and an antibody specific for human Bcl-2 that failed to crossreact with mouse Bcl-2. This approach avoided false positives owing to cloning of cDNAs encoding Bcl-2.

9. When using the λEX-lox system, phage can be conveniently converted to plasmids by using the loxP-cre recombination system. After isolation of single plaques, infect an *E. coli* host strain that expresses P1 cre recombinase (BM25.8). Plate *E. coli* BM25.8 on M9 plates containing 1 μg/mL thiamine, 34 μg/mL chloramphenicol, and 50 μg/mL kanamycin. Pick a single colony into LB medium containing 10 m*M* MgSO$_4$, 0.2% maltose, 50 μg/mL kanamycin, 34 μg/mL chloramphenicol,

and incubate at 37°C overnight. Mix phage in SM buffer with 100 μL of prepared BM25.8 cells and then plate cells on LB ampicillin or carbenicillin plates with top agarose, and incubate overnight at 37°C to obtain colonies.

References

1. Sikela, J. M. and Hahn, W. E. (1987) Screening an expression library with a ligand probe: isolation and sequence of a cDNA corresponding to a brain calmodulin-binding protein. *Proc. Natl. Acad. Sci. USA* **84**, 3038–3042.
2. Gershoni, J. M. (1988) Protein-blot analysis of receptor-ligand interactions. *Biochem. Soc. Trans.* **16**, 138,139.
3. Fazleabas, A. T. and Donnelly, K. M. (1992) Characterization of insulin-like growth factor binding proteins by two-dimensional polyacrylamide gel electrophoresis and ligand blot analysis. *Anal. Biochem.* **202**, 40–45.
4. Hasegawa, Y., Hasegawa, T., Yokoyama, T., Kotoh, S., Tsuchiya, Y., and Kurimoto, F. (1992) Western ligand blot assay for human growth hormone-dependent insulin-like growth factor binding protein (IGFBP-3): the serum levels in patients with classical growth hormone deficiency. *Endocrinol. Jpn.* **39**, 121–127.
5. Grissom, F., Rivero-Crespo, F., Lindgren, B., and Hall, K. (1993) Ligand blot analysis: validation of quantitative capabilities and utilization for measurement of truncated insulin-like growth factor regulation of Hep-G2 insulin-like growth factor binding protein-1 production. *Anal. Biochem.* **212**, 412–420.
6. Martínez-Ramírez, A. C., González-Nebauer, S., Escriche, B., and Real, M. D. (1994) Ligand blot identification of a *Manduca sexta* midgut binding protein specific to three *Bacillus thuringiensis* CrylA-type ICPs. *Biochem. Biophys. Res. Commun.* **201**, 782–787.
7. Fields, S. and Song, O. (1989) A novel genetic system to detect protein–protein interactions. *Nature* **340**, 245,246.
8. Kaelin, W. G., Jr., Krek, W., Sellers, W. R., DeCaprio, J. A., Ajchenbaum, F., Fuchs, C. S., Chittenden, T., Li, Y., Farnham, P. J., Blanar, M. A., Livingston, D. M., and Flemington, E. K. (1992) Expression cloning of a cDNA encoding retinoblastoma-binding protein with E2F-like properties. *Cell* **70**, 351–364.
9. Helin, K., Lees, J. A., Vidal, M., Dyson, N., Harlow, E., and Fattaey, A. (1992) A cDNA encoding a pRB-binding protein with properties of the transcription factor E2F. *Cell* **70**, 337–350.
10. Buyse, I. M., Shao, G., and Huang, S. (1995) The retinoblastoma protein binds to RIZ, a zinc-finger protein that shares an epitope with the adenovirus E1A protein. *Proc. Natl. Acad. Sci. USA* **92**, 4467–4471.
11. Blackwood, E. M. and Eisenman, R. N. (1991) Max: a helix-loop-helix zipper protein that forms a sequence-specific DNA-binding complex with myc. *Science* **251**, 1211–1217.
12. Ayer, D. E., Kretzner, L., and Eisenman, R. N. (1993) Mad: a heterodimeric partner for max that antagonizes myc transcriptional activity. *Cell* **72**, 211–222.
13. Blanar, M. A. and Rutter, W. J. (1992) Interaction cloning: identification of a helix-loop-helix zipper protein that interact with c-Fos. *Science* **256**, 1014–1018.

14. Takayama, S., Sato, T., Krajewski, S., Kochel, K., Irie, S., Millan, J. A., and Reed, J. C. (1995) Cloning and functional analysis of BAG-1: a novel Bcl-2-binding protein with anticell death activity. *Cell* **80,** 279–284.

15. Yang, E., Zha, J., Jockel, J., Boise, L. H., Thompson, C. B., and Korsmeyer, S. J. (1995) Bad, a heterodimeric partner for Bcl-XL and Bcl-2, displaces Bax and promotes cell death. *Cell* **80,** 285–291.

16. Vinson, C. R., LaMarco, K. L., Johnson, P. F., Landschulz, W. H., and McKnight, S. L. (1988) *In situ* detection of sequence-specific DNA binding activity specified by a recombinant bacteriophage. *Genes Dev.* **2,** 801–806.

17. Singh, H., Clerc, R. G., and LeBowitz, J. H. (1989). Molecular cloning of sequence-specific DNA binding proteins using recognition site probes. *BioTechniques* **7,** 252–261.

18. Defeo-Jones, D., Huang, P. S., Jones, R. E., Haskell, K. M., Vuocolo, G. A., Hanobik, M. G., Huber, H. E., and Oliff, A. (1991) Cloning of cDNAs for cellular proteins that bind to the retinoblastoma gene product. *Nature* **352,** 251–254.

19. Shan, B., Zhu, X., Chen, P.-L., Durfee, T., Yang, Y., Sharp, D., and Lee, W.-H. (1992) Molecular cloning of cellular genes encoding retinoblastoma-associated proteins: identification of a gene with properties of the transcription factor E2F. *Mol. Cell Biol.* **12,** 5620–5631.

20. Ren, R., Mayer, B. J., Cicchetti, P., and Baltimore, D. (1993) Identification of a ten-amino acid proline-rich SH3 binding site. *Science* **259,** 1157–1161.

21. Su, L.-K., Vogelstein, B., and Kinzler, K. W. (1993) Association of the APC tumor suppressor protein with catenins. *Science* **262,** 1734–1737.

22. Gu, W., Bhatia, K., Magrath, I. T., Dang, C. V., and Dalla-Favera, R. (1994) Binding and suppression of the myc transcriptional activation domain by p107. *Science* **264,** 251–254.

23. Macgregor, P. F., Abate, C., and Curran, T. (1990) Direct cloning of leucine zipper proteins: Jun binds cooperatively to the CRE with CRE-BP1. *Oncogene* **5,** 451–458.

24. Cheng, H.-J. and Flanagan, J. G. (1994) Identification and cloning of ELF-1, a developmentally expressed ligand for the Mek4 and Sek receptor tyrosine kinases. *Cell* **79,** 157–168.

16

Yeast Two-Hybrid Library Screening

Ian G. Cowell

1. Introduction

The two-hybrid system was originally devised by Fields and Song *(1)* as a protein interaction detection system in yeast. Subsequently, it has been employed in many laboratories as a means of screening cDNA and genomic fusion libraries for protein interaction partners *(2–8)*. The method relies on the fact that transcription factors, such as the yeast GAL4 factor, consist of separable DNA-binding and transcriptional regulatory domains; the former being required to direct the latter to appropriate promoters where transcriptional activation is effected, usually by direct or indirect interaction of the activation domain with the general transcription machinery. The essence of the two-hybrid system is the in vivo reconstitution of a functional transcriptional activator from two interacting polypeptides, one fused to a sequence-specific DNA-binding domain and the other to a potent transcriptional activation domain. Interaction is detected when the reconstituted transcription factor activates a reporter gene (*see* Fig. 1). DNA-binding domain and activation domain fusion proteins are expressed from plasmid DNAs. The DNA-binding and transcriptional activation components are usually derived from the yeast GAL4 transcription factor (*see* Figs. 1 and 2), although alternatives have been used by some workers *(6,9,10)*. Where the GAL4 DNA-binding domain is used, yeast strains in which the wild-type GAL4 gene has been deleted are employed. For the purposes of identifying new protein partners, DNA encoding the polypeptide for which partners are sought (bait polypeptide) is ligated into a yeast shuttle vector to create a DNA-binding domain–bait fusion (Fig. 2A). Library cDNA is ligated into a second shuttle vector to create an activation domain-tagged cDNA library (*see* Fig. 2B). In the original method as proposed by Chien et al. *(2)*, the bait (GAL4 DNA-binding domain) construct and activation domain fusion library were cotransformed into a yeast strain containing an integrated *LacZ* gene

From: *Methods in Molecular Biology, Vol. 69: cDNA Library Protocols*
Edited by: I. G. Cowell and C. A. Austin Humana Press Inc., Totowa, NJ

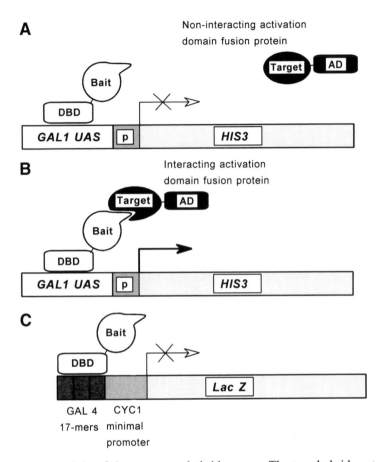

Fig. 1. The principle of the yeast two-hybrid system. The two-hybrid system is a method of detecting specific protein–protein interactions between two polypeptides in vivo. The "bait" and "target" polypeptides are fused to the DNA-binding and transcriptional activation domains of the transcription factor GAL4, respectively (other DNA-binding and activation domains have been used). In vivo interaction of "bait" and "target" polypeptides results in formation of a functional activator and activation of GAL4-responsive reporter genes. Yeast strains used for two-hybrid analysis, such as YPB2 and HF7c, contain two integrated GAL4-responsive reporter genes *HIS3* and *LacZ*. Activation of the *HIS3* reporter gene confers the ability to grow on media lacking histidine. **(A)** In the situation where bait and target proteins do not interact, the bait polypeptide is targeted to the promoter of the *HIS3* reporter gene through specific protein–DNA interactions between the GAL4 DNA-binding domain (DBD) and the GAL1 UAS, but the activation domain is not recruited to the promoter, resulting in no transcription of the *HIS3* gene. **(B)** Where there is a functional interaction between bait and target polypeptides, a functional activator is formed on the GAL1 UAS; *HIS3* expression ensues, resulting in histidine prototrophy. **(C)** The *LacZ* reporter gene in strains, such as YPB2 or HF7c, is driven by an artificial GAL4-dependent promoter. Interaction of bait and target polypeptides results in expression of β-galactosidase activity, which is easily assayed.

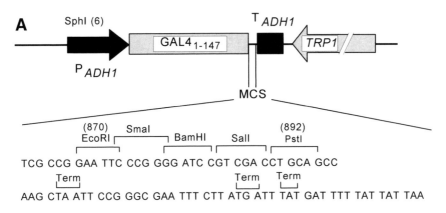

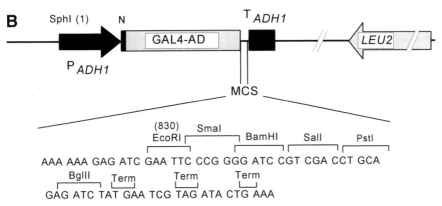

Fig. 2. Cloning sites in the DNA-binding domain fusion vector pGBT9 and activation domain vector pGAD424. (A) The region of pGBT9 around the segment encoding the DNA-binding domain of GAL4 (residues 1–147). P_{ADH1}, *ADH1* promoter; MCS, multicloning site; T_{ADH1}, *ADH1* transcriptional terminator. The *TRP1* auxotrophic marker gene is also indicated. The first TCG codon in the following sequence corresponds to residue 147 of GAL4. Restriction sites are indicated as are termination codons (Term). (B) The region of pGAD424 around the segment encoding the GAL4 activation domain (residues 768–881). The first GAG codon in the sequence corresponds to codon 881 of GAL4. The segment designated N indicates the SV40 large T nuclear localization sequence engineered into the N-terminus of GAL4 *(2)*.

driven by a GAL4-responsive promoter *(2)*. Yeast-containing plasmids encoding interacting polypeptides were scored by the presence of β-galactosidase activity (blue colonies in the presence of the chromogenic substrate X-Gal). The method was subsequently developed to facilitate the screening of large mammalian cDNA libraries *(11)*. This innovation employs a *his3* test yeast strain (that is auxotrophic for histidine) that contains, in addition to the *lacZ* reporter, an integrated *HIS3* gene that is transcribed at a significant level only

Table 1
10X Dropout Solution

Component	mg/L
L-Isoleucine	300
L-Valine	1500
Adenine hemisulfate	200
L-Arginine-HCl	200
L-Histidine-HCl	200
L-Leucine	1000
L-Lysine-HCl	300
L-Methionine	200
L-Phenylalanine	500
L-Threonine	2000
L-tryptophan	200
L-Tyrosine	300
Uracil	200

in the presence of the reconstituted activator (*see* Fig. 2). Yeasts transformed with the bait plasmid and the activator library are plated on minimal plates lacking histidine. Only *HIS3* colonies are then carried forward for further analysis. The genotype of two commonly used yeast reporter strains is given in Note 1.

2. Materials

2.1. Maintenance of Yeast Strains

1. A solution of 40% glucose (dextrose) in water: Autoclave to sterilize.
2. Yeast, peptone, dextrose (YPD) medium per liter: 20 g peptone, 10 g yeast extract. Adjust pH if necessary to 5.8–6.5. Add 15 g agar for plates. Autoclave and add 50 mL of glucose (dextrose) from a sterile 40% stock.
3. YPDA: Prepare YPD as in item 2, and after autoclaving add adenine sulfate to 40 mg/L from a filter-sterilized stock.
4. Yeast nitrogen base (Difco, Detroit, MI, cat. no. 0919-15-3).
5. Amino acids: L-tryptophan, L-histidine-HCl, L-arginine-HCl, L-methionine, L-tyrosine, L-leucine, L-isoleucine, L-lysine-HCl, L-phenylalanine, L-glutamic acid, L-aspartic acid, L-valine, L-threonine, L-serine (all from Sigma, St. Louis, MO).
6. Adenine sulfate (Sigma, cat. no. A2196).
7. Uracil.
8. 10X dropout solution: Add the components in items 5 to 7 in the proportions given in Table 1, omitting the appropriate ingredient (*see* Section 3.1.). Dissolve in MilliQ water, and autoclave to sterilize. Store for up to 1 yr at 4°C.
9. Synthetic (SD) medium and SD agar (per liter of medium): dissolve 6.7 g Difco yeast nitrogen base in 850 mL MilliQ deionized water, and add 20 g agar (omit

the agar for liquid media). Autoclave at 120°C, at 15 lbs for no more that 15 min. Add 100 mL of the appropriate 10X dropout solution (Table 1) and 50 mL 40% glucose. For agar medium, pour plates on the thick side, and leave on the bench the right way up with lids on for 2–3 d to dry. Plates may then be wrapped and stored for up to several months at 4°C. If no short autoclave cycle is available for SD agar, make up the nitrogen base and agar in 425 mL of water each, autoclave separately, combine while still hot, and then add 100 mL 10X dropout solution and 50 mL 40% glucose.

2.2. Yeast Phenotype Verification

1. Two-hybrid yeast strains (see Note 1).
2. SD plates lacking trp, leu, his, or ura (see Sections 2.1. and 3.1.).
3. YPDA and YPD plates.
4. Disinfectant-free toothpicks.

2.3. Yeast Transformation and Library Screening

2.3.1. Transformation with Bait Plasmid

1. Fresh plate of (verified phenotype) yeast, such as HF7c (see Note 1).
2. YPD (see Section 2.1.).
3. Autoclaved 500-mL centrifuge pots, Beckman JA-10 or equivalent.
4. Sterile water.
5. 50% Polyethylene glycol (PEG) 4000: Make 200–500 mL in water and autoclave.
6. 10X TE: 100 mM Tris-HCl, pH 7.8, 10 mM EDTA. Half fill a number of glass universal bottles and autoclave.
7. Lithium acetate: 1M prepared in water. Adjust to pH 7.5 with dilute acetic acid and autoclave.
8. 1X TE/LiAc: Mix 1 mL of 10X TE and 1 mL of 1M LiAc in a 30-mL sterile universal, and add 8 mL sterile water.
9. PEG/LiAc: Mix 1 mL of 10X TE, 1 mL of 1M LiAc, 8 mL 50% PEG in a 30-mL sterile universal.
10. Dimethyl sulfoxide (DMSO).
11. 10 mg/mL Sonicated salmon sperm DNA.
12. 9-cm SD plates without tryptophan (–T plates); see Section 2.1.

2.3.2. Yeast Transformation with Activation Domain Library

1. Amplified activation domain library (see Note 3).
2. 2.5 L YPD medium (see Section 2.1.).
3. 1.5 L –TL synthetic medium (see Section 2.1. and 3.1.).
4. 250 mL –TLH synthetic medium (see Section 2.1. and 3.1.).
5. 20 mL 1X TE/LiAc (see Section 2.3.1.).
6. 140 mL PEG/LiAc. Prepare fresh by combining in a sterile container: 14 mL 10X TE, 14 mL 1M LiAc, 112 mL 50% PEG.
7. Sterile water.
8. Several 9-cm –TL plates.
9. 10 mg/mL Sonicated salmon sperm DNA.

2.3.3. Library Plating

1. Several 9-cm –TL plates.
2. At least 50 15-cm –THL plates (see Note 4).
3. 1M 3-Aminotriazole (3-AT) made up in water. Prepare 10–20 mL and filter sterilize (see Note 4).

2.4. Further Analysis and Elimination of False Positives

2.4.1. Gridding Out His⁺ Transformants

1. 9-cm –THL plates.
2. 7-cm Filter paper circles (Whatman no. 1), imprinted with a grid of 1-cm squares. Number in pencil, wrap in an aluminum envelope, and autoclave to sterilize.
3. 14-cm Sterile filter papers.
4. Sterile toothpicks.

2.4.2. β-Galactosidase Filter Assays

1. 5-Bromo-4-chloro-3-indolyl-β-D-galactopyranoside (X-Gal): dissolved in N,N-dimethylformamide (DMF) at 20 mg/mL. Store in 1-mL aliquots in 1.5-mL microfuge tubes at –20°C.
2. Z buffer: 16.1 g/L disodium hydrogen phosphate · $7H_2O$, 5.5 g/L sodium dihydrogen phosphate, $1H_2O$, 0.75 g/L potassium chloride, 0.25 g/L magnesium sulfate · $7H_2O$. Make 1 L. Autoclave to sterilize.
3. Working X-Gal solution: 10 mL Z buffer, 167 µL X-gal stock, 27 µL β-mercapto-ethanol.

2.4.3. Curing Transformant of the Bait Plasmid

1. 500 mL –L (SD) medium (see Section 2.1.).
2. –L plates (see Section 2.1.).
3. –LT plates (see Section 2.1.).

3. Methods

It is advisable for workers unfamiliar with yeast manipulations to read Sherman's article, "Getting Started with Yeast," in *Methods in Enzymology (12)*. This and other contributions in the volume provide additional information beyond the scope of this chapter.

3.1. Maintenance of Yeast Strains and Transformants

To select for and maintain yeast plasmid transformants, yeast strains with characteristic auxotrophic markers are grown on synthetic media lacking the cognate nutrient. The required enzymatic activity is then provided by a plasmid-borne marker gene. Common transformation markers include *trp1*, *leu2*, *his3*, and *ura3* affecting tryptophan, leucine, histidine, and uracil biosynthesis, respectively. SD consists of a minimal medium or nitrogen base plus glucose combined with a "dropout solution," which contains amino acid and nucleotide supplements (*see* Table 1). One or more of the dropout solution components are omitted to make a selective medium. In this chapter, SD lacking leucine is referred to as –L medium, that lacking tryptophan as –T medium, that lacking both leucine and tryptophan as –TL medium, and so on.

3.2. Yeast Phenotype Verification

Before embarking on transformation and library screening, it is worthwhile checking the phenotype of the yeast strain to be used. The author has used the strain HF7c obtained commercially from Clontech (*see* Note 1), but a number of other strains have been used by other workers (*see* Note 1). HF7c is characterized by *trp1*, *leu2*, and *his3* auxotrophic markers and contains a second *HIS3* gene driven by the GAL4-responsive *GAL1* promoter. This strain also contains an inserted *LacZ* reporter gene driven by an artificial GAL4-responsive promoter.

1. Using a sterile inoculating loop, scrape some cells from the top of a frozen glycerol stock (*see* Note 5) and streak onto a YPDA plate. Incubate at 30°C for 2–3 d until colonies about 1–2 mm across have formed.
2. Take four or five well-isolated colonies, and using sterile toothpicks, streak some of the cells from each colony onto a sector of a second YPD or YPDA plate, and in parallel onto –Ura, –Trp, –Leu, and –His synthetic medium plates and on –H plates containing 5 m*M* 3-AT (*see* Note 4).
3. Incubate at 30°C for 3–4 d. Cells will grow on YPD plate and on –U plates (if the strain being used is Ura⁺), but should not grow on either –T or –L plates. Some slow growth may occur on –His plates owing to leaky expression of the *HIS3* reporter gene, but no growth should occur on the plates containing 3-AT.
4. Colonies displaying the expected phenotype should be used as the stock for library screening in Section 3.3.

3.3. Yeast Transformation and Library Screening

Construction of the bait plasmid encoding the DNA-binding domain fusion is not considered here in detail. DNA-binding domain fusion vectors are discussed in Note 2 and Table 2. The following are written assuming the use of a bait and activation library plasmids carrying *trp1* and *leu2* selection markers, respectively (*see* Tables 2 and 3), and a *trp1*, *leu2* reporter strain, such as HF7c (*see* Table 4) carrying a *HIS3* reporter gene, allowing preliminary screening for positive two-hybrid interaction by selection for histidine prototrophy.

3.3.1. Transformation with Bait Plasmid (see Note 2)

This protocol is derived from the methods of Schiestl and Gietz *(15)* and Gietz et al. *(16)*.

Table 2
DNA-Binding Domain Fusion Vectors Used in the Two-Hybrid System

Plasmid	Size	DNA-binding domain	Selection	Useful cloning sites	Other characteristics	Refs.
pMA424	12 kb	GAL4	HIS3	EcoRI, BamHI, SalI		(9)
pGBT9[a]	5.4 kb	GAL4	TRP1	EcoRI, SmaI, BamHI, SalI, PstI		(11)
pAS2[b]	8.5 kb	GAL4	TRP1	NdeI, NcoI, SmaI, BamHI, SalI, PstI	Contains CYH^{S}2 gene conferring cycloheximide sensitivity on transformed cells	(4,13)
pBTM116	5.4 kb	LexA	TRP1	EcoRI, SmaI, BamHI, SalI, PstI		(11)

[a]Available commercially from Clontech, licensed from the Research Foundation of the State University of New York.
[b]Available commercially from Clontech, licensed from Baylor College.

Table 3
Activation Domain Fusion Vectors Used in the Two-Hybrid System

Plasmid	Size, kb	Activation domain	Selection	Useful cloning sites	Other characteristics	Refs.
pGAD2F	13	GAL4[a]	LEU2	BamHI	pGAD1F, 2F, and 3F contain the BamHI site in three different frames with respect to the activation domain	(2)
pGAD10	6.6	GAL4[a]	LEU2	BglII, XhoI, BamHI, EcoRI		(11)
pGAD424[b]	6.6	GAL4[a]	LEU2	EcoRI, SmaI, BamHI, SalI, PstI, BglII		(9,11)
pJG4-5		B42	TRP1	EcoRI, XhoI, HindIII	Fusion protein expression is inducible with galactose, reducing potential toxicity problems	(6)
pVP16	8.1	VP16	LEU2	BamHI, NotI		(9)
pACT	7.65	GAL4[a]	LEU2	BglII, EcoRI, BamHI	Automatic excision from the λ vector λ-ACT	(4)

[a]A nuclear localization signal was engineered at the N termini of the GAL4 activation domain in the GAD series (2) and –Act (4).
[b]Available commercially from Clontech, licensed from the Research Foundation of the State University of New York.

Table 4
Yeast Reporter Strains Used for Two-Hybrid Library Screening[a]

Strain	Mating type	Transformation markers	Reporter genes	Refs.
YPB2	MAT**a**	trp1, leu2	GAL1$_{UAS}$-LEU2$_{TATA}$-HIS3, (GAL 17-mers)$_3$-CYC1$_{TATA}$-LacZ	(11)
CTY1	MATα	trp1, leu2	GAL1-LacZ, GAL1-HIS3	(11)
Y185	MAT**a**	trp1, leu2	GAL1-LacZ, GAL1-HIS3	(11)
Y190	MAT**a**	trp1, leu2	GAL1-LacZ, GAL1-HIS3	(4)
HF7c[b]	MAT**a**	trp1, leu2	GAL1$_{UAS-TATA}$-HIS3, (GAL 17-mers)$_3$-CYC1$_{TATA}$-LacZ	(13)
CG-1945[b,c]	MAT**a**	trp1, leu2	GAL1$_{UAS-TATA}$-HIS3, (GAL 17-mers)$_3$-CYC1$_{TATA}$-LacZ	(14)
L40	MAT**a**	trp1, leu2	(LexAop)$_4$-HIS3, (LexAop)$_8$-LacZ	(7,9)
Y187d	MATα	trp1, leu2, his3	GAL1-LacZ	(13)

[a]Only strains with a two-hybrid activated reporter gene conferring a means of selection for positive interaction are listed.
[b]Available commercially from Clontech, licensed from the Research Foundation of the State University of New York.
[c]Used in conjunction with DNA-binding domain vector pAS2 encoding the CYC^{S2} gene, loss of the bait construct can be selected for by plating on plates containing cycloheximide.
[d]Available commercialy from Clontech, licensed from Baylor University, used in yeast mating; *see* Note 12.

1. Use a single colony of phenotype-verified reporter strain, such as HF7c (*see* Table 4), to inoculate 10 mL of YPD. Shake overnight at 30°C.
2. Measure the OD_{600} of the culture, which should be about 1.5–2, and dilute the overnight culture into 50 mL YPD at 30°C to give a final OD_{600} of 0.2.
3. Incubate for a further 3 h at 30°C.
4. Pellet the cells by centrifuging at 1000g for 3 min using a bench top centrifuge.
5. Resuspend the cells in 25 mL of sterile water, and pellet as above.
6. Gently resuspend the cells in 0.6 mL of TE/LiAc.
7. For each transformation, add to a sterile 1.5-mL microfuge tube: 0.5 μg plasmid DNA, and 100 μg sonicated salmon sperm DNA. Mix and add 100 μL of yeast cell suspension from step 6.
8. Mix gently, but thoroughly, and add 600 μL of PEG/LiAc solution. Vortex to mix, and place the tubes in a 30°C incubator for 30 min. Mix gently several times during this incubation.
9. Add 70 μL of DMSO to each tube, mix gently, and place the tubes in a 42°C water bath for 15 min.
10. Briefly chill on ice, and pellet cells by spinning for 5 s in a microfuge.
11. Aspirate the supernatant, resuspend the cells in 500 μL of sterile TE, and spread the cells onto appropriate selective plates (Trp minus, –T for most bait plasmid/reporter strain combinations). Spread 100–200 μL/9-cm plate or 200–400 μL/15-cm plate.
12. Incubate at 30°C for 2–4 d for colonies to appear.
13. Keep stocks of the transformed strain on selective medium plates. Use one of the Trp⁺ colonies to prepare competent cells for large-scale transformation as described in Section 3.3.2. It is also advisable to make a glycerol stock of the transformed strain (*see* Note 4).

Before going any further, it is advisable to carry out some preliminary tests with the bait construct as described in Note 6.

3.3.2. Yeast Transformation with Activation Domain Library

This protocol is derived from the methods of Schiestl and Gietz *(15)* and Gietz et al. *(16)*.

1. Inoculate one of the Trp⁺ colonies from Section 3.3.1. into 5 mL selective –T medium. Incubate at 30°C overnight with shaking.
2. Inoculate the whole 5-mL culture into 300 mL of prewarmed –T medium in a 1-L flask, and shake for 24 h at 30°C (*see* Note 7).
3. Measure the OD_{600}, which should be approx 1.2–1.5, and add sufficient culture to each of two 2.5-L flasks each containing 500 mL of YPD prewarmed to 30°C to attain an OD_{600} of 0.3.
4. Shake at 30°C for 4 h.
5. Pellet the cells by centrifugation at 2000g for 5 min using a Beckman JA-10 or similar rotor.
6. Decant the supernatant and resuspend the cell pellet in 500 mL water.
7. Pellet the cells as in step 5, decant the supernatant, resuspend the washed cells in 20 mL TE/LiAc solution, and transfer to a sterile 250-mL flask.

8. Add, premixed, 1 mL denatured salmon sperm DNA (10 mg/mL) and 500 μg of activation library plasmid DNA, and mix (*see* Note 3).
9. Add 140 mL of PEG/LiAc solution and mix again.
10. Incubate at 30°C for 30 min with shaking.
11. Transfer to a sterile 2-L flask, and add 17.6 mL of DMSO. Swirl to mix while adding the DMSO.
12. Stand the flask in a 42°C water bath for 6 min to heat-shock the cells, swirling occasionally.
13. Transfer the flask to a water bath at 20°C to cool.
14. Pellet the cells at 2000*g*. Decant the supernatant and gently resuspend the cells in 200 mL YPD.
15. Pellet the cells as in step 14, and resuspend in 1 L of the same medium prewarmed to 30°C, and shake at that temperature for 1 h.
16. Remove 1 mL of the cell suspension from step 15, wash the cells once in –TL medium, and resuspend in 1 mL of the same medium. Plate 1, 10, and 100 μL of the cell suspension on –TL plates to determine the primary transformation efficiency (*see* Note 8).
17. Pellet the remainder of the cells as in step 5, and wash with 500 mL –TL medium.
18. Resuspend the cell pellet in 1 L warm –TL medium, and incubate with gentle shaking for 4 h.
19. Pellet the cells as in step 5, and wash twice in 100 mL –TLH medium.
20. Resuspend the final pellet in 10 mL of the same medium.

3.3.3. Library Plating

1. To estimate the final transformation and plating efficiency, dilute 10 μL of the cell suspension from Section 3.3.2. into 1 mL of TE, and spread 1, 10, and 100 μL onto single 9- or 15-cm –TL plates (*see* Note 8). This represents one millionth, one one hundred thousandth, and one ten thousands of the transformation, respectively.
2. Spread 100 μL of the cell suspension from step 20 in Section 3.3.2. on each of a series of –THL plates (or –THL plates containing 3-AT; *see* Note 4). This requires at least 50 plates to screen enough colonies to cover all of the primary transformants (*see* Note 9)!
3. Leave the plates right side up on the bench with lids on for 1–2 h for any surplus liquid to dry. Transfer plates, lid-side-down to a 30°C incubator.
4. His+ colonies should appear after 2–4 d, although up to 10 d may be required for weak positives to appear.

3.4. Further Analysis and Elimination of False Positives

False positives are a big problem in two-hybrid screening. Assuming the bait construct has already been tested to check that it lacks activation potential alone (*see* Note 6), false positives may arise through fortuitous binding of the activation domain fusion to the test promoter, nonspecific binding to the bait, or through one of a number of less well-understood routes. The first step after His+ colonies appear is to test them for β-galactosidase activity (*see* Section

3.4.3.). The *LacZ* reporter in strain HF7c, for example, is driven by a different GAL4-responsive promoter than the *GAL1-HIS3* reporter. His$^+$, β-galactosidase$^+$ colonies are then carried forward for further analysis. Since it is possible for more than one activation domain fusion plasmid to be present in each β-galactosidase-positive colony, such colonies should be streaked onto –TL plates to allow segregation and then retested for β-galactosidase activity. The next stage is to check that reporter gene activity requires both activation domain and DNA-binding domain constructs. This is achieved by curing the tentative positives of the DNA-binding domain plasmid and retesting the resulting Trp$^-$ cells for β-galactosidase activity and/or His prototrophy. Only transformants that are Trp$^-$, Leu$^+$, and β-galactosidase$^-$ are candidates for genuine positives. A rapid method for curing of the bait plasmid has been devised by Elledge and colleagues *(13)* using the bait vector pAS2 (*see* Table 2), which carries a gene conferring cycloheximide sensitivity on resistant reporter strains, such as Y190 *(13)* or CG-1945. Plating on –L plates containing cycloheximide then selects for bait plasmid loss (*see* Note 12).

Further testing then involves retransformation of the Trp$^-$, Leu$^+$, and β-galactosidase$^-$ activation domain fusion-containing strain with the original bait plasmid along with a series of controls, including the original bait vector and one or more different fusion constructs, which would not be expected to interact with the same target. An ideal control would be a mutant bait domain, known to be inactive in some way. The retransformed strains would then be tested again for β-galactosidase activity or His prototrophy. As a faster alternative to transformation, these plasmids can be introduced using yeast mating methods (*see* Note 12). Those positives that are β-galactosidase-positive when retransformed with the bait plasmid, but not with the controls, are candidates for real positives! The final step is to isolate the activation domain plasmid DNA and to confirm the protein–protein interaction by independent means.

3.4.1. Transferring Colonies to Filter Papers for β-Galactosidase Assays

For β-galactosidase filter assays, colonies are transferred onto filter paper circles, leaving most of each colony in place for later analysis. If there are a large number of positives, it is easier to transfer cells from each colony directly onto sterile filter papers as described in Section 3.4.1.1. Otherwise, it is probably best to grid the colonies out as described in Section 3.4.1.2.

3.4.1.1. PREPARING REPLICA LIFTS OF POSITIVE COLONIES

1. Lay a sterile 14-cm Whatman filter paper directly on top of each 15-cm plate from the two-hybrid screen, and make sure there are no air bubbles. Make several orientation marks through the filter paper and agar using a syringe needle dipped in India ink.

2. Using blunt-ended forceps, carefully remove each filter and place directly onto a new –THL plate, colony side up.
3. Incubate the original plates at 30°C for 24 h to allow regeneration of the colonies, and then store the plates at 4°C until required.
4. Incubate the new plates at 30°C until colonies are clearly visible on the filter papers.
5. Proceed with β-galactosidase filter assays as described in Section 3.4.2.

3.4.1.2. GRIDDING POSITIVE COLONIES

1. Lay a series of 7-cm gridded filter papers (*see* Section 2.4.1.) directly onto a set of –THL plates.
2. Using sterile toothpicks, grid His⁺ colonies from Section 3.3. onto the plates from step 1, and onto a second set of –THL plates, using the same pattern for both sets. The second set of plates will be the master set of tentative positives.
3. Incubate at 30°C for 2–4 d until colonies or small blobs of cells have appeared on the filter papers.

3.4.2. β-Galactosidase Filter Assay

1. Using flat-ended forceps, carefully lift the filters from the agar plates one at a time, and float colony-side-up in a pool of liquid nitrogen. After a few seconds, submerge the filter for 10 s, and then remove from the liquid nitrogen and lay colony side up on a tissue paper to thaw (*see* Note 10).
2. Once thawed, carefully place the filter colony-side-up on a second filter paper soaked in freshly prepared Z buffer containing X-Gal (*see* Section 2.4.2.).
3. Lift the filter sandwich colony-side-up into a 9-cm Petri dish, replace the lid, and incubate at 30°C for 30 min to several days (*see* Note 11).
4. β-Galactosidase-positive colonies can be matched with the colonies on the master plates from their grid positions or orientation marks, and can be carried forward for further analysis.

3.4.3. Curing of Bait Plasmid (see also Ref. 13 and Note 12)

1. For each β-galactosidase-positive transformant, set up an overnight culture in 2 mL –L medium. Use 1 drop of the overnight culture to inoculate a further 2 mL of –L medium, and incubate overnight again.
2. Take 100 μL of the second overnight culture, dilute to 1 mL with sterile water, and plate out 50–100 μL of the culture onto a series of –L plates.
3. Using sterile toothpicks, grid out 20–30 colonies from each plate onto –L and –TL plates.
4. Transformants that grow on –L, but fail to grow on –LT plates are cured of the bait plasmid (which carries a *TRP* marker gene). These transformants should be retested for β-galactosidase activity, and those that are now negative in this assay should be saved for further verification (*see* Note 12).

3.4.4. Recovery of Plasmid DNA from Transformed Yeast

Further analysis and identification of cDNA inserts require isolation of activation domain plasmid DNA. Only plasmid DNA heavily contaminated with genomic DNA and other impurities can easily be isolated from yeast. Therefore, plasmid DNA is shuttled into a suitable *Escherichia coli* strain for preparation. Usefully, *E. coli* strains, such as HB101, are leucine auxotrophs owing to the *leuB* mutation, which can be complemented by the yeast *LEU2* gene, and so activation domain plasmids with *LEU2* as their transformation marker can be selected for by plating transformed *E. coli* cells on minimal plates lacking leucine. A method for the recovery from yeast of plasmid DNA suitable for transformation of *E. coli* can be found in ref. *17*. Electroporation or high-efficiency chemical means should be used for *E. coli* transformation with this material.

4. Notes

1. Two of the most commonly used reporter strains for two-hybrid library screening are YPB2 *(11)* (*MATa, ura3-52, his3-200, ade2-101, lys2-801, trp1-109, leu2-3, 112, gal4-542, gal80-358, LYS2::GAL1$_{UAS}$-LEU2$_{TATA}$-HIS3, URA3::(GAL4 17-mers)$_3$-CY1C$_{TATA}$-LacZ*) and HF7c *(14)* (*MATa, ura3-52, his3-200, ade2-101, lys2-801, trp1-109, leu2-3, 112, gal4-542, gal80-358, LYS2::GAL1$_{UAS,TATA}$-HIS3, URA3::(GAL4 17-mers)$_3$-CYC1$_{TATA}$-LacZ*) (*see* Table 4) both constructed for use with GAL4 DNA-binding domain bait plasmids. The *HIS3* reporter gene in HF7c reportedly has a lower level of background expression (*see also* Note 4). For bait constructs using the LexA DNA-binding domain, the strain L40 has been described *(7,9)* (partial genotype: *MATa, trp1, leu2, his3, LYS::lexA-HIS3, URA3::lexA-LacZ*). Some of the other yeast strains that have been used for two-hybrid library screening are listed in Table 4 and ref. *11*.

2. Table 2 lists some DNA-binding domain fusion plasmids (bait plasmids). The important features of each of these plasmids are as follows:
   a. A yeast transformation marker (*TRP1* in most cases);
   b. A DNA segment encoding sequence-specific DNA-binding domain (the DNA-binding domain from the yeast transcription factor GAL4 in most cases) driven by a constitutive promoter, such as the ADH1 promoter, and placed immediately upstream of a series of restriction sites for cloning of insert cDNAs;
   c. The origin of replication from the yeast 2 µ plasmid; and
   d. The *Bla* gene for ampicillin resistance in *E. coli* and an *E. coli* origin.
   The plasmid pAS2 *(13)* possesses a number of additional features (*see* Table 2). The sequence of the multiple cloning sites in pGBT9 *(11)* is shown in Fig. 2A; the multicloning sites of pGBT9 and pAS2 *(13)* are identical between the *Sma*I and *Pst*II sites, but the pAS2 MCS also contains *Nde*I and *Nco*I sites (containing in-frame ATG codons) just upstream of the *Sma*I site (*see* ref. *13*).

3. The quality of the activation domain-tagged cDNA library is obviously of central importance for successful two-hybrid screening. It is outside the scope of this chapter to describe in detail the construction of such libraries beyond the following points. First, two approaches can be made in the construction of activation domain-tagged plasmid libraries. The first approach is to generate a relatively large amount of cDNA from the target tissue or cell line. Sufficiently large libraries may then be generated by direct ligation into the activation domain vector

(*see* ref. *3*, for example). Following plasmid transformation by electroporation, transformed *E. coli* are plated onto a large number of 15-cm LB-amp plates (approx 50–100) ideally at about 1×10^5 colonies/plate. After overnight incubation, cells are collected by scraping them from the plates and plasmid DNA prepared by standard alkaline lysis *(18)*. In the second method, described by Durfee et al. *(4)*, a library is constructed in the λ-phage vector λ-ACT, which contains an embedded copy of the two-hybrid activation domain-tagged plasmid pACT. In vivo excision allows the direct generation of the pACT activation domain library, which may be amplified as described. Whichever method is used for activation domain-tagged library construction, it is argued that random primed first-strand cDNA synthesis is preferable, since this eliminates biases toward C-terminal sequences. Rather than generating libraries themselves, many workers will obtain libraries from other workers or from commercial sources (Clontech has on catalog a number of activation domain libraries constructed largely in pGAD10). Plasmid libraries obtained commercially or from other workers will usually require amplification as described above prior to yeast transformation (for which 500 μg of plasmid DNA are required/transformation). Caution should be exercised here, and *E. coli* transformation should aim to generate at least as many colonies as the estimated original complexity of the library. A high-efficiency transformation procedure, such as electroporation, should be used, plating the results of several independent transformations containing 10–50 ng of library DNA each on a large number of 15-cm plates.

4. The *HIS3* reporter gene in such strains as YPB2 (*see* Table 4) is leaky, resulting in slow growth on –HTL plates in the absence of GAL4 activity. This can produce a significant background when the transformed two-hybrid library is plated. Background growth can be eliminated by the addition of 5–20 m*M* 3-amino-triazole (3-AT) to plates. The reporter strain HF7c *(14)* contains a slightly different *HIS3* reporter (*see* Table 4) and is reported not to suffer in this way.

5. Glycerol stocks of yeast strains are prepared by adding 0.5 mL of YPDA or appropriate SD medium containing 20% glycerol to 0.5 mL of a midlog culture. The glycerol-containing culture can then be stored in cryogenic storage vials at –80°C. Glycerol stocks remain viable in excess of 1 yr at this temperature.

6. Prior to library screening, the bait construct should be tested for fortuitous transcriptional activation properties in the absence of interacting activation domain-tagged plasmid. Perform a series of small-scale transformations as in Section 3.3.1. using:
   a. The bait construct alone;
   b. The bait construct and the activation domain vector from which the library to be screened was constructed; and
   c. The empty bait vector alone.
   Plate out on appropriate selective plates, and test for β-galactosidase activity as in Sections 3.4.2. and 3.4.3. In addition, grid the transformants on plates lacking histidine to check for fortuitous activation of the HIS3 reporter gene. None of the transformants should give significant reporter activity.

7. The cells should be at approximately midlog phase at this point. If the OD_{600} is below 1.2, continue incubation until the optical density falls into this range.

8. The primary transformation efficiency using this method should be in excess of 1×10^4 colonies/µg of plasmid DNA. The protocol given here should therefore give >5 million transformants. The protocol includes a 4-h recovery period. This time can be optimized for different strains. Since some growth will occur during this period, a second series of aliquots should be plated after this recovery period to determine the plating efficiency.

9. This is a large number of plates, but do not be tempted to plate more cells per plate, since growth of His⁺ colonies is inhibited by too dense plating. Even at 100 µL/plate, each plate will receive 50,000 transformants.

10. Freeze thawing allows the cells to be permeable.

11. Strong signals may appear in hours, whereas weak positives will take at least overnight for β-galactosidase activity to be apparent.

12. Two significant enhancements to the basic method as described here that speed up false-positive elimination have been described by Harper et al. *(13)*. First, curing of putative positive colonies of bait plasmid is facilitated by using the DNA-binding domain vector pAS2 for the bait construct and a tetracycline-sensitive yeast strain, such as Y190 *(15)* or CG-1945 *(see* Table 4). Plasmid pAS2 carries the *CYH2* gene conferring cycloheximide sensitivity on suitable resistant strains. Using pAS2-derived bait plasmids and one of the aforementioned strains, segregants that have spontaneously lost the pAS2– derivative can be selected by picking Leu⁺, Trp⁺, His⁺, LacZ⁺ colonies into 200 µL of sterile water and spreading 100 µL of the cell suspension onto each of two –L plates containing cycloheximide at 1 µg/mL. The resulting Trp⁺, leu⁻, colonies can be tested for *LacZ* expression as in Section 3.4.2. The second enhancement involves mating the **a** strain Trp⁺, leu⁻ colonies described above with α mating type yeast, such as Y187 *(13)* carrying either the original DNA-binding domain bait plasmid or one of a number of controls, such as the empty DNA-binding domain vector or unrelated DNA-binding domain fusions. Resulting **a**/α diploids selected on –TL plates now carrying both plasmids are again tested for β-galactosidase activity and/or ability to grow on –TLH plates. Original positives that produce diploid colonies on both –TL and –TLH plates (and that are positive for β-galactosidase) only when mated with α cells containing the DNA-binding domain bait plasmid and not any of the controls are strong candidates for genuine positive interactions. This procedure eliminates the necessity of isolating plasmid DNAs and retransforming to test for false positives caused by nonspecific interactions of the activation domain fusion *(see* Section 3.4.). To produce the diploids, pick a single colony of each cured Trp⁺, Leu⁻ two-hybrid positive into a microfuge tube containing 500 µL of YPD, and add a single colony of the appropriate α mating type transformant. Incubate overnight with vigorous shaking, and plate 50 µL of each culture on –TL and –TLH plates.

References

1. Fields, S. and Song, O. (1989) A novel genetic system to detect protein–protein interactions. *Nature* **340,** 245,246.
2. Chien, C. T., Bartel, P. L., Sternglanz, R., and Fields, S. (1991) The two-hybrid system: a method to identify and clone genes for proteins that interact with a protein of interest. *Proc. Natl. Acad. Sci. USA* **88,** 9578–9582.

3. Hannon, G. J., Demetrick, D., and Beach, D. (1993) Isolation of the Rb-related p130 through its interaction with Cdk2 and cyclins. *Genes Dev.* **7**, 2378–2391.
4. Durfee, T., Becherer, K., Chen, P. L., Yeh, S. H., Yang, Y., Kilburn, A. E., Lee, W. H., and Elledge, S. J. (1993) The retinoblastoma protein associates with the protein phosphatase type 1 catalytic subunit. *Genes Dev.* **7**, 555–569.
5. Hardy, C. F. J., Sussel, L., and Shore, D. (1992) A RAP1-interacting protein involved in transcriptional silencing and telomere length regulation. *Genes Dev.* **6**, 801–814.
6. Gyuris, J., Golemis, E., Chertkov, H., and Brent, R. (1993) Cdi1, a human G1 and S phase protein phosphatase that associates with Cdk2. *Cell* **75**, 791–803.
7. Vojtek, A. B., Hollenberg, S. M., and Cooper, J. A. (1993) Mammalian ras interacts directly with the serine/threonine kinase Raf. *Cell* **74**, 205–214.
8. Dunaief, J. L., Strober, B. E., Guha, S., Khavari, P. A., Ålin, K., Luban, J., Begemann, M., Crabtree, G. R., and Goff, S. P. (1994) The retinoblastoma protein and BRG1 form a complex and cooperate to induce cell cycle arrest. *Cell* **79**, 119–130.
9. Hollenberg, S. M., Sternglanz, R., Cheng, P. F., and Weintraub, H. (1995) Identification of a new family of tissue-specific basic helix-loop-helix proteins with a two-hybrid system. *Mol. Cell Biol.* **15**, 3813–3822.
10. Dalton, S. and Triesman, R. (1992) Characterisation of SAP-1, a protein recruited by serum response factor to the c-fos serum response element. *Cell* **68**, 597–612.
11. Bartel, P. L., Chien, C. T., Sternglanz, R., and Fields, S. (1993) Using the two hybrid system to detect protein–protein interactions, in *Cellular Interactions in Development: A Practical Approach* (Hartley, D. A., ed.), Oxford University Press, Oxford, UK, pp. 153–179.
12. Sherman, F. (1991) Getting started with yeast. *Methods Enzymol.* **194**, 3–21.
13. Harper, J. W., Adami, G. R., Wei, N., Keyomarsi, K., and Elledge, S. J. (1993) The p21 Cdk-interacting protein Cip is a potent inhibitor of G1 cyclin-dependent kinases. *Cell* **75**, 805–816.
14. Feilotter, H. E., Hannon, G. J., Ruddell, C. J., and Beach, D. (1994) Construction of an improved host strain for two hybrid-screening. *Nucleic Acids Res.* **22**, 1502,1503.
15. Schiestl, R. H. and Gietz, R. D. (1989) High efficiency transformation of intact cells using single-stranded nucleic acids as a carrier. *Curr. Genet.* **16**, 339–346.
16. Gietz, D., St. Jean, A., Woods, R. A., and Schiestl, R. H. (1992) Improved method for high efficiency transformation of intact yeast cells. *Nucleic Acids Res.* **20**, 1425.
17. Hoffman, C. S. and Winston, F. (1987) A ten-minute DNA preparation from yeast efficiently releases autonomous plasmids for transformation of *Escherichia coli*. *Gene* **57**, 267–272.
18. Sambrook, J., Frisch, E. F., and Maniatis, T. (1989) *Molecular Cloning: A Laboratory Manual*. Cold Spring Harbor Laboratory, Cold Spring Harbor, NY.

17

Signal Sequence Trap

Expression Cloning Method for Secreted Proteins and Type 1 Membrane Proteins

Kei Tashiro, Toru Nakano, and Tasuku Honjo

1. Introduction

Intercellular signaling and cell adhesion are among the most critical mechanisms for development and maintenance of multicellular organisms. Although a large number of molecules involved in signaling or adhesion have been cloned, there still remain many unknown molecules that are important in these functions. Most of the molecules involved in signaling or adhesion are secreted or membrane-anchored proteins. Many of these proteins contain a signal sequence or leader peptide in the N-terminal of their premature form *(1)*. The traditional strategy for cloning these genes requires a functional assay to detect the specific function of each molecule. To establish a general cDNA cloning method for growth factors, hormones, neuropeptides, their receptors, and adhesion molecules, we have developed a new cloning strategy for selecting cDNA fragments encoding N-terminal signal sequences *(2)*. This method, called Signal Sequence Trap, turns out to be an efficient method to isolate 5'-cDNA fragments from secreted or transmembrane proteins. We have already obtained a number of cDNA clones of putative growth factors, receptors, or adhesion molecules *(2–5)*. Signal Sequence Trap can clone not only secreted proteins, GPI-anchored proteins, and plasma membrane proteins, but also proteins located in endoplasmic reticulum (ER), Golgi apparatus (GA), and lysosome.

The epitope-tagging expression plasmid vector used in Signal Sequence Trap is the pcDLSRα-Tac(3') vector *(2,6)*, which directs the cell-surface expression of Tac (human CD25, α chain of the human interleukin-2 receptor)

From: *Methods in Molecular Biology, Vol. 69: cDNA Library Protocols*
Edited by: I. G. Cowell and C. A. Austin Humana Press Inc., Totowa, NJ

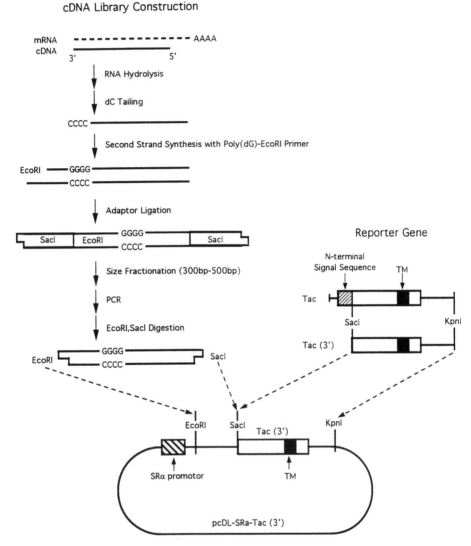

Fig. 1. Schematic diagram of signal sequence trap.

fusion proteins when inserts with N-terminal signal sequences are cloned in-frame, in the correct orientation *(7)* (*see* Fig. 1). The Tac epitope-tagged fusion protein expressed on plasma membranes is easily detected with anti-human CD25 antibodies *(8)*. The first strand of cDNA is synthesized with random hexamers, deoxycytosine (dC) tails are added at the 3'-end of the first-strand cDNA, and second-strand synthesis is carried out with an *Eco*RI-linker primer that contains polydeoxyguanosine (dG). The product is ligated with a

Step I: Preparation for First Screening

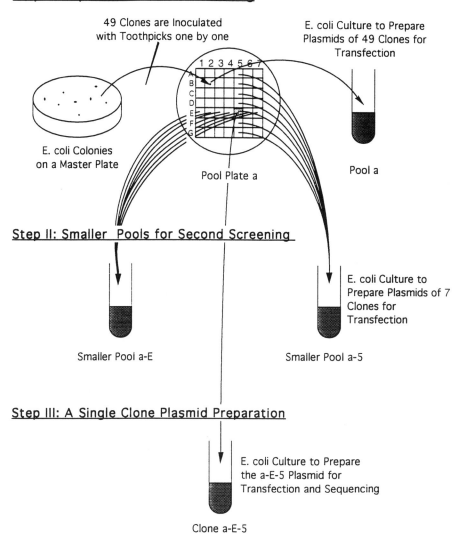

Fig. 2. Schematic view of sib-screening.

*Sac*I adapter, size-fractionated to 300–500 bp by agarose gel electrophoresis (*see* Note 1), and amplified by polymerase chain reaction (PCR) *(9)*. The amplified fragments are digested with *Eco*RI and *Sac*I, size-fractionated by agarose gel electrophoresis again, and inserted into the pcDL-SRα-Tac(3') vector in the same orientation with the Tac cDNA. Using the resulting expression library, sib-screening should be done (*see* ref. *10* and Note 2) (*see* Fig. 2).

After *Escherichia coli* transformation, 49 individual colonies are plated on a 9-cm agar plate in a matrix format (seven rows by seven lines) and assigned to one pool. COS-7 cells are transfected with plasmid DNAs of each pool *(11)*, and fusion proteins expressed on the cell surface are microscopically detected by immunostaining with antihuman CD25 antibodies. If a pool contains a positive clone plasmid, 14 smaller pools consisting of seven individual clones in each row or line of the matrix are tested until a single positive clone can be determined. The nucleotide sequence information of the positive clones is used to identify hydrophobic amino acid residues, which are the core of N-terminal signal sequence. Also, a homology search in the database is done to compare the sequence with known proteins. With stringent hybridization condition, the full-length cDNA clones can be obtained by using trapped cDNA fragment encoding the N-terminal signal sequence to probe an oligo(dT)-primed cDNA library.

2. Materials

2.1. First-Strand Synthesis

1. PolyA⁺ RNA extracted from the cells or tissues of interest.
2. Random hexamer DNA (Takara, Kyoto, Japan).
3. DEPC-treated H_2O.
4. 5X Reverse transcription buffer: $0.25M$ Tris-HCl, pH 8.3, $0.375M$ KCl, 15 mM $MgCl_2$.
5. $0.1M$ Dithiothreitol (DDT).
6. 5 mM dNTP: 5 mM dATP, 5 mM dTTP, 5 mM dCTP, 5 mM dGTP.
7. Super Script II, (Gibco BRL, Gaithersburg, MD), 200 U/µL.
8. $0.5M$ EDTA.
9. $1N$ NaOH.
10. $2M$ Tris-HCl, pH 7.4.
11. $1N$ HCl.
12. TE: 10 mM Tris-HCl, pH 8.0, 0.1 mM EDTA.
13. $10M$ Ammonium acetate.
14. 10 µg/µL Glycogen.
15. Phenol/CIAA: 25:24:1 phenol equilibrated with $0.1M$ Tris-HCl, pH. 8.0/chloroform:isoamyl alcohol.
16. Ethanol.

2.2. dC-Tailing and Second-Strand Synthesis

1. 1 mM dCTP
2. 5X Reverse transcription buffer: $0.25M$ Tris-HCl, pH 8.3, $0.375M$ KCl, 15 mM $MgCl_2$ (for dC-Tailing, final concentration is 1/2X, not 1X).
3. Terminal deoxynucleotidyl transferase (Stratagene, La Jolla, CA), 20 U/mL.
4. TE: 10 mM Tris-HCl, pH 8.0, 0.1 mM EDTA.

5. 3M Sodium acetate.
6. Phenol/CIAA: 25:24:1 phenol equilibrated with 0.1M Tris-HCl, pH 8.0/chloroform:isoamyl alcohol.
7. 10 µg/µL Glycogen.
8. Ethanol.
9. 25 µM ESLG primer: 5'-GCGGCCGCGAATTCTGACTAACTGAC(G)$_{17}$.
10. 5X Reverse transcription buffer: 0.25M Tris-HCl, pH 8.3, 0.375M KCl, 15 mM MgCl$_2$.
11. 5 mM dNTP: 5 mM dATP, 5 mM dTTP, 5 mM dCTP, 5 mM dGTP.
12. 0.1M DTT.
13. 200 U/µL Super script II (Gibco BRL, Gaithersburg, MD).

2.3. Sac I Adapter Ligation and Size Selection

1. 25 µM SSEH oligomer DNA: 5'-CCGCGAGCTCGATATCAAGCTTGTAC.
2. 10X Kination buffer: 700 mM Tris-HCl, pH 7.6, 100 mM MgCl$_2$, 50 mM DTT.
3. 5 mM dATP.
4. 10 U/µL T4 Polynucleotide kinase: (New England Biolabs).
5. Phenol/CIAA: 25:24:1 phenol equilibrated with 0.1M Tris-HCl, pH 8.0/chloroform/isoamyl alcohol.
6. 3M Sodium acetate.
7. Ethanol.
8. 25 µM LLHES oligomer DNA: 5'-GAGGTACAAGCTTGATATCGAGC TCGCGG.
9. 10X Ligation buffer: 500 mM Tris-HCl, pH 7.8, 100 mM MgCl$_2$, 100 mM DTT, 10 mM adenosine triphosphate (ATP), 250 µg/mL BSA.
10. 400 U/µL T4 DNA ligase (New England Biolabs).
11. Agarose.
12. 1X TAE: 0.04M Tris-acetate, 1 mM EDTA.
13. 10 mg/mL Ethidium bromide.
14. 6X Loading buffer: 0.25% bromophenol blue, 30% glycerol in H$_2$O.
15. DNA size markers.
16. The GeneClean II Kit (BIO 101 Inc., La Jolla, CA), containing 6M sodium iodite, Glassmilk and New Wash.
17. TE: 10 mM Tris-HCl, pH 8.0, 0.1 mM EDTA.

2.4. PCR Amplification

1. Thermal cycler.
2. 10X PCR buffer: 100 mM Tris-HCl, pH 8.3, 500 mM KCl, 15 mM MgCl$_2$ 0.01% (w/v) gelatin.
3. 5 mM dNTP: 5 mM dATP, 5 mM dTTP, 5 mM dCTP, 5 mM dGTP.
4. 25 µM ESL oligomer DNA: 5'-GCCGCGAATTCTGACTAACTGAC.
5. 25 µM LLHES oligomer DNA:5'-GAGGTACAAGCTTGATATCGAGC TCGCGG.
6. 5 U/µL *Taq* DNA polymerase (Takara).

7. Mineral oil.
8. Agarose.
9. 1X TAE: 0.04M Tris-acetate, 1 mM EDTA.
10. 10 ng/mL Ethidium bromide.
11. 6X Loading buffer: 0.25% bromophenol blue, 30% glycerol in H_2O.
12. DNA size markers.

2.5. Insert cDNA Digestion with Sac I and Eco RI

1. 10X Low-salt restriction enzyme buffer: 100 mM Tris-HCl, pH 7.5, 100 mM $MgCl_2$, 10 mM DTT.
2. 10 U/μL *Sac*I.
3. 10X Medium-salt restriction enzyme buffer: 100 mM Tris-HCl, pH 7.5, 100 mM $MgCl_2$, 10 mM DTT, 500 mM NaCl.
4. 10 U/μL *Eco*RI.
5. Agarose.
6. 1X TAE: 0.04M Tris-acetate, 1 mM EDTA.
7. 10 ng/mL Ethidium bromide.
8. 6X Loading buffer: 0.25% bromophenol blue, 30% glycerol in H_2O.
9. DNA size markers.
10. TE: 10 mM Tris-HCl, pH 8.0, 0.1 mM EDTA.

2.6. Vector Preparation

1. 10X Low-salt restriction enzyme buffer: 100 mM Tris-HCl, pH 7.5, 100 mM $MgCl_2$, 10 mM DTT.
2. pcDL-SRα-hRAR(5')-Tac(3'): a negative control plasmid; it has 5'-portion of human retinoic acid receptor, which does not bear N-terminal signal sequence *(2)*. It can be obtained from Tasuku Honjo (Department of Medical Chemistry, Kyoto University, Kyoto, Japan).
3. 10 U/μL *Sac*I.
4. 10X Medium-salt restriction enzyme buffer: 100 mM Tris-HCl, pH 7.5, 100 mM $MgCl_2$, 10 mM DTT, 500 mM NaCl.
5. 10 U/μL *Eco*RI.
6. Agarose.
7. 1X TAE: 0.04M Tris-acetate, 1 mM EDTA.
8. 10 ng/mL Ethidium bromide.
9. 6X Loading buffer: 0.25% bromophenol blue, 30% glycerol in H_2O.
10. DNA size markers.
11. TE: 10 mM Tris-HCl, pH 8.0, 0.1 mM EDTA.

2.7. Vector-Insert Ligation

1. 10X Ligation buffer: 500 mM Tris-HCl, pH 7.8, 100 mM $MgCl_2$, 100 mM DTT, 10 mM ATP, 250 μg/mL BSA.
2. 400 U/μL T4 DNA ligase (New England Biolabs).

2.8. Transformation

1. XL-1 Blue competent cells (Stratagene).
2. SOC medium.
3. LB agar plates containing 100 μg/mL ampicillin.

2.9. Making Pool Plate and DNA Preparation

1. LB agar plates containing 100 μg/mL ampicillin.
2. Terrific broth containing 100 μg/mL ampicillin *(10)*.
3. Autoclaved toothpicks.
4. GTE buffer: 50 mM glucose, 25 mM Tris-HCl, pH 8.0, 10 mM EDTA (pH 8.0).
5. 0.2N NaOH, 1% sodium dodecyl sulfate (SDS).
6. 5M Potassium acetate.
7. 5M Lithium chloride.
8. Isopropanol.
9. Ethanol.
10. 100 μg/μL RNase A.
11. Phenol/CIAA: 25:24:1 phenol equilibrated with 0.1M Tris-HCl, pH 8.0/chloroform/isoamyl alcohol.
12. 10 μg/μL Glycogen.
13. 3M Sodium acetate.
14. pcDL-SRα-hRAR(5')-Tac(3').
15. pcDL-SRα-hG-CSF(5')-Tac(3'): A positive control plasmid; it has a 5'-portion of human G-CSF, which has an N-terminal signal sequence *(1)*. It can be obtained from Tasuku Honjo.

2.10. Transfection into COS-7 Cells

1. COS-7 cells (can be obtained from American Type Culture Collection [ATCC], Rockville, MD).
2. Phosphate-buffered saline (PBS) (–).
3. 0.05% (w/v) Trypsin-0.02% (w/v) EDTA-PBS (–).
4. Dulbecco's modified Eagle's medium (DMEM) (+): supplemented with 10% (v/v) fetal calf serum.
5. DMEM (–).
6. 1M Tris-HCl (pH 7.4).
7. 10 mg/mL DEAE-dextran.
8. 10 mM Chloroquine.
9. 10% (v/v) dimethyl sulfoxide (DMSO).

2.11. Cell Surface Immunostaining with Anti-Tac

1. Fluorescein (FITC)-conjugated antihuman CD25 (Dako, Carpinteria, CA).
2. PBS (–).
3. Fetal calf serum.
4. 0.02% (w/v) EDTA in PBS (–).

2.12. Making Smaller Pools and Identifying Positive Clones

Materials used here are listed in Section 2.9–2.11.

2.13. Hydropathy Analysis and Database Search

1. SRA primer 5'-TTTACTTCTAGGCCTGACG.
2. TAC primer 5'-CCATGGCTTTGAATGTGGCG.
3. Nucleic acid and protein analysis software (Gene Works, IntelliGenetics, Mountain View, CA).
4. A personal computer accessible to the databases that have the information of Genbank, EMBL, DDBJ, and Swiss Prot.

3. Methods

Section 3.1., describes the 5' terminal-enriched cDNA library construction for Signal Sequence Trap. Section 3.2. contains the protocol of sib-screening with anti-Tac Immunostaining. Section 3.3. describes types of analysis.

3.1. SST-cDNA Library Construction

3.1.1. First-Strand Synthesis and RNA Hydrolysis

1. Mix, with caution, 1 μg of polyA$^+$ RNA and 10 ng of random hexamer in DEPC-treated H$_2$O in a final volume of 11 μL (*see* Notes 3–5).
2. Gently stir and quick spin.
3. Incubate at 70°C for 10 min, and quickly chill on ice.
4. Add 4 μL of 5X reverse transcription buffer, 2 μL of 0.1M DTT, and 2 μL of 5 mM dNTP mix solution, and gently mix (*see* Note 6).
5. Immediately add 1 μL of Super Script II (200 U/μL), gently stir, and quick spin.
6. Incubate at 49°C for 60 min.
7. Add 1 μL of 0.5M EDTA, and gently mix.
8. Add 39 μL of H$_2$O, 15 μL of 1N NaOH, gently mix, and quick spin.
9. Incubate at 70°C for 20 min.
10. Add 4 μL of 2M Tris-HCl, pH 7.4, 14.7 μL of 1N HCl, 115 μL of TE, 140 μL of 10M ammonium acetate, 1 μL of 10 μg/μL glycogen, vortex gently, and quick spin.
11. Add 350 μL of phenol/CIAA and vortex.
12. Spin to separate the phases, and then add 2.5 vol of ethanol to the aqueous phase.
13. Put on dry ice for 20 min, and centrifuge at the full speed for 15 min.
14. Rinse the pellets with 950 μL of 75% ethanol and air-dry.
15. Dissolve the single-stranded cDNA in 20 μL of TE.

3.1.2. dC-Tailing and Second-Strand Synthesis

1. Denature cDNA at 70°C for 5 min, and chill on ice.
2. Mix 20 μL of first-strand cDNA with 6 μL of 1 mM dCTP and 3 μL of 5X reverse transcription buffer (*see* Notes 7 and 8).
3. Add 1 μL of terminal deoxynucleotidyl transferase (20 U/μL), gently stir, and quick spin.

4. Incubate at 37°C for 10 min, and then at 70°C for 15 min.
5. Add 170 μL of TE, 20 μL of 3*M* sodium acetate, vortex vigorously for 1 min, add 200 μL of phenol/CIAA, and vortex vigorously for 1 min (*see* Notes 1 and 9).
6. Spin to separate the phases, then add 1 μL of 10 μg/μL glycogen, 40 μL of 3*M* sodium acetate, and 2.5 vol of ethanol to the aqueous phase.
7. Set on dry ice for 20 min, and centrifuge at the full speed for 15 min.
8. Rinse the pellets with 950 μL of 75% ethanol and air-dry.
9. Dissolve in 37 μL of H_2O.
10. Add 2 μL of ESLG primer (25 μ*M*), 10 μL of 5X reverse transcription buffer, 2 μL of 5 m*M* dNTP, and 6 μL of DTT, gently vortex, and quick spin (*see* Note 6).
11. Incubate at 70°C for 10 min, and chill on ice.
12. Add 3 μL of Super script II (200 U/μL), gently vortex, and quick spin.
13. Incubate at 42°C for 30 min.
14. Incubate at 75°C for 15 min, and chill on ice until used in Section 3.1.3. (*see* Note 9).

3.1.3. Sac*I* Adapter Ligation and Size Selection

1. Incubate 84 μL of SSEH (25 μ*M*) at 70°C for 5 min, and then chill on ice.
2. Add 10 μL of 10X Kination buffer, 5 μL of 5 m*M* dATP, mix, and spin.
3. Add 2 μL of T4 polynucleotide kinase (10 U/μL), mix, and quick spin.
4. Incubate at 37°C for 30 min.
5. Incubate at 70°C for 20 min.
6. Add 100 μL of phenol/CIAA, and vortex.
7. Spin to separate the phases, and then add 10 μL of 3*M* sodium acetate and 2.5 vol of ethanol to the aqueous phase.
8. Set on dry ice for 20 min, and centrifuge at the full speed for 15 min.
9. Rinse the pellets with 950 μL of 75% ethanol, and air-dry.
10. Dissolve the DNA in 50 μL of H_2O.
11. Measure the OD_{260}, and adjust the DNA concentration to 25 μ*M*.
12. Mix 80 μL of 5'-phosphorylated SSEH (25 μ*M*), and 80 μL of LLEH (25 μ*M*).
13. Incubate at 90°C for 5 min, and slowly cool down to 22°C, 10 μL of this solution should be used in the next step. You can keep the annealed *Sac*I adapter at –20°C. You need not repeat previous steps every time.
14. Mix 15 μL of double-stranded cDNA, 9 μL of annealed *Sac*I adapter, and 3 μL of 10X ligation buffer, vortex, and quick spin.
15. Add 3 μL of T4 DNA ligase (400 U/μL), mix, and quick spin.
16. Incubate at 16°C for 4–16 h.
17. Incubate at 65°C for 5 min, and chill on ice.
18. Make a 1.6% agarose/TAE gel containing 30 ng/mL ethidium bromide, and load the 25 μL of ligation solution with 6 μL of 6X loading buffer. Load DNA size markers, and be sure to keep the latter at least one slot away from the cDNA.
19. Run on the gel, and cut out required fractions (e.g., 300–450, 450–650, 650–900 bp) (*see* Note 1), put each agarose block into Eppendorf tubes, and weigh them. Agarose gel blocks in one tube should be <300 μg. If a gel is >300 μg, divide into two or more tubes (*see* Note 10).

20. Add 3 vol of NaI stock solution.
21. Incubate at 55°C for 5 min and vortex. You will see the agarose gel melt and become almost transparent.
22. Add 10 μL of Glassmilk, and set on ice for 5 min.
23. Wash pellet three times with New Wash.
24. Elute DNA into 20 μL of TE. Call these DNA solution "original PCR template solutions."

3.1.4. PCR Amplification

1. Make a series of dilutions of the original PCR template solutions (e.g., 1/10, 1/100, 1/1000) to check the quality of the library.
2. Mix 1 μL of the original or a diluted PCR template solution, 2.5 μL of 10X PCR buffer, 2 μL of 5 mM dNTP, 1 μL of ESL (25 μM), 1 μL of LLHES (25 μM), 17.2 μL of H$_2$O, and 0.3 μL of *Taq* DNA polymerase (5 U/μL), the total volume will be 25 μL (*see* Note 11.). Vortex and quick spin.
3. Overlay a drop of mineral oil.
4. Start the thermal cycle on a program consisting of 94°C for 4 min, followed by 25 cycles of denaturation for 30 s at 94°C, annealing for 1 min at 49°C, and synthesis for 3 min at 72°C, and then followed by synthesis for 5 min at 72°C.
5. Make a 1.6% agarose/TAE gel. Load the 8 μL of each PCR reaction products mixed with 2 μL of 6X loading buffer.
6. Take a photo and make sure smears of the DNA appear not only in the reaction using the original PCR template solution, but also in the reaction using the diluted templates (*see* Note 12).
7. Cut out the agarose gel containing DNA smear generated from the original PCR template solution and recover the DNA as described in Section 3.1.3.
8. Dissolve the DNA in 8 μL of H$_2$O.

3.1.5. Insert cDNA Digestion with Sac I and Eco RI

1. Mix 8 μL of recovered DNA solution, 1 μL of 10X low-salt restriction enzyme buffer, and 1 μL of *Sac*I (10 U/μL).
2. Incubate at 37°C for 3 h.
3. Add 2 μL of 10X medium salt restriction enzyme buffer, 7 μL of H$_2$O, and 1 μL of *Eco*RI.
4. Incubate at 37°C for 3 h, then incubate at 65°C for 5 min, and quickly chill on ice.
5. Run on the 1.6% agarose/TAE gel, observe the smear of DNA and recover the DNA as described in Section 3.1.3., steps 19–24.
6. Dissolve in 8 μL of TE.

3.1.6. Vector Preparation

1. Add 3 μL of 10X low-salt restriction enzyme buffer, 3 μg of negative control plasmid, pcDL-SRα-hRAR(5')-Tac(3') (*see* Note 13), H$_2$O to 27 μL, mix, and quick spin. Add 3 μL of *Sac*I, mix, and quick spin.

2. Incubate at 37°C for 3 h.
3. Add 10X medium salt restriction enzyme buffer, 21 μL of H$_2$O, and 3 μL of *Eco*RI, mix, and quick spin.
4. Incubate at 37°C for 3 h, then incubate at 65°C for 5 min, and quickly chill on ice.
5. Run on the 1.0% agarose/TAE gel. You will find two bands, the 291-bp band representing the retinoic acid receptor insert and the 4.5-kb digested vector band.
6. Recover the digested vector DNA as described in Section 3.1.3.
7. Dissolve in 15 μL of TE.
8. Estimate the DNA concentration using 1 μL of DNA solution to run on the 0.9% agarose/TAE gel.

3.1.7. Vector-Insert Ligation

1. Add 1 μL of insert DNA prepared in Section 3.1.5., 100 ng of digested vector prepared in Section 3.1.6., 1 μL of 10X ligation buffer, H$_2$O to 9 μL, and 1 μL of T4 DNA ligase (400 U/μL), mix, and quick spin. As a control without insert DNA, add 100 ng of digested vector prepared in Section 3.1.6., 1 μL of 10X ligation buffer, H$_2$O to 9 μL, and 1 μL of T4 DNA ligase, mix, and quick spin.
2. Incubate at 16°C for 4–16 h, and then store the ligation reaction solutions at –20°C until required for transformation.

3.1.8. Transformation

1. Thaw a 100-μL aliquot of XL-1 Blue competent cell on ice, and transfer to a Falcon 2159 tube.
2. Add 1 μL of ligation solution, and set on ice for 30 min.
3. Heat-shock at 42°C for 30 s, and then immediately replace on ice.
4. Add 1.1 mL of SOC medium, and shake at 37°C for 40 min.
5. Spread 100 μL of above mixture on 12 LB agar plates containing 100 μg/mL ampicillin.
6. Transformation efficiencies are usually in the range 0.5 × 10^5–1 × 10^6 transformants/μg of DNA.
7. Check the insert size by miniprep DNA, *Sac*I-*Eco*RI digestion, and run on the 1.6% agarose/TAE gel.

3.2. Sib-Screening with Anti-Tac Immunostaining

3.2.1. Making Pool Plates and DNA Preparation (see Note 2)

1. Draw seven rows by seven lines matrix format on the bottom surface of LB agar plates containing 100 μg/mL ampicillin as shown in Fig. 2, and assign a pool plate number. It is advisable to check the system by screening a small scale, such as 1000 clones. When 1000 clones are going to be screened, prepare 20 pool plates and 20 2-mL cultures of terrific broth containing 100 μg/mL ampicillin in a Falcon 2159 tube (*see* Fig. 2). One person can inoculate 1000–3000 clones on LB agar plates in the 7 × 7 format in 6 h, and can prepare 20–60 pools of plasmids on the next day.

2. Forty-nine individual colonies are picked up and inoculated onto both individual squares on the pool plate and in a 2-mL terrific broth culture, and assigned to one pool as shown in Fig. 2.
3. Set the pool plates at 37°C overnight, and let a single colony grow in a square. Forty-nine individual colonies should appear on each pool plate.
4. Shake the 2-mL culture tube at 37°C overnight.
5. Transfer 1.2 mL of the full growth bacterial culture to an Eppendorf tube, and spin at the full speed for 30 s. Remaining cultures should be kept at 4°C.
6. Resuspend the cell in 100 µL of GTE buffer (*see* Note 14).
7. Add 200 µL of 0.2N NaOH, 1% SDS, and invert several times.
8. Add 150 µL of 5M potassium acetate, and invert 10 times.
9. Add 450 µL of 5M lithium chloride, and invert 10 times
10. Put on ice for 5 min, and centrifuge at the full speed for 5 min.
11. Transfer supernatant to a new Eppendorf tube.
12. Add 600 µL of isopropanol.
13. Put on dry ice for 5 min, and centrifuge at the full speed for 5 min.
14. Rinse the pellets with 75% ethanol, and air-dry.
15. Dissolve the pellets in 190 µL of H_2O.
16. Add 10 µL of RNase A (100 µg/µL).
17. Incubate at 37°C for 30 min.
18. Add 200 µL of phenol/CIAA, and vortex well.
19. Centrifuge at the full speed for 5 min to separate phases and save the aqueous phase for the next step.
20. Add 1 µL of glycogen (10 µg/µL), 20 µL of 3M sodium acetate, and 450 µL of ethanol, and put on dry ice for 20 min.
21. Spin at the full speed for 15 min, rinse the pellet with 1 mL of 75% ethanol, and air-dry.
22. Dissolve in 50 µL of H_2O. Store at –20°C.
23. Using 2 µL of above DNA solution, and check the DNA concentration by running on the 1% agarose/TAE gel. At least 200 ng of DNA are needed in one transfection procedure.
24. Prepare pcDL-SRα-hRAR(5')-Tac(3') and pcDL-SRα-hG-CSF(5')-Tac(3') DNA using the same procedure as described in steps 4–23.

3.2.2. Transfection into COS-7 Cells

1. Mix 500 ng of pcDL-SRα-hRAR(5')-Tac(3') and H_2O to 25 µL, and assign it to "Negative Control." Mix 10 ng of pcDL-SRα-hG-CSF(5')Tac(3'), 490 ng of pcDL-SRα-hRAR(5')-Tac(3'), and H_2O to 25 µL, and assign it to "Positive Control."
2. Harvest exponentially growing COS-7 cells by trypsinization and replate six-well culture plate 7×10^4 cells/well 24 h before transfection. Add 2 mL of DMEM(+), and incubate at 37°C in a humidified incubator in an atmosphere of 5% CO_2.
3. Remove the medium from the cells by aspiration, and wash twice with 2 mL of DMEM(–).

4. Add 605 μL of the DNA/DEAE-dextran solution consisting of 532 μL of DMEM(–), 30 μL of 1M Tris-HCl (pH 7.4), 12 μL of DEAE-dextran (10 mg/mL), 6 μL of chloroquine (10 mM), and 25 μL of plasmid solution prepared in Section 3.2.1.

5. Return the cells to the incubator, and incubate for 3 h.

6. Remove the solution, add 1 mL of 10% (v/v) DMSO, and set at room temperature for 2 min.

7. Remove the solution, and wash with 1.5 mL of DMEM(–) once and with 1.5 mL of DMEM(+) once.

8. Add 2 mL of DMEM(+), return to the incubator, and incubate for 44–72 h.

3.2.3. Cell Surface Immunostaining with Antihuman CD25

1. Mix 50 μL of FITC-conjugated antihuman CD25 and 950 μL of PBS(–) supplemented with 1% (v/v) FCS. Set on ice (*see* Note 15).

2. Remove the DMEM(+) from the COS-7 cells, and wash with 2 mL of PBS(–) twice.

3. Add 1 mL of 0.02% (w/v) EDTA.

4. Scrape off the COS-7 cells with cell scraper, suspend the cells with a pipetman p1000, and transfer the cells into Eppendorf tubes.

5. Spin at 8500g for 5 s, and remove the supernatant.

6. Add 20 μL of diluted FITC-conjugated antihuman CD25 prepared in step 1, and suspend well.

7. Set on ice for 20 min. Shield from light.

8. Tap the tube, add 800 μL of PBS(–) supplemented with 1% (v/v) FCS, spin at 8500g for 5 s, and remove the supernatant.

9. Tap the tube, add 800 μL of PBS(–) supplemented with 1% (v/v) FCS, suspend the cells with a pipetman p1000, spin at 8500g for 5 s, and remove the supernatant.

10. Resuspend the COS-7 cells in 6–10 μL of PBS(–) supplemented with 1% (v/v) FCS.

11. Observe with the fluorescent microscope.

12. In COS-7 cells transfected with "Positive Control" DNA, margins of 1 in 20–500 cells are glittering intensely. In other words, more than 10 surface staining positive cells are detected in 1 μL of "Positive Control" DNA transfected cell suspension prepared in Section 3.2.2., step 1. Only pools, in which COS-7 cells are transfected and that show surface fluorescence as strong as the cells transfected with the positive control plasmid should be judged as "positive." For unknown reasons, a limited number of weak surface-stained cells can be sometimes observed in cells transfected with "Negative Control" DNA. Therefore, comparison in intensity and numbers with controls is needed to determine if a pool or clone is positive or negative. Typically, 2–8 positive clones are trapped in 1000 clones (20 pools) (*see* Note 5).

3.2.4. Making Smaller Pools and Identifying Single Positive Clones

1. Pick up seven individual colonies in a row or a line on the pool plate, and inoculate in 2 mL of terrific broth, containing 100 μg/mL of ampicillin. Fourteen sets of smaller pool consisting seven colonies should be prepared for one positive pool.

2. Prepare DNA from 14 sets of 2-mL cultures for one positive pool as described in Section 3.2.1., steps 4–22.
3. Transfect COS-7 cells with smaller pools as described in Section 3.2.2., and stain the cells with FITC-conjugated antihuman CD25 as described in Section 3.2.3.
4. One single positive clone can be obtained by determining which one in the smaller pool of seven rows and which one in the smaller pool of seven lines is positive (*see* Note 16).
5. Prepare DNA of one clone assumed to be positive, transfect COS-7 cells with this DNA, and stain the transfected COS-7 cells with anti-Tac.

3.3. Analysis

3.3.1. Hydropathy Analysis and Database Search

1. Determine the base sequence of positive clones in both directions, using SRA primer and Tac primer (*see* Note 7).
2. Make sure there is a start codon followed by an open reading frame fused with Tac (3'). Check if the base-sequence near the ATG fits with Kozak's rule *(12)*, or there are one or more in-frame stop codons upstream of the start codon. Either is enough (*see* Note 17).
3. Draw the hydropathy profiles of deduced amino acid sequences of positive clones, compare the shape of hydropathy profiles with that of authentic N-terminal signal sequences, and make sure the putative N-terminal regions are as hydrophobic as authentic ones (*see* Note 17) *(13,14)*.
4. Calculate to make sure there is a reasonable cleavage site by signal peptidase by von Heine's method using Gene Works program (*see* Note 17) *(14)*.
5. Compare the sequence information with the databases in both DNA and protein levels for homology using searching programs, such as BLAST or FASTA.
6. Using obtained cDNA fragments as probes, you may check the RNA expression and may screen an oligo(dT)-primed cDNA library to obtain full-length clones. Since a full coding region can never be trapped by Signal Sequence Trap, the screening for a full-length clone is always needed.
7. Draw the hydropathy profiles of deduced amino acid sequences of full-length clones, and make sure the N-terminal regions are hydrophobic (*see* Note 18).
8. Compare the full-length sequence information with database in both DNA and protein levels for homology.
9. Check whether the deduced proteins have ER or GA retention signals *(15)*.

4. Notes

1. The 300–500 bp fraction should be tested first for two reasons:
   a. Longer cDNAs, which contain the whole coding regions, do not generate fusion proteins because of the appearance of the stop codons.
   b. In fractions shorter than 200 bp, too many artifacts appear in our experience.
2. The sib-screening method described here is faster than FACS sorting followed by hirt fraction plasmid recovery.

3. cDNA construction should be started with 200 ng–5 μg of polyA⁺ RNA. Contamination with rRNA will result in increasing the clone number you have to screen.

4. To enrich 5'-end of cDNA, <20 ng of random hexamer primer should be used for 1 μg of polyA⁺ RNA.

5. Probability of the positive clones is typically 1/100–1/500 but can be 1/80–1/1500, depending on quality of polyA⁺ RNA, random hexamer/polyA⁺ RNA ratio, elongation of first-strand cDNA synthesis, size fraction, and cell source.

6. The elongation of cDNA can be monitored by adding [α-^{32}P] dCTP and applying it on an alkaline agarose gel.

7. If dC-tailing is too long, there will be some difficulties in determination of base sequences, because poly(dG) longer than 25 bases prevents sequencing reaction. 1/2X reverse transcription buffer, which contains as low as 1.5 mM MgCl$_2$, tends to give better results than tailing buffer containing cacodylate *(16)*, although cacodylate has been recommended for use for a long time *(10,17,18)*. If there is still trouble in base sequencing, another sequencing primer, SLG9 (20-mer) 5'-GACTAACTGACGGGGGGGGG can be tried.

8. The combination of dA-tailing and poly(dT) primer can be used instead of the combinations of dG-tailing and poly(dC) primer. Poly(dT) does not prevent the sequencing reaction so severely as poly(dG).

9. Cutting at the middle of coding region of cDNA by shearing with ultrasound sonication may bring more efficient fusion protein generation, although sometimes, recovery of DNA after sonication is rather poor. Sonication steps can be added as follows:

   a. Add H$_2$O to the double-stranded cDNA prepared in Section 3.1.2., step 14 to the final volume of 400 μL.

   b. Sonicate 5–10 times for 30 s at setting 5, continuous output, 100% duty cycle in a 15-mL disposable tube, chilling with ice-cold water, using SONICATOR (HEAT-SYSTEM-ULTRASONICS, Farmingdale, NY). Before shearing cDNA, you should practice sonication using such DNA fragments as λ-phage DNA digested with *Hin*dIII, and determine the settings and time.

   c. Transfer to an Eppendorf tube, do phenol/CIAA extraction, ethanol precipitation, rinse, and dissolve DNA in 7 μL of H$_2$O.

   d. Add 1 μL of 10X second-strand buffer, 1 μL of 2 mM dNTP, vortex, and quick spin.

   e. Add 1 μL of T4 DNA polymerase (3 U/μL), stir gently, and quick spin.

   f. Incubate at 37°C for 10 min, then vortex violently, and spin.

   g. Incubate at 75°C for 15 min, quickly chill on ice, and go to Section 3.1.4.

10. In Section 3.1.3., steps 19–24, DNA recovery method using Geneclean II kit is recommended because the loss of around 300 bp DNA fragments was minimum in our hands.

11. Instead of *Taq* DNA polymerase, PFU DNA polymerase (Stratagene) or Deep Vent DNA Polymerase (New England Biolab), may be useful to avoid mutation during PCR.

12. When smearing of DNA fragments is observed in the 1/1000 diluted reaction, even if it is very faint, there will be little chance to trap exactly the same clones generated by PCR.

13. For vector preparation, negative control plasmid, pcDL-SRα–hRAR(5')-Tac(3') should be used.

14. The Wizard miniprep kit (Promega, Madison, WI) may be used, instead of steps 6–22 of Section 3.2.1.

15. The optimal concentrations of antibodies should be titrated before use. Alternatively, the combination of anti-Tac monoclonal antibodies (MAb) (anti-Tac ascitis can be obtained from T. A. Waldman (NIH Bethesda) or T. Uchiyama (Kyoto University, Kyoto, Japan) and FITC-conjugated AffinPure goat antimouse IgG (H + L) (1 mg/mL) (Jackson ImmunoReseach Laboratories, West Grove, PA) can be used.

16. In 10% of positive pools obtained in the first screening, no positive clones appear in the secondary screening for unknown reasons. Sometimes more than one clone appears from one positive pool in the secondary screening.

17. In the fraction (300–500 bp), 15–30% of the anti-Tac surface-stained positive clones do not match the three criteria described in Section 3.3.1., steps 2–4. The ratio of artificial clones depends on the stringency of the judgment of anti-Tac staining *(4)*.

18 Results of computer homology search or cloning of full-length cDNA shows that 10–30% of the clones which match the three criteria (Section 3.3.1., steps 2–4) do not encode N-terminal, but the middle of the coding regions. Some of them code putative transmembrane regions. The ratio depends on the stringency of the judgment of the existence of N-terminal signal sequence in Section 3.3.1., step 3 *(4)*.

Acknowledgments

We are grateful to Taku Uchiyama for providing anti-Tac ascitis and Yutaka Takebe for the pcDLSRα plasmid.

References

1. Sabatini, D. D., Kreibich, G., Morimoto, T., and Adesnik, M. (1982) Mechanisms for the incorporation of proteins in membranes and organelles. *J. Cell Biol.* **92**, 1–22.

2. Tashiro, K., Tada, H., Heilker, R., Shirozu, M., Nakano, T., and Honjo, T. (1993) Signal sequence trap: a cloning strategy for secreted proteins and type I membrane proteins. *Science* **261**, 600–603.

3. Nakamura, T., Tashiro, K., Nazarea, M., Nakano, T., Sasayama, S., and Honjo, T. (1995) The murine lymphotoxin-β receptor cDNA: isolation by the signal sequence trap and chromosomal mapping. *Genomics* **30**, 312–319.

4. Shirozu, M., Tada, H., Tashiro, K., Nakamura, T., Lopez, N. D., Nazarea, M., Hamada, T., Sato, T., Nakano, T., and Honjo, T. (1996) Characterization of a novel secreted and membrane proteins isolated by the signal sequence trap method. *Genomics,* in press.

5. Hamada, T., Tashiro, K., Tada, H., Inazawa, J., Shirozu, M., Shibahara, K., Nakamura, T., Nakano, T., and Honjo, T. (1996) Isolation and characterization of a novel secretory protein, stromal cell-derived factor-2 (SDF-2) by the signal sequence trap method. *Gene,* in press.

6. Takebe, Y., Seiki, M., Fujisawa, J., Hoy, P., Yokota, K., Arai, K., Yoshida, M., and Arai, N. (1988) SR alpha promoter: an efficient and versatile mammalian cDNA expression system composed of the simian virus 40 early promoter and the R-U5 segment of human T-cell leukemia virus type 1 long terminal repeat. *Mol. Cell Biol.* **8,** 466–472.

7. Nikaido, T., Shimizu, A., Ishida, N., Sabe, H., Teshigawara, K., Maeda, M., Uchiyama, T., Yodoi, J., and Honjo, T. (1984) Molecular cloning of cDNA encoding human interleukin-2 receptor. *Nature* **311,** 631–635.

8. Uchiyama, T., Broder, S., and Waldman, T. A. (1981) A monoclonal antibody (anti-Tac) reactive with activated and functionally mature human T cells. I. Production of anti-Tac monoclonal antibody and distribution of Tac (+) cells. *J. Immunol.* **126,** 1393–1397.

9. Saiki, R. K., Scharf, S., Faloona, F., Mullis, K. B., Horn, G. T., Erlich, H. A., and Arnheim, N. (1985) Enzymatic amplification of beta-globin genomic sequences and restriction site analysis for diagnosis of sickle cell anemia. *Science* **330,** 1350–1354.

10. Sambrook, J., Fritsch, E. F., and Maniatis, T. (1989) *Molecular Cloning: A Laboratory Manual,* Cold Spring Harbor Laboratory, Cold Spring Harbor, NY.

11. Gluzman, Y. (1981) SV40-transformed simian cells support the replication of early SV40 mutants. *Cell* **23,** 175–182.

12. Kozak, M. (1989) The scanning model for translation: an update. *J. Cell Biol.* **108,** 229–241.

13. Kyte, J. and Doolittle, R. F. (1982) A simple method for displaying the hydropathic character of a protein. *J. Mol. Biol.* **157,** 105–132.

14. von Heine, G. (1986) A new method for predicting signal sequence cleavage sites. *Nucleic Acids Res.* **14,** 4683–4690.

15. Nilsson, T. and Warren, G. (1994) Retention and retrieval in the endoplasmic reticulum and the Golgi apparatus. *Curr. Opinion Cell Biol.* **6,** 517–521.

16. Schuster, D. M., Buchman, G. W., and Rashtchian, A. (1993) A simple and effective method for amplification of EDNA ends using 5' RACE. *Focus* **14,** 46–52.

17. Grosse, F. and Manns, A. (1993) Terminal deoxyribonucleotidyl transferase (EC 2. 7. 7. 31). *Methods Mol. Biol.* **16,** 95–105.

18. Ausubel, F. U., Breut, R., Kingston, R. E., Moore, D. D., Seidman, J. G., Smith, J. A., and Struhl, K. (1987) *Current Protocols in Molecular Biology.* Greene Publishing and Wiley-Interscience, New York.

18

Isolation of Genetic Suppressor Elements (GSEs) from Random Fragment cDNA Libraries in Retroviral Vectors

Andrei V. Gudkov and Igor B. Roninson

1. Introduction

Cellular phenotypes resulting from decreased function of a specific gene are manifested as recessive, since they are suppressed in the presence of a normal allele of the corresponding gene. The powerful gene-transfer techniques, which have played a key role in the studies of dominant phenotypes, are not readily applicable to recessive traits, since expression of a recessive allele does not generally affect the cellular phenotype. In haploid organisms, random gene disruption can be used as a general method of cloning recessive genes, but this approach is limited to genes that are not essential for cell growth and is not, at least presently, applicable to diploid cells. This explains why studies of recessive genes of higher eukaryotes (e.g., tumor suppressor genes) are in many cases lagging far behind the analysis of dominant genes (e.g., oncogenes).

The identification and functional analysis of recessive genes in mammalian cells have been boosted by the ability to select genetic suppressor elements (GSEs) that induce the desired phenotype by suppression of specific genes *(1–7)*. GSEs are short (<500 bp) cDNA fragments that produce a phenotype when expressed in the cells; this phenotype is usually opposite to that of the full-length cDNA from which they are derived. GSEs inhibiting recessive genes behave as dominant selectable markers in gene-transfer protocols and can therefore serve as tools for studying recessive mechanisms.

There are two types of GSE: antisense-oriented GSEs encoding efficient inhibitory antisense RNA molecules and sense-oriented GSEs encoding functional protein domains that interfere with the protein function in a dominant

From: *Methods in Molecular Biology, Vol. 69: cDNA Library Protocols*
Edited by: I. G. Cowell and C. A. Austin Humana Press Inc., Totowa, NJ

fashion. GSEs are isolated by preparing an expression library containing randomly fragmented DNA of the gene or genes targeted for suppression, introducing this library into the appropriate recipient cells, selecting cells with the desired phenotype, recovering the inserts from the expression vectors contained in the selected cells, and testing the recovered sequences for functional activity. Both sense- and antisense-oriented GSEs can be isolated through this procedure; the orientation, sequence, and mechanism of action of the GSEs are determined only after their biological activity has been established.

Principles and applications of the GSE methodology, as well as the results obtained with the use of this approach have been summarized elsewhere *(1)*. Briefly, the random fragment selection strategy was first applied in a prokaryotic system and used for the isolation of GSEs from the DNA of phage λ, by selecting GSEs for the ability to protect *Escherichia coli* from λ-induced lysis *(2)*. This approach was then extended to mammalian cells in a study where a set of GSEs were isolated from the cDNA encoding human topoisomerase II (topo II), by the ability of such GSEs to confer resistance to etoposide, a topo II-interacting drug *(3)*. In the latter study, a protocol was developed for preparing a random fragment library in the form of a mixture of recombinant retroviruses capable of highly efficient gene transfer into different mammalian cells. GSE selection was also carried out from random fragment libraries of several other single genes (e.g., p53 *[4]* and BCL-2 *[5]*), leading in all cases not only to the generation of new genetic tools, but also to new insights into the mechanisms of function and regulation of the studied genes.

Construction of GSE libraries from more complex targets, such as the mixture of multiple cDNA clones *(6)* or total cellular cDNA *(7)*, has extended the applications of GSE methodology from characterization of already known genes to identification of new genes whose inactivation is associated with various selectable phenotypes. GSEs isolated from such a library are most likely to induce their effect by inhibiting the gene from which they are derived. GSEs therefore can be used as probes for cloning full-length cDNA for the genes whose suppression caused the selected phenotypes. Several retroviral libraries carrying random fragments of normalized cellular mRNA sequences were constructed and used for isolation of GSEs inducing resistance to cytotoxic drugs or abnormal cell growth properties. This approach, for example, allowed us to demonstrate the involvement of the motor protein kinesin in drug sensitivity and senescence *(7)*.

Thus, the GSE strategy provides a general method for cloning genes involved in complex cellular phenotypes and for identifying functional domains of individual gene products. Isolated GSEs may be used as tools to study the corresponding gene function not only in cell culture but, in perspective, in the whole organism, if introduced into the genome of transgenic animals. Phenotypic changes or developmental disorders associated with genetically introduced GSEs may indicate the normal function of the genes targeted by such GSEs.

As a methodology, the GSE approach is still young and open to improvements. Some stages of library preparation, delivery, or selection are flexible and may be done in several alternative ways (we have tried to describe some of the alternatives in Section 4.). Thus, the protocols presented herein reflect our as yet limited experience in GSE development and may not necessarily provide the optimal experimental design.

2. Materials

2.1. GSE Libraries from Individual cDNAs

2.1.1. Random Fragmentation of a cDNA Insert by DNase I

1. GeneClean kit for DNA purification from agarose gels (BIO 101, Midwest Scientific, St. Louis, MO).
2. DNase I (Sigma, St. Louis, MO): 1 mg/mL in $0.01N$ HCl; keep frozen at $-70°C$ in small aliquots.
3. 10X DNase I buffer: 500 mM Tris-HCl, pH 7.5, 100 mM MnCl$_2$, 1 mg/mL bovine serum albumin (BSA); prepare fresh and keep on ice.

2.1.2. Filling the Ends of DNase I-Generated Fragments

1. T4 DNA polymerase (New England Biolabs, Beverly, MA).
2. Klenow fragment of *E. coli* DNA polymerase (New England Biolabs).
3. 10X T4 polymerase buffer: 500 mM NaCl, 100 mM Tris-HCl, pH 7.9, 100 mM MgCl$_2$, 50 mg/mL BSA, 10 mM dithiothreitol (DTT).
4. Stock solution of dNTPs, 2.5 mM each.

2.1.3. Adapter Ligation, Size Fractionation and Polymerase Chain Reaction (PCR) Amplification

1. Two oligonucleotides annealed to form a double-stranded adapter (Fig. 1): store frozen in annealing buffer: 10 mM Tris-HCl, pH 7.5, 7 mM MgCl$_2$, 100 mM NaCl.
2. T4 DNA ligase (New England Biolabs).
3. 5X Ligation buffer: 250 mM Tris-HCl, pH 7.8, 100 mM MgCl$_2$, 50 mM DTT, 5 mM adenosine triphosphate (ATP), 125 mg/mL BSA.
4. *Taq* DNA polymerase (Promega, Madison, WI) or equivalent enzyme.
5. 10X PCR buffer: 500 mM KCl, 100 mM Tris-HCl (pH 9.0 at 25°C).
6. dNTP 10X stock solution, 2.5 mM each.
7. Sense-oriented adaptor oligonucleotide (Fig. 1), water solution, 1 mg/mL; keep frozen at $-70°C$ in small aliquots.

A

```
5'-TCTCTAGATCGATCAGTCAGTCAGGATG...insert...
3'-ATAAGAGATCTAGCTAGTCAGTCAGTCCTAC
          Cla I
```

```
          ...insert...CATCCTGACTGACTGATCGATCTAGAGAATA-3'
                      GTAGGACTGACTGACTAGCTAGATCTCT-5'
                                     Cla I
```

B

```
5'-AATCATCGATGGATGGATGG...insert...CCATCCATCCATCGATGATTAAA-3'
3'-AAATTAGTAGCTACCTACCTACC        GGTAGGTAGGTAGCTACTAA-5'
     Cla I                                       Cla I
```

C

```
5'-AAACGAATTCACAATGGATGGATGG...insert...TAGTTAGTTAGGATCCTGC-3'
3'-    GCTTAAGTGTTACCTACCTACC         ATCAATCATTCCTAGGACGAAA-5'
     EcoRI                                       BamHI
```

D

```
5'-NNNN[restr. site][Pu]NNATGGNATGGNATGG...insert...
```

Fig. 1. Adapter sequences used for different GSE libraries. **(A)** Adapter used for topo II random fragment library *(3)*; contains *Cla*I cloning site, single ATG codon in sense, and three stop codons in all reading frames in antisense orientation. Forms relatively long inverted repeats at the ends of inserts. **(B)** Adapter with *Cla*I cloning site and three ATG codons in all reading frames; was used for large GSE library from normalized NIH 3T3 cDNA *(7)*; does not contain stop codons, which are provided by vector sequences. **(C)** Asymmetrical adapters used for p53 library *(4)*; one contains *Eco*RI cloning site and three ATG codons, and another one contains three stop codons and *Bam*HI site; allows for oriented cloning. In this and in the previous adapters, ATG contexts do not perfectly match Kozak's initiator codon consensus. **(D)** Optimized, though never experimentally tested, adapter sequence in which ATG codons are put in perfect Kozak's consensus for initiator codons.

2.1.4. Cloning and Amplification of the Library

1. Centricon-100 filters (Amicon, Beverly, MA, catalog no. 4212).
2. Purified vector plasmid DNA (pLNCX, pLXSN *[8]* or other retroviral vectors).
3. Restriction enzymes and their 10X reaction buffers.
4. Calf intestinal alkaline phosphatase (New England Biolabs).
5. 10X Alkaline phosphatase buffer: 500 mM NaCl, 100 mM Tris-HCl, pH 7.9, 100 mM MgCl$_2$, 10 mM DTT.
6. L-broth (per liter): 10 g tryptone, 5 g yeast extract, 10 g sodium chloride. Autoclave.

7. L-agar: L-broth containing 1.5% bacto-agar. Autoclave to melt agar and sterilize.
8. Ampicillin.

2.2. GSE Libraries from Normalized Total Cellular cDNA

2.2.1. cDNA Synthesis

1. cDNA synthesis system, such as Gibco BRL (Grand Island, NY; catalog no. 8267SA) or equivalent kit.
2. PolyA$^+$ RNA, 2–10 µg, water solution 0.2–1 mg/mL (keep at –70°C).
3. 5'-Phosphorylated random hexanucleotides, water solution, 1 mg/mL.
4. α-[^{32}P]-dCTP at 5000 Ci/mmol (Amersham, Arlington Heights, IL).

2.2.2. Filling the Ends, Ligation with Adapter, and PCR Amplification

Refer to the materials listed in Section 2.1.3.

2.2.3. Normalization of cDNA

1. 2X Hybridization solution: 0.6*M* Na-phosphate, pH 7.5, 2 m*M* EDTA, 0.2% SDS.
2. Hydroxylapatite for nucleic acid chromatography (Bio-Rad, Richmond, CA).
3. 0.6*M* Phosphate buffer, pH 7.2.
4. Centricon-100 (Amicon).

2.2.4. Cloning

1. Highly efficient competent bacterial cells for electroporation (Invitrogen, San Diego, CA, INVaF' or TOP10 Library Size Electroporation Kits).
2. L-agar, ampicillin.

2.3. Screening of GSE Libraries

2.3.1. GSE Library Delivery

1. Packaging cell lines: BOSC23 *(9)* (ecotropic) and BING *(10)* (amphotropic).
2. Dulbecco's modified Eagle's medium (DMEM) tissue-culture media with 10% newborn or fetal calf serum.
3. Target cells and appropriate culture media.
4. Solutions for calcium-phosphate transfection *(9)*: 2X HBS: 50 m*M* HEPES, pH 7.05, 10 m*M* KCl, 12 m*M* dextrose, 280 m*M* NaCl, 1.5 m*M* Na$_2$HPO$_4$; 1*M* CaCl$_2$; 25 m*M* chloroquine.
5. CsCl-quality purified retroviral vector plasmids.
6. Polybrene (hexadimethtine bromide; Sigma), 4 mg/mL sterile water stock solution.
7. Geneticin (G418) (Gibco BRL) or other appropriate selective agent.

2.3.2. Virus Rescue from Selected Cells

1. PCR solutions (*see* Section 2.1.3.).
2. PCR primers.
3. Standard cloning reagents (purified vectors, restriction enzymes, etc.) *(11)*.

2.3.3. Control of Library Representativity

1. PCR solutions.
2. Primers.
3. Reagents for standard Southern blot hybridization *(11)*.

3. Methods
3.1. GSE Libraries from Individual cDNAs

GSE libraries from individual cDNAs are useful for structure/functional analysis of specific genes whose suppression leads to selectable changes in cellular phenotype. GSEs serve as efficient genetic tools allowing suppression of the gene's function in various cellular systems. In addition, the identification of sense-oriented GSEs is a method of mapping functional protein domains. Examples of successful GSE isolation from single cDNA libraries include topo II *(3)*, p53 *(4)*, and BCL-2 *(5)*.

3.1.1. Random Fragmentation of a cDNA Insert by DNase I

1. Excise cDNA insert from the plasmid using appropriate restriction enzyme(s), and purify the insert from agarose gel by GeneClean procedure or any other alternative method (e.g., phenol extraction from low-melting agarose or electroelution *[11]*).
2. Keep the isolated DNA in TE buffer at a concentration of 0.1–1 mg/mL. As a sufficient amount for further procedures, we recommend preparing at least 10 μg of the insert.
3. To estimate the conditions of DNase I digestion, first treat several 0.5-μg aliquots of isolated DNA inserts with DNase I (freshly prepared 1:1000 dilution of the stock solution in reaction buffer) at 16°C in freshly prepared DNase I buffer *(see* Note 1) for different time periods (from 5 s to 1 min). Stop the reactions by adding 0.5M EDTA to the final concentration of 25 mM and estimate the degree of DNA degradation by electrophoresis in 1.5% agarose gel next to a 123-bp ladder (Gibco BRL) or other appropriate size marker.
4. Use the reaction conditions leading to the desired size range *(see* Note 2) to scale up the reaction proportionally to the amount of DNA. Remove DNase I by phenol–chloroform extraction and ethanol precipitation (here and in Section 3.1.2. where this procedure is mentioned, do two phenol–chloroform extractions followed by one chloroform extraction, add EDTA to 20 mM, precipitate with 2.5 vol of ethanol, and wash DNA precipitate twice with 70% ethanol to remove traces of organic solvents).

3.1.2. Filling the Ends of DNase I-Generated Fragments

1. Dissolve DNA in T4 DNA polymerase buffer containing 0.25 mM of each of dNTPs (approx 10 μL for 1 μg DNA).
2. Add T4 DNA polymerase and Klenow fragment of *E. coli* DNA polymerase (5 U/μg each), and incubate at 37°C for 1 h.

3. Stop reaction by adding EDTA to 25 m*M*.
4. Purify DNA by phenol–chloroform extraction.

3.1.3. Adapter Ligation, Size Fractionation, and PCR Amplification

1. Mix equimolar amounts of sense and antisense strands of the adaptor (5–10 µg total) in 50 µL of annealing buffer. Put the mixture into a water bath preheated to 90°C, and allow to cool down slowly to room temperature.
2. Mix 1 µg of DNase-treated cDNA insert with 1 µg of annealed adapter (Fig. 1; *see* Note 3) in 10 µL reaction buffer with ATP and 1 µL of T4 ligase.
3. Incubate the reaction at room temperature for at least 1 h.
4. Load the ligation mixture on 1.5% agarose gel. After electrophoresis, cut out DNA fragments of the desired size range (*see* Note 2), and elute them from agarose using any appropriate method.
5. If necessary, amplify the resulting mixture of fragments by PCR using "sense" strand of the adapter as a primer for both ends. Before PCR, incubate the mixture with *Taq* DNA polymerase for 10 min at 55°C to eliminate nicks between the adapter and cDNA fragments. Before doing large-scale PCR, estimate in a pilot set of reactions the minimal number of cycles necessary to reach the plateau (under our conditions, this number varied from 8–16 cycles). To avoid artifacts, do not exceed this value in large-scale PCR (*see* Note 4).

3.1.4. Cloning and Amplification of the Library

1. Purify the PCR product by phenol–chloroform extraction, and centrifuge through Centricon-100 filters to remove primers and short artifactual PCR products.
2. Estimate the quantity of your cDNA preparation, and treat it with the desired restriction enzyme(s) using ≥50 U of enzyme/µg of DNA for 2–16 h.
3. Purify cDNA fragments by phenol-chloroform extraction and centrifugation through Centricon-100.
4. Digest vector DNA (*see* Note 5) with the appropriate restriction enzyme(s), and (in case of using a single restriction enzyme) treat it with alkaline phosphatase according to manufacturer's recommendations.
5. Before large-scale ligation, determine optimal ligation conditions using different ratios of insert and vector DNAs, and subsequently testing transformation efficiency and the proportion of recombinant clones among the transformants (our standard ligation conditions: molar ratio 5:1, DNA concentration 10–50 ng/mL).
6. Once optimal conditions are selected, do a large-scale ligation and transformation of competent bacterial cells, and plate them on L-agar with an appropriate antibiotic. Based on our experience, an extensive set of GSEs can be derived from single-gene cDNA libraries containing about 5000–10,000 recombinant clones/kb of cDNA.
7. Wash off the bacterial colonies from agar by thoroughly scraping the surface with a spatula, using several milliliters of L-broth/plate.

8. To amplify the library, allow the resulting bacterial suspension to grow for 3–4 h in L-broth with the appropriate antibiotic (about 1 L of broth/30,000 colonies).
9. Isolate and purify plasmid DNA. For purification, use either CsCl density centrifugation or any other alternative method that allows highly purified plasmid DNA to be used for calcium-phosphate transfection.

3.2. GSE Libraries from Normalized Total Cellular cDNA

Highly complex GSE libraries of random fragments of total cellular mRNA are expected to contain suppressors of practically all cellular genes. Such libraries can therefore be used for the identification and cloning of new genes whose suppression alters the cell phenotype. A total polyA$^+$ RNA preparation is a mixture of unequally represented sequences, some very abundant and some extremely rare. In order to increase the probability of isolating GSEs from rare cDNAs, a polyA$^+$ RNA-derived cDNA preparation is subjected to a normalization procedure, providing for uniform abundance of different DNA sequences. Assuming that the mammalian genome contains at least 10,000 genes, and that the incidence of GSE-containing clones in a single-gene cDNA library is on the order of 1/100–1/500 (in our experience), the minimal complexity of a mammalian GSE library has to be at least 10^7 independent recombinant clones.

3.2.1. cDNA Synthesis

To achieve equal representation of 5'- and 3'-mRNA sequences in a random fragment cDNA preparation, cDNA synthesis is carried out on fragmented mRNA using random oligonucleotides as primers (*see* Note 6).

1. To determine the conditions for RNA fragmentation, treat 3–5 aliquots of the polyA$^+$ RNA (each aliquot contains 50 ng of RNA in 10 μL) in a boiling water bath for 2–15 min.
2. Use treated RNA samples for pilot cDNA synthesis using a commercially available cDNA synthesis kit according to the manufacturer's protocol for double-stranded cDNA synthesis, using random hexanucleotides at a concentration of 500 ng/mL as primers. Monitor the size of the first strand of cDNA by labeling it with α-[^{32}P]-dCTP and subjecting it to electrophoresis under denaturing conditions *(11)*.
3. For large-scale cDNA synthesis, use the conditions of mRNA degradation under which most of the product ranges between 200 and 1,000 bp (*see* Note 7). Synthesize at least 100 ng of cDNA.

3.2.2. Filling the Ends, Ligation with Adapter, and PCR Amplification

Proceed as described for the single-gene library protocol (*see* Section 3.2.1.). PCR amplification is an essential step in a complex cDNA library preparation, since it is necessary to generate enough cDNA for normalization (20 μg) and cloning (10–20 μg).

3.2.3. Normalization of cDNA

This is a modification of the procedure of Patanjali et al. *(12).*

1. Denature 20 µg of cDNA by boiling for 5 min in 25 µL of TE buffer, followed by immediate cooling on ice.
2. Add 25 µL of 2X hybridization solution, and divide the mixture into four aliquots in Eppendorf tubes, 12.5 µL each.
3. Add one to two drops of mineral oil to each sample to avoid evaporation, and place the tubes into a 68°C water bath for annealing.
4. Freeze down one tube every 12 h.
5. After the last time-point, dilute each of the annealing mixtures with water to a final volume of 500 µL and subject DNA to hydroxylapatite (HAP) chromatography (*see* Note 8).
6. Put into Eppendorf tubes HAP suspension equilibrated with 0.01*M* phosphate-buffered saline (PBS) so that the volume of HAP pellet is approx 100 µL.
7. Preheat tubes with HAP and all the solutions used below, and keep them at 65°C. Remove the excess of PBS, and add diluted annealing solution. Mix thoroughly, but gently, by shaking in a 65°C water bath.
8. Leave the tubes in the water bath until HAP pellet is formed (a 15-s centrifugation to collect the pellet is okay; do not exceed 1000*g* in the microcentrifuge to avoid damage of HAP crystals).
9. Carefully replace the supernatant with 1 mL of preheated 0.01*M* PBS, and repeat the process.
10. To elute single-stranded DNA (ssDNA), suspend the HAP pellet in 500 µL of PBS at the single-strand elution concentration determined as described in Note 8 (e.g., 0.16*M*), collect the supernatant, and repeat the process. Combine supernatants, and remove traces of HAP by centrifugation.
11. Concentrate ssDNA by centrifugation, and wash it three times using 1 mL of water on Centricon-100.
12. Amplify isolated ssDNA sequences by PCR with the sense primer from the adapter, using a minimal number of cycles to obtain 10 µg of the product (it usually requires about 20 PCR cycles to synthesize 10 µg of PCR product in 12–20 50-µL reactions using approx ¹/₁₀ of the isolated ssDNA).
13. Check that the size of the PCR product remains within the desired range (200–500 bp) (*see* Notes 2, 3, and 7).
14. Test the normalization quality by Southern or slot-blot hybridization with ³²P-labeled probes for high, moderate- and low-expressing genes using 0.3–1.0 µg of normalized cDNA/lane (slot) (Fig. 2). We use β-actin and β-tubulin cDNAs as probes for high-expressing genes, c-*myc* and topo II cDNAs for moderate, and c-*fos* cDNA for low-expressing genes.
15. Compare cDNA isolated after different annealing times with the original unnormalized cDNA. Make sure that the probes are of a similar size and specific activity. Use best-normalized ssDNA fraction, which produces the most uniform signal intensity with different probes, for large-scale PCR amplification to

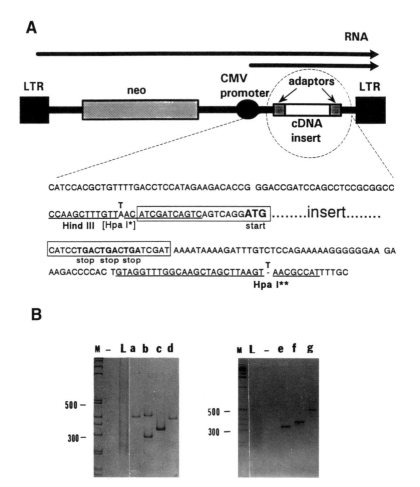

Fig. 2. (A) Structure of an integrated provirus containing an insert of topo II cDNA insert. Adapter-derived sequences are bracketed, with translation initiation and termination codons shown in boldface. Sequences used as PCR primers for isolation and recloning of the inserts are underlined; changes in the primer sequences leading to inactivation (*) or creation (**) of the indicated restriction sites are shown above the vector sequence. (B) PCR amplification of proviral inserts from DNA of HeLa cells after infection with the retroviral library (lane L) and from DNA of HeLa cell clones obtained after etoposide selection (lanes a–g). The sense strand of the adapter shown in Fig. 1A was used as the PCR primer. Too many PCR cycles used for amplification of the library cDNA inserts (lane L) in the left gel resulted in the artifactual increase in the size of PCR products owing to a concatemer formation. PCR products were separated in polyacrylamide gel and stained with ethidium bromide. Size markers are indicated in base pairs.

synthesize at least 20 µg of the product for cloning. Avoid increasing the number of PCR cycles to increase the yield of the product. Instead, use more ssDNA template to obtain the desired amount by scaling up the number of PCR reactions or the reaction volume.

3.2.4. Cloning

In order to construct a plasmid library with a complexity $\geq 10^7$ clones, it is necessary to prepare large amounts of vector and insert DNAs, to optimize carefully the conditions of ligation and transformation, and to have competent cells with very high transformation efficiency. We have obtained adequate results using the InVitrogen transformation kit, which provides competent cells for electroporation with an efficiency of 10^9 transformants/µg of plasmid DNA. This kit also includes competent cells for chemical transformation that allow one to select optimal ligation conditions using minimal amounts of material in pilot transformations. It is usually enough to have 5–10 µg of restriction enzyme-digested inserts and 20–30 µg of digested and alkaline phosphatase-treated vector.

1. Estimate optimal vector-to-insert ratio in the ligation mixture according to the manufacturer's (InVitrogen) recommendations.
2. Determine the proportion of recombinant clones after transformation (this should be no less than 60%) by PCR analysis of 20–40 individual colonies, using the sense strand of the adapter as the PCR primer.
3. Do large-scale ligation under predetermined conditions, and purify DNA from the ligation mixture by phenol–chloroform extraction and ethanol precipitation. Dissolve the DNA in TE buffer at a concentration of 0.3–1 µg/µL. Divide ligated DNA into several batches (approx 1 µg of DNA in each), and electroporate competent cells using one batch at a time.
4. Plate bacterial suspension containing $2–5 \times 10^6$ transformed cells on L-agar with the appropriate antibiotic, at no more than 10^3 colonies/cm^2.
5. To maintain the representativity of the library, isolate and purify plasmid DNA directly from bacteria grown on agar plates, rather than from bacteria grown in suspension. Estimate the quantity of DNA obtained after each transformation, and keep each batch of plasmid DNA separately.

3.3. GSE Library Screening

Successful isolation of biologically active GSEs from random fragment cDNA libraries requires careful and accurate tuning of the selection conditions that depend on the nature of the selectable phenotype, cell type, and so on. The examples of several selection schemes are given in our papers *(5–10)*. Here we describe some general considerations and essential experimental steps involved in any GSE selection protocol.

It is important to keep in mind that since most GSEs are relatively weak and because the levels of transgene expression vary widely in a retrovirally transduced cell population, one cannot expect to see the selectable phenotypic alterations in every cell that carries a GSE. Therefore, in order to be selected, every element in the GSE library should be delivered to numerous (at least 30–100) cells. This is best achieved by increasing the number of infected cells and by reaching a very high efficiency of viral transduction.

The relatively weak effect of most GSEs also means that the phenotypic alterations caused by the GSEs may not be strong enough to allow the application of selection conditions that would be so stringent as to eliminate all spontaneous background (as determined with the same recipient cells infected with an insert-free retroviral vector). Although in some cases we were able to select GSEs under zero-background conditions, it is prudent to assume before the start of a new selection that GSEs would only work under the conditions that give a selection background of 10^{-5} to 10^{-4} or even higher.

It is often difficult to combine both requirements (a large number of library-infected cells and relatively mild selection conditions) in the design of the selection. This problem is overcome by the possibility of performing multiple rounds of selection. After each round, all the GSEs and "passenger" cDNA inserts integrated in the selected cells are recovered and reintroduced into fresh recipient cells. The cycle of selection and recovery is repeated until the number of clones becomes small enough to provide a clear increase in selectability of library-transduced cells over the control. At this time, the selected elements can be individually tested for the GSE activity. An example of a two-round selection is the isolation of GSEs conferring resistance to etoposide from a normalized GSE library of NIH 3T3 cDNA *(7)*.

3.3.1. GSE Library Delivery

1. Plate the ecotropic (for mouse or rat recipient cells) or amphotropic (for recipient cells from other mammals) packaging cells (Note 9) at 6×10^6 cells/100-mm plate. (The number of plates with packaging cells should match the number of plates with the target cells that will be used for infection and selection. This number is calculated from the library size, the requirements of the selection procedure, and the predetermined efficacy of retroviral infection.)
2. The next day, transfect each plate with 10 μg of plasmid DNA using standard calcium-phosphate procedure with modifications suggested by W. Pear for BOSC23 and BING cells *(see* ref. *11* and Note 10).
3. Prior to transfection, change media on packaging cells by adding 9 mL of fresh DMEM with 10% serum. Dissolve plasmid DNA in 375 μL of sterile water, add 125 μL of 1*M* CaCl$_2$, and add 0.5 mL 2X HBS by bubbling. Immediately (within 1–2 min) add this solution to the cells.

4. Plate 2×10^6 target cells/150-mm plates.
5. Collect media from packaging cells 24 and 48 h after transfection, filter it through 0.45-mm filters, dilute twofold with fresh media, add polybrene to 4 µg/mL, and put on target cells for at least 2 h. Rock plates gently several times during this period. Initiate selection as soon as possible, but not earlier than 48 h after the last infection. In most cases, the target cells do not require G418 selection (or other selection depending on the nature of the selectable marker in the vector) (*see* Note 10). A small subpopulation of the infected cells, however, may be subjected to G418 treatment to determine the efficiency of infection.

3.3.2. Virus Rescue from Selected Cells

Although we have used several procedures to rescue the integrated virus from infected cells (*see* Note 11), only one method so far has proven to be reliable for all the different types of target cells. This method involves PCR amplification of the cDNA inserts, followed by recloning into the retroviral vector. The same procedure is used both to generate a GSE-enriched library at intermediate rounds of selection and to derive retroviral clones carrying potential GSEs for individual testing. Primers for this PCR correspond to vector sequences flanking the inserts. It is important to have sense primer sequence partially overlapping with the beginning of the adapter. This allows the avoidance of PCR artifacts caused by annealing of adapter-derived inverted repeats flanking the insert. One of several possible primer designs that allows oriented cloning of cDNA fragments back into the same vector is given in our work on topo II GSE isolation (ref. *3*; Fig. 3).

3.3.3. Control of Library Representativity

This is important at different stages of library delivery and screening. It is based on the analysis of PCR-generated population of cDNA inserts by gel electrophoresis and, in the case of the total cDNA library, by Southern hybridization with probes for several cellular genes. As long as good representativity is maintained in the library, both the electrophoretic pattern and the hybridization profiles should show a smear within the GSE size range. Loss of representativity (which occurs during the selection as a normal and desired event, but may also arise as an unfortunate consequence of uneven infection or unequal growth of bacteria during library amplification) is accompanied by the appearance of discrete bands on the gel or in Southern blots. An alternative method is based on PCR using several pairs of primers, one of which is uniform and corresponds to vector sequence, and another representing internal sequences of certain cDNA sequences that are present in the GSE library. Until representativity is preserved, each of these pairs generates multiple bands in PCR.

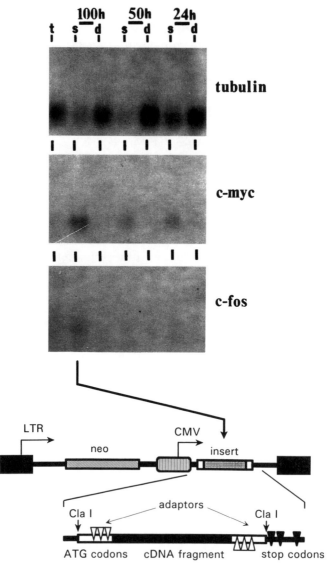

Fig. 3. Construction of normalized cDNA library. (Upper) Normalization of cDNA fragments from NIH 3T3 cells. Total cDNA (lane t) was reannealed for the indicated periods of time, and the single-stranded (lane s) and double-stranded (lane d) fractions were analyzed by Southern hybridization with the indicated probes. (Lower) Structure of an integrated provirus containing a cDNA fragment in LNCX vector.

3.3.4. Testing Individual Clones for GSE Activity

This is the final and most difficult stage of screening. Although the exact protocol for such testing depends on the specific features of the selection procedure used, there are some general observations and rules that are applicable to most of the GSEs.

GSEs are not conventional dominant selectable markers: in many cases, their biological activity is relatively weak and can only be demonstrated under mild selection conditions. This is particularly true for GSEs isolated from total cDNA libraries, rather than from individual genes. Do not necessarily expect that a GSE would have a stronger activity than that of the enriched virus population at the end of the selection. It is often easier to obtain a retroviral population that is capable of inducing the desired phenotype than to prove the activity of individual elements.

In most cases, GSEs are unable to induce the desired phenotype in every transduced cell. It is therefore important to deliver the putative GSEs into a large number of target cells (10^5 to 10^6) and to compare it with an equally large population of control cells. Avoid analyzing cell clones: it is often impossible to determine the relative impact of clonal variability and of the GSE effect on the phenotype.

Make sure that both experimental and control cells undergo exactly the same conditions of treatment. Even minor discrepancies in the treatment conditions (e.g., differences in the number of packaging cells taken for transfection with control and experimental plasmids, number of virus infections, etc.) may significantly jeopardize the result. To avoid difficulties in reproducing biological effects of GSEs during individual testing, it is necessary to maintain the same assay conditions throughout the whole process of selection (the stock of media, the lot of serum, the same generation of target cells, etc.).

Since the GSE enrichment during selection is often easier to demonstrate than direct GSE activity, it can be used as an assay for GSE confirmation: a mixture of control vector-transduced cells with the GSE-carrying cells (with the excess of control cells) may be subjected to the selection procedure in the expectation that cells with the GSE would be preferentially selected.

Some of the GSEs may have a detrimental effect on the cell growth. Even if such effect is very weak, cells with low GSE expression will gradually overgrow in the population, thus diluting the GSE-induced phenotype. Furthermore, spontaneous silencing of the GSE-expressing transcription unit may also occur over time. We therefore strongly recommend the investigator to analyze cell phenotypes immediately after GSE transduction (within several days). In most cases, G418 selection is not necessary for GSE testing and may even lead to a decrease in the GSE activity.

In some cases, weak GSE effects may be enhanced by an increase in GSE expression. This can be achieved by introducing multiple GSE copies into the target cells through repeated infection *(7)*. Furthermore, when only a single GSE is available for the given gene (as is usually the case with the GSEs selected from total cDNA libraries), subsequent isolation of full-length cDNA sequence and screening of a new GSE library prepared from the cloned cDNA is likely to yield more active GSEs.

4. Notes

1. In the presence of Mn^{2+}, DNase I makes mostly double-strand breaks, whereas Mg^{2+} stimulates single-strand break formation *(11)*. It is important to prepare DNase I-buffer with Mn^{2+} just before mixing the reaction and to keep it in tightly closed tubes in an ice bath. The solution may have a light brown color.
2. For GSE library preparation, we usually use random fragments whose size varies from 80–500 bp. No correlation between the GSE size and activity was found within this size range. PCR amplification of cDNA fragments after ligation with adapter (*see* Note 3) may lead to a substantial decrease in the average size of the fragments owing to more efficient amplification of shorter fragments. This problem can be resolved by separate isolation and processing of cDNA fragments of two size ranges (e.g., 80–200 bp and 200–500 bp), which can be combined at the ligation stage.
3. Adapter design is determined by the vector and the restriction map of the cDNA target (if known). Generally, adapters should provide the insert with the cloning sites that are absent or rare in the insert, with translation initiator codon(s) at the beginning, and stop codons at the end of the cDNA fragment. The adapters also have a protruding end at one side to ensure directional ligation. Several suggested adapter designs are shown in Fig. 1. Adapters may have the initiator codon in one or in all three open reading frames. Initiator and stop codons (in different orientations) may be included into the same adapter sequence (Fig. 1A). Alternatively, initiator and stop codons may be separated between two different adapters (Fig. 1C). The latter design allows one to have shorter adaptor sequences and to use different cloning sites for directional recloning. Finally, adapters need not carry any of the initiation or termination codons if they are provided in the vector (Fig. 1B). The role of the adapter in such a case is reduced to supplying the cDNA fragments with uniform sequence tags for PCR amplification and cloning. Adapter sequence is a subject for future optimization and testing (Fig. 1D). For single-gene GSE libraries, we recommend the adaptor design shown in Fig. 1C, keeping in mind that the restriction sites for cloning should not be present within the target cDNA.
4. Ligation with adapters supplies cDNA fragments with inverted repeats. At the plateau of the PCR reaction, when the primer concentration is low and the concentration of the product is high, these repeats may link different cDNA fragments by crossannealing. This leads to formation of artifactual concatemeric PCR

products (*see* Fig. 3A). This problem does not occur if the proper number of cycles and primer concentrations are maintained.

5. We are primarily using two retroviral vectors constructed in A. D. Miller's laboratory (Fred Hutchinson Cancer Research Center, Seattle, WA): pLNCX and pLXSN *(12)*. Both vectors contain two promoters, one of which drives the expression of the *neo* (G418 resistance) gene. GSEs are expressed from the cytomegalovirus promoter in LNCX (*see* Fig. 2) and from the retroviral (Moloney murine leukemia virus) LTR in the LXSN vector. Both vectors have been successfully used for preparation and screening of single-cDNA GSE libraries. Note that GSEs in pLXSN are transcribed as a part of long proviral RNA whose secondary structure may in some cases interfere with the function of short antisense-oriented elements.

6. cDNA preparations initiated from oligo(dT) primer are known to underrepresent 5'-end mRNA sequences. Similarly, cDNAs synthesized on full-length mRNAs using random primers underrepresent 3'-end mRNA sequences. Fragmentation of mRNA allows us to reach more equal 5'- and 3'-end sequence representation if cDNA synthesis is done from a random primer. Random primers should be phosphorylated to ensure successful ligation with an adapter.

7. The presence of cDNA fragments longer than the desired size range prior to PCR is no reason for concern: The average length of the final products will be significantly decreased as a result of PCR amplification that favors shorter fragments. To preserve longer fragments in the library (400–600 bp), it is recommended to isolate them from the gel after PCR amplification of normalized ssDNA and to amplify them separately. This fraction is added to the bulk of the DNA preparation before cloning.

8. Before HAP chromatography, phosphate buffer used for DNA elution must be calibrated. Prepare serial dilutions of a $1M$ stock of PBS ranging from $0.01–0.5M$ concentration, and determine PBS concentrations at which ssDNA and double-stranded DNA (dsDNA) are eluted from HAP. As a test DNA, use any ^{32}P-labeled PCR product generated under the conditions of asymmetrical PCR and containing a mixture of ssDNA and dsDNA. It is important to keep the temperature of the tubes with HAP at 68°C all the time, especially during ssDNA elution. At lower temperatures, ssDNA may form a net of fragments linked through annealed adapter regions that can behave as partially dsDNA and be retained in HAP. An example of cDNA normalization is shown in Fig. 3.

9. In our experience, ecotropic BOSC 23 *(9)* and amphotropic BING *(10)* packaging cell lines, developed by W. S. Pear in the laboratory of D. Baltimore (Rockefeller University) from human 293 cells, provide by far the highest retroviral titer (at least 10^6/mL) on transient transfection. The titer is maintained for 2 d, allowing two viral infections of target cells from one transfection. However, growth of some target cells (e.g., mouse embryo fibroblasts) is negatively affected by the media conditioned by BOSC23 or BING cells. Therefore, the following steps are recommended.

a. Dilute virus-containing media at least twofold with fresh media;

b. Expose target cells to the virus for 2–4 h with subsequent replacement with fresh media; and

c. Make sure that control cells are treated with an insert-free virus under exactly the same conditions.

In the cases where 293-derived cell supernatants have an adverse effect, one can also consider using NIH 3T3-derived packaging cells, such as GP+E86 (ecotropic) and GP+envAm12 (amphotropic) *(13)*, which are not efficient enough to ensure delivery of complex libraries, but can be used for smaller single-gene libraries *(3,4)*.

10. Transfection protocol for BOSC23 and BING cells suggested by their creators contains several differences from the conventional calcium-phosphate DNA transfection protocol *(11)*:

a. Cells should be about 80% confluent prior to transfection; otherwise significant cell death occurs after precipitate is added;

b. DNA precipitate is added to the cells immediately after mixing 2X HBS with DNA-CaCl$_2$ solution; and

c. It is also recommended by W. Pear *(11)* to add 25 mM of chloroquine into media prior to transfection and keep cells with precipitate and chloroquine for 7–11 h; then media are replaced with a chloroquine-free medium, which is used as a source of the virus 48 h after transfection. (We usually skip chloroquine treatment and media replacement and do infection twice: 24 and 48 h after transfection.)

11. Fifty to 100% infection of most target cells can be reached after optimizing the conditions for infection. This allows one to avoid the expensive and time-consuming G418 selection of target cells. In addition, it makes it possible to do GSE selection right after the delivery of the GSE library, before loss of complexity may occur owing to variations in cell growth rate or transcriptional inactivation of proviruses.

12. The alternative methods for provirus rescue from the selected cells are cell fusion with murine NIH 3T3-derived packaging cells *(13)* and long-range PCR of the entire integrated provirus, rather than just the cDNA insert. To rescue the virus by fusion, plate into a 100-mm plate a mixture of 4 × 10^6 ecotropic packaging cells GP+E86 *(13)* with 2 × 10^6 selected cells. Allow the cells to attach and spread (5–12 h), wash them three times with PBS to remove media with serum completely and load 1 mL of 10% polyethylene glycol (PEG, mol wt 6000, Sigma) for 1 min. Dilute and remove PEG, wash cells twice with PBS, and add fresh media. Collect media with rescued virus 24 and 48 h after PEG treatment, add polybrene, and infect amphotropic packaging cells, line GP + envAm12 *(13)*. Titer of rescued virus is usually quite low (up to 10^3). Select infected cells on G418 adding the drug 24 h after last infection. Virus produced by the resulting cell population usually has titer up to 10^5 and may be used for infection of fresh target cells and secondary screening. Some GSE selections can be performed right on these packaging cell lines producing the library (for example, the first

round of selection of etoposide-resistant Gses was done on the mixture of eco-
and amphotropic packaging cells "ping-ponging" the NIH 3T3 GSE library *[7]*).
In this case, virus rescue step may be omitted since selected cells produce virus
by themselves. Both virus rescue by fusion and long-range PCR are faster and
less laborious than the main procedure, which involves recloning, but their efficacy has varied greatly among different target cell types and individual investigators. Thus, we would not yet recommend these procedures for rescuing
complex mixtures of cDNA inserts until further optimization.

Acknowledgments

The GSE methodology development was supported by National Cancer
Institute grants R37CA40333, RO1CA39365, RO1CA56736, RO1CA62099
(I. B. Roninson) RO1CA60730, R21CA62045, RO3TW00475 (A. V. Gudkov),
a grant from Ingenex, a US Army grant DAMD17-94-J-4038, a Faculty Research
Award from the American Cancer Society (I. B. Roninson), and American
Cancer Society (Illinois Division), grants 91-58 and 93-64 (A. V. Gudkov).
The authors thank all the members of their laboratories who contributed to
the development of the GSE protocols and the investigators who provided vec-
tors, cell lines, and other materials used in this work.

References

1. Roninson, I. B., Gudkov, A. V., Holzmayer, T. A., Kirschling, D. J., Kazarov, A.
   R., Zelnick, C. R., Mazo, I. A., Axenovich, S. A., and Thimmapaya, R. (1995)
   Genetic suppressor elements: new tools for molecular oncology. *Cancer Res.*
   55, 4023–4028.
2. Holzmayer, T. A., Pestov, D. G., and Roninson, I. B. (1992) Isolation of dominant
   negative mutants and inhibitory antisense RNA sequences by expression selection of random DNA fragments. *Nucleic Acids Res.* **20,** 711–717.
3. Gudkov, A. V., Zelnick, C., Kazarov, A. R., Thimmapaya, R., Suttle, D. P., Beck,
   W. T., and Roninson, I. B. (1993) Isolation of genetic suppressor elements, inducing resistance to topoisomerase II-interactive cytotoxic drugs, from human
   topoisomerase II cDNA. *Proc. Natl. Acad. Sci. USA* **90,** 3231–3235.
4. Ossovskaya, V. S., Mazo, I. A., Strezoska, Z., Chernova, O. B., Chernov, M. V.,
   Stark, G. R., Chumakov, P. M., and Gudkov, A. V. (1996) Use of genetic suppressor elements to dissect distinct biological effects of separate p53 domains. *Proc.
   Natl. Acad. Sci USA* **93,** in press.
5. Tarasewich, D. G., Schott, B., Xuan, Y., Chang, B., Salov, S., Holzmayer, T. A.,
   and Roninson, I. B. (1996) Isolation of a genetic element inhibiting BCL-2
   expression and sensitizing tumor cells to chemotherapeutic drugs. In preparation.
6. Pestov, D. G. and Lau, L. F. (1994) Genetic selection of growth inhibitory
   sequences in mammalian cells. *Proc. Natl. Acad. Sci. USA* **91,** 12,549–12,553.
7. Gudkov, A. V., Kazarov A. R., Thimmapaya, R., Mazo, I. A., Axenovich, S., and
   Roninson, I. B. (1994) Isolation of genetic suppressor elements from a retroviral

normalized cDNA library: identification of a kinesin associated with drug sensitivity and senescence. *Proc. Natl. Acad. Sci. USA* **91,** 3744–3748.

8. Miller, A. D. and Rosman, G. J. (1989) Improved retroviral vectors for gene transfer and expression. *BioTechniques* **7,** 980–986.

9. Pear, W. S., Nolan, G. P., Scott, M. L., and Baltimore, D. (1993) Production of high titer helper-free retroviruses by transient transfection. *Proc. Natl. Acad. Sci. USA* **90,** P8392–8396.

10. Pear, W. S., Scott, M. L., Baltimore, D., and Nolan, G. (1995) Production of high-titer helper-free retroviruses with an amphotropic host range by transient transfection. In preparation.

11. Sambrook, J., Fritsch, E. F., and Maniatis, T. (1989) *Molecular Cloning. A Laboratory Manual.* Cold Spring Harbor Laboratory, Cold Spring Harbor, NY.

12. Patanjali, S. R., Parimoto, S., and Weissman, S. M. (1991) Construction of a uniform abundance (normalized) cDNA library. *Proc. Natl. Acad. Sci. USA* **88,** 1943–1947.

13. Markowitz, D., Goff, S., and Bank, A. (1988) Construction and use of a safe and efficient amphotropic packaging cell line. *Virology* **167,** 400–406.

19

Expression and Preparation of Fusion Proteins from Recombinant λgt11 Phages

Sheng-He Huang and Ambrose Jong

1. Introduction

The phage λgt11 system has become increasingly popular for expression of cDNAs or genomic DNAs either in phage plaques or in bacteria lysogenized with recombinant phages *(1,2)*. It offers the advantages of high cloning efficiency, high-level expression, the relative stability of β-galactosidase fusion proteins, and simple approaches to purify the fusion proteins. After the desired clone is detected and purified, it is often necessary to obtain preparative amounts of recombinant protein specified by the fusion of the foreign sequence to the carboxyl-terminus of β-galactosidase in λgt11 expression system. The conventional method for preparing fusion proteins from the recombinant λgt11 clones involves production of phage lysogens in *Escherichia coli* strain Y1089 followed by inducing *lacZ*-directed fusion protein expression with isopropyl-β-D-thiogalactopyranoside (IPTG) *(1)*. This method has two limitations: it is time-consuming, and phage lysogeny occurs at a low frequency. We have previously described a method for making fusion proteins from LB agar plates containing *E. coli* Y1090 infected with a high concentration of recombinant λgt11 phages (up to 5 × 10⁶ PFU/150 × 15-mm plate) *(3)*. A liquid culture method for preparing fusion proteins from *E. coli* Y1090 infected with the λgt11 clones has previously been described *(4)*. More recently, some improvements have been made on the plate method by repeating induction and elution *(5)*. Although the liquid culture method allows the recovery of only 0.2–1% of total proteins *(6,7)*, this method generally yields 5–10% of expressed protein in solution, that is, most lysed cells are trapped in the agar and the expressed proteins are recovered in a small volume of inducing solution, resulting in a higher final concentration of protein. More than 200 μg of fusion protein can be obtained from one plate.

From: *Methods in Molecular Biology, Vol. 69: cDNA Library Protocols*
Edited by: I. G. Cowell and C. A. Austin Humana Press Inc., Totowa, NJ

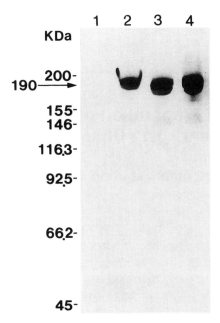

Fig. 1. Induction of the 190-kDa ERP72 fusion protein with various concentrations of IPTG (lane 1, no IPTG; lane 2, 2.5 m*M*; lane 3, 5 m*M*; lane 4, 10 m*M*). Fusion protein was detected by Western blot analysis with anti-ERP72 antibody *(8)*.

Currently, the plate method appears to be a simple and efficient way to express and prepare fusion proteins from recombinant λgt11 phages (Figs. 1 and 2). This chapter describes how the β-galactosidase fusion proteins can be made and isolated from the recombinant λgt11 phages with the plate method.

2. Materials

2.1. Plating Bacteria

1. Bacterial strain: Y1090 strain is deficient in the *lon⁻* protease. In *lon⁻* cells, β-galactosidase fusion proteins can accumulate at much higher levels than in wild-type cells.
2. Phage: λgt11.
3. LB/ampicillin medium: 10 g bacto-tryptone, 5 g yeast extract, 5 g NaCl/L. After autoclaving and cooling, add 100 mg/L ampicillin from a 100 mg/mL stock solution prior to use.
4. LB plates: 1.5% agar in LB containing 10 m*M* $MgCl_2$ and 100 µg/mL ampicillin.
5. LB top agar: 0.7% agar in LB containing 10 m*M* $MgCl_2$ and 100 µg/mL ampicillin.
6. 1*M* $MgSO_4$ (autoclave).
7. Phage buffer (SM): 50 m*M* Tris-HCl, pH 7.5, 100 m*M* NaCl, and 10 m*M* $MgCl_2$ or $MgSO_4$.

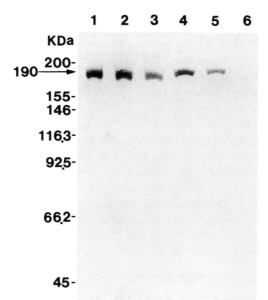

Fig. 2. Repeated induction and elution of the 190-kDa ERP72 fusion protein from agar plates. *E. coli* Y1090 cells were infected and plated by the procedure described in Section 3. Plates were incubated for 3 h at 42°C, and 5 mL of 5 m*M* IPTG in 0.5X LB containing 10 m*M* MgSO₄ added to each plate. Incubation was carried out at 37°C for 3 h (lane 1), and the supernatant saved. Induction and elution were repeated five times with 1-h intervals between inductions (lanes 2–6). Fusion proteins recovered after each induction was analyzed by Western blotting.

2.2. Fusion Protein Expression

1. 1*M* IPTG: 240 mg in 1 mL H₂O. Store frozen.
2. 0.5X LB containing 10 m*M* MgSO₄.

2.3. Detection of Fusion Proteins

1. 5X Polyacrylamide gel electrophoresis (PAGE) sample buffer: 15% β-mercaptoethanol, 15% sodium dodecyl sulfate (SDS), 1.5% bromophenol blue, and 50% glycerol.
2. 10% Polyacrylamide gels containing 0.2% SDS.

2.4. Preparing Crude Fusion Proteins

1. TEP buffer: 0.1*M* Tris-HCl, pH 7.5, 10 m*M* EDTA, and 1 m*M* phenylmethylsulfonyl fluoride (PMSF).
2. Saturated or solid (NH₄)₂SO₄.

2.5. Immunoaffinity Purification of Fusion Proteins

1. Immunoadsorbent: ProtoSorb *lacZ* Adsorbent (Promega, Madison, WI).
2. Necessary buffers *(6)*.

3. Methods
3.1. Plating Bacteria

1. Grow host cells: Streak the Y1090 on LB agar plates containing 100 µg/mL ampicillin. After growing this plate overnight at 37°C, store it at 4°C. Inoculate the plating cells from a single colony in LB containing 100 µg/mL ampicillin for 8–18 h at 37°C.
2. Infect the cells with phages: Up to 5×10^6 recombinant phages/plate (150 × 15 mm) can be used. Mix 700 µL of the Y1090 culture with up to 200 µL of the phage in SM. Allow the phage to adsorb the cells for 15–30 min at room temperature.
3. Plate cells and phages: Add 7.5 mL of LB top agar (at 45–50°C) to the infected cells, pour onto an LB agar + ampicillin plate (at room temperature), and plate evenly over the surface of the plate. Use slightly dry plates (2 d old) so that better adhesion of the top agar to the bottom agar will be obtained.
4. Grow plaques: Allow the top agar of the plates to solidify for 30 min at room temperature. Incubate the plates at 42°C for 3–3.5 h.

3.2. Fusion Protein Induction

1. Add 5–10 mL of 5 mM IPTG in 0.5X LB containing 10 mM MgSO$_4$ to each plate.
2. Incubate at 37°C for 3 h. Recover the supernatant.
3. Repeat steps 1 and 2, except that the incubation time is reduced to 1–2 h. Pool the eluate from each induction.

3.3. Detection of Fusion Proteins

1. Transfer 24 µL of the eluate to a microcentrifuge tube. Spin at 12,000g for 5 min to remove the cellular debris and agar, and then transfer the supernatant to a fresh tube.
2. Add 6 µL of 5X PAGE sample buffer to the tube, and boil for 5 min. Analyze samples on a 10% polyacrylamide gel and visualize proteins with Coomassie blue, or detect the fusion protein on Western blot with specific antibodies.

3.4. Preparing Crude Fusion Proteins

1. Centrifuge the pooled eluates at 15,000g for 10 min to get rid of the debris and save the supernatant.
2. To the supernatant, add either solid ammonium sulfate or 3 vol of saturated ammonium sulfate to 75% saturation, and stir at 4°C for 20 min.
3. Centrifuge at 15,000g for 20 min, discard the supernatant, and redissolve the pellet in cold TEP buffer at approx 20 mg/mL.

3.5. Immunoaffinity Purification of Fusion Proteins (6)

After diluting the crude fusion proteins with 50 mM Tris-HCl buffer, pH 7.3, to about 4 mg/mL total protein, it is convenient to purify the fusion protein by the prepared immunoaffinity column (ProtoSorb *lacZ* Immmunoaffinity Adsorbent) available from Promega.

4. Notes

1. In some cases, allowing expression at 30°C rather than at 37°C could help stabilize expressed fusion proteins.
2. Low amount of IPTG and phages may be used if expression is too high. The expression level of fusion proteins can be estimated by immunoblotting analysis on a plaque lift along with dot blotting a series of known amounts of the antigen on a membrane disk, which is also used to make a plaque lift. The lower detection limits of alkaline phosphatase- and peroxidase-conjugated second antibodies are 20–50 and 200–500 pg of antigens, respectively *(1)*.
3. In certain cases where the fusion protein recovery is extremely low (i.e., <5 μg fusion protein/plate), the yield may be improved by a couple of ways. Some fusion proteins trapped in the top agar can be extracted with TEP buffer and precipitated with ammonium sulfate as above. The recovery may be increased by including a fractionation step (such as gel filtration) prior to immunoaffinity purification.
4. Fusion proteins may be purified by alternative ways. If the immunoaffinity column is not available, the large size of the fusion protein is suitable for preoperative SDS-PAGE and gel-filtration chromatography.

Acknowledgments

This work was supported by the grants GM 39436 and GM 48492 from the National Institutes of Health.

References

1. Mierendorf, R. C., Percy, C., and Young, R. A. (1987) Gene isolation by screening gt11 libraries with antibodies. *Methods Enzymol.* **152,** 458–469.
2. Young, R. A. and Davis, R. W. (1991) Gene isolation with λgt11 system. *Methods Enzymol.* **194,** 230–238.
3. Huang, S. H., Tomich, J., Wu, H. P., Jong, A., and Holcenberg, J. (1989) Human deoxycytidine kinase: sequence of cDNAs and Analysis of Expression in cell lines with and without enzyme activity. *J. Biol. Chem.* **264,** 14,762–14,768.
4. Runge, S. W. (1992) Rapid analysis of λgt11 fusion proteins without subcloning or lysogen induction. *BioTechniques* **12,** 630–631.
5. Huang, S. H. and Jong, A. (1994) Efficient induction and preparation of fusion proteins from recombinant λgt11 clones. *Trends in Genetics* **10,** 183.
6. Promega (1991) *Promega Protocols and Applications Guide*, 2nd ed. Madison, WI.
7. Singh, H., Clerc, R. G., and LeBowitz, J. H. (1989) Molecular Cloning of sequence-specific DNA binding proteins using recognition site probes. *BioTechniques* **7,** 252–261.
8. Huang, S. H., Gomer, C., Sun, G. X., Wong, S., Wu, C., Liu, Y. X., and Holcenberg, J. (1992) Molecular characterization of a 72-kD human stress protein: a homologue to murine ERP72. *FASEB J.* **6,** A1670.

20

Computer Analysis of Cloned Sequences

Paul R. Caron

1. Introduction
1.1. Goals of Analyses

Analysis of the data generated from cDNA sequences, an important step in the final stages of any sequencing project, can provide insights into gene structure and function as well as help to direct the experimental approaches to obtaining the sequence of a full-length clone even when only preliminary data are available. The types of analyses that should be performed will depend on the immediate goal, and may evolve from trying to identify whether a clone is genuine or just an artifact to the stage where one would like to draw conclusions about function and domain structure of a previously unknown gene product encoded by the cDNA clone. This chapter provides an overview of the types of analyses that should be performed to extract the most information from a sequence. No attempt is made to review the current status of sequence analysis software, but particular programs are used to illustrate the general approaches that should be taken.

1.2. Software Choices

There are currently sequence analysis software programs for each of the major computer operating systems. Although many of these software packages have very useful functions, the exponential growth of the sequence databases necessitates the use of centralized computing facilities, which are able to handle the large volume of data associated with the sequence databases as well as provide the computing power necessary to search these data rapidly. These resources may be databases and programs accessible to the users at individual sites or network-based program servers. This chapter discusses both approaches, using the programs that are part of the Genetics Computer Group's (GCG's) (Madi-

From: *Methods in Molecular Biology, Vol. 69: cDNA Library Protocols*
Edited by: I. G. Cowell and C. A. Austin Humana Press Inc., Totowa, NJ

son, WI) Wisconsin Sequence Analysis Package™ for most of the examples
(1). Section 4. contains pointers for accessing network-based searching services. Although many other implementations of most of the functions in the
GCG package exist, this package is widely available and provides a consistent
interface to a multitude of sequence analysis programs. This package is available for VMS and a number of UNIX operating systems. This chapter specifically discusses using the UNIX command-line interface to GCG release 8.0.
Section 4. contains a summary of the differences between the UNIX and VMS
implementations. No discussion of any graphical interface to the GCG programs or any similar programs is presented.

This chapter presents the general methodology needed to characterize a
recently cloned sequence. It is impossible in this brief summary to cover
thoroughly all of the options for the programs discussed; those interested in
more detailed information should consult *Methods in Molecular Biology*, vol.
24 in this series *(2)* as well as the original program documentation. Comparisons between the various databases' searching techniques can be found in
Altschul et al. *(3)*. Throughout this chapter, text in **bold** should be entered
by the user, <return> is used to denote pressing the "return" or "enter" key;
<ctrl-d> signifies pressing "d" while holding down the "control" key; and the
user should replace <filename> with the name of his or her file.

2. Materials

Most of the programs discussed in this chapter are part of release 8.0 of the
Wisconsin Sequence Analysis Package available from the GCG. The examples
were run under IRIX 5.0 on a Silicon Graphics computer. However, the commands should be virtually identical under all of the UNIX operating systems
currently supported by GCG. Translation of the commands to VMS is straight-
forward and is discussed in Section 4. A VT100 terminal connection to the host
enables users with a variety of desktop computers and workstations to use the
programs similarly. Unless specifically stated, the default values were used for
all optional parameters in the examples presented.

Macaw *(4)* and Nentrez are available from the National Center for Biotechnology Information (NCBI) in both Mac and Windows versions. Macaw version 2.0.5 and Nentrez version 3.015 were obtained by anonymous ftp from
ncbi.nlm.nih.gov. Local copies of the sequence databases were obtained from
the NCBI and reformatted with the tools provided with the GCG package.
Remote databases were accessed from the NCBI. Direct access to the internet
is required to use programs, such as Nentrez or Network Blast, both of which
also require preregistration with the NCBI. Access to these programs is also
available without preregistration through the World Wide Web (WWW) at
http://www.ncbi.nlm.nih.gov.

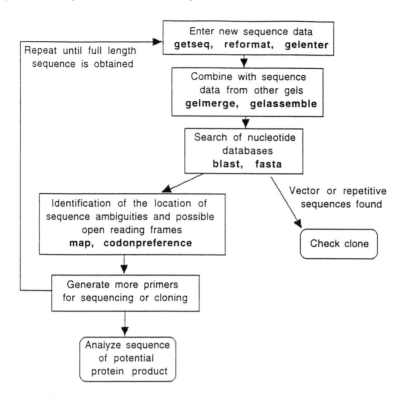

Fig. 1. Initial characterization of nucleotide sequence data.

3. Methods

The sequence analysis problem can be divided into two phases as depicted in Figs. 1 and 2. The first phase involves obtaining a full-length cDNA with high confidence in the reliability of the nucleotide sequence. In the second phase, one attempts to predict what this cDNA encodes, and how much can be extrapolated about the structure and function of the presumed protein product. The flowcharts presented in these two figures display the general approaches to the problems, whereas specific programs are discussed in the text.

3.1. Data Entry/Fragment Assembly

Sequence data can be entered into a computer file either automatically from a DNA sequencing system or manually using custom software or a word processor. The DNA sequence should be saved either directly in a GCG format or as a plain text file. The file containing the sequence data can be transferred directly to the system running GCG and "reformatted," or for short sequences, the sequence data can be "cut and pasted" into GCG-compatible files. Throughout this chapter, it is assumed that files are in a compatible format for subse-

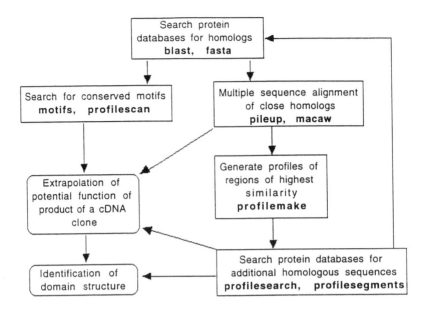

Fig. 2. Identification of structure and function of protein product.

quent analysis. Type **reformat** <filename> to format a noncompliant text file into a GCG-compatible format. Tools are also available to convert files directly from GenBank, EMBL, FASTA, PIR, Staden, and IntelliGenetics formats.

3.1.1. Entering a New Sequence Directly into a File

Type **getseq** <return>, followed by the name of a file in which to store the sequence. Paste the sequence, followed by <ctrl-D>. Nucleic acid sequence can be represented by any valid IUPAC symbols.

3.1.2. Assembling Multiple Fragments

Often sequence data are obtained from multiple gels that result from sequencing-related clones and the use of primers originating on both strands of the DNA. Depending on the size and complexity of the sequence project, it may be possible to assemble these data into a single consensus sequence using the gel reading software or a word processor. At other times, it will be necessary to use fragment assembly software to derive a consensus sequence. Details for using the GCG fragment assembly tools can be found in Chapter 2 of *Methods in Molecular Biology*, vol. 24, as well as in the GCG program manual.

3.2. Restriction Site Mapping

The location of commercially available restriction sites within a sequence are valuable landmarks that can be used to compare locations within a sequence

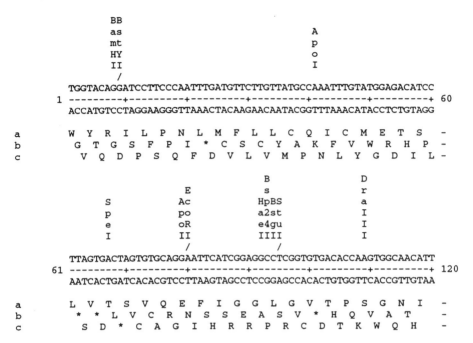

Fig. 3. Example of the output from **map**. The command **map -six** was used to locate the position of restriction enzyme cleavage sites in the sequence. Amino acid translations in the three forward reading frames is shown. Asterisks indicate the location of stop codons.

with DNA fragment sizes obtained while cloning. Identification of new sites in the sequence would provide additional handles that can be used to manipulate the clone further.

3.2.1. Generating a Linear Restriction Map

The **map** command will generate a linear restriction map of a nucleotide sequence and can display the translated amino acid sequence below it. This type of display allows one to visualize the correspondence of the nucleotide sequence with open reading frames and stop codons. An example of the output of this command, in which the restriction enzyme list was limited to enzymes with six or more nucleotides in their recognition site, is shown in Fig. 3. Files other than the default can be used as a source of enzyme or factor recognition specificities and amino acid translation schemes by command line specification.

3.2.2. Generating a List of Restriction Sites

A tabular list of the location of restriction enzyme sites can be generated by the command **mapsort**. Similar to the **map** command, the file that defines cleavage specificity can be defined at the command line.

3.3. Finding Open Reading Frames

Viewing the translated amino acid sequence generated by the **map** command is useful in identifying stop codons in a newly sequenced region and locating regions in the cDNA sequence that contain potential frame shifts. Searching protein sequences against all reading frames of a DNA sequence using programs, such as **blast**, can also identify possible frame shifts. Other statistical analyses of the DNA sequence can also be used to identify the correct translation frame(s).

3.3.1. Codonpreference

Each organism has a set of codons that are used preferentially when expressing proteins of relatively high abundance. This bias can be used to locate the coding vs noncoding reading frames of a sequence. The program **codonpreference** uses the information contained in codon usage tables for a given organism to identify regions of the sequence that most likely encode protein *(5)*. Codon usage tables for a variety of organisms from *Escherichia coli* to human are available in the GCG package. Tables of any specificity can be generated using the program **codonfrequency**. *See* Section 4. for details on configuring the GCG package for graphic output before running **codonpreference**.

3.3.2. Testcode

An additional measure of the potential of a region to encode a protein can be obtained by analyzing the distribution of nucleotides in every third position. The degeneracy of the genetic code allows this position to have much more variability than the others while conserving the amino acid sequence. The program **testcode** uses a statistical analysis of the nucleotide sequence to predict the location of coding vs noncoding regions *(6)*.

3.3.3. Translate

The protein sequence can be extracted from the cDNA sequence once the proper open reading frame has been identified using the **translate** program. This program requires that one know the nucleotide number of the starting and ending codon.

3.4. Homology with Other Proteins/Genes

3.4.1. Local vs Network Searching

Probably the most important question to ask when obtaining a new sequence is "What is this sequence similar to?" One would like to know if this is a homolog of, or identical to, a previously identified sequence. Obviously, one would like to perform a thorough search using the most recent and complete databases. Unfortunately, with the exponential growth of these data and daily new releases, maintaining an up-to-date local copy of these databases requires much time and storage space. Thus, to make the most efficient use of resources, there has been a shift toward using network servers to perform the searches. Searching local databases has advantages in terms of complete control of data security, response time, and unlimited number of queries.

3.4.2. Blast

Blast searches are very fast searches used to find local segments of homology; no gaps in the aligned segments are allowed *(7)*. The GCG implementation of **blast** allows searching of local databases as well as remote databases using servers at NCBI. This program can be used to query nucleotide sequences against nucleotide sequence databases, or the query sequence can be translated into amino acid sequences to search protein sequence databases. Protein query sequences can likewise be used to search either the protein or nucleotide databases. Starting with a nucleotide query sequence, blast searching of nucleotide sequences will find very closely related sequences, whereas searching of protein sequences will potentially identify more distant homologs. This is apparent in the results of **blast** searches using a sample sequence in Fig. 4. An estimate of the probability of a match is calculated based on the highest scoring match, but caution should be used when interpreting this number owing to repetitive sequences and contamination by vector sequences.

3.4.3. Fasta

The **fasta** program allows sequence searching for regions of local homology; unlike **blast**, gaps are allowed *(8)*. This program is more sensitive than **blast** when searching nucleotide databases. Another advantage is that the GCG version of **fasta** is able to search local sequence databases in the format used by the other GCG programs. The **blast** program requires that local databases be reformatted before searching, which increases the amount of storage space processing time required to maintain up-to-date data.

3.5. Retrieving Homologous Sequences and Descriptions

3.5.1. Fetch

Copies of sequences of interest that are stored in a local copy of a sequence database can be obtained using the **fetch** program. Typing **fetch** followed by the accession number or sequence ID will create a file containing the complete database entry. The command **fetch -outf=term:** or **fetch -outf=gcgstdout** will retrieve the sequence entry to the terminal window.

```
                                                              Smallest
                                                                Sum
A                                                    High    Probability
Sequences producing High-scoring Segment Pairs:      Score   P(N)      N

gb|U07681|HSU07681   Human NAD(H)-specific isocitrate deh...   383   1.6e-35   2
gb|U07980|BTU07980   Bos taurus NAD+-dependent isocitrate...  329   1.7e-29   2
emb|X67310|TBBISODEH S.tuberosum mRNA for beta-isopropylm...  179   6.2e-06   1
gb|M33099|BACIPMD    B.coagulans 3-isopropylmalate dehydr...  110   0.9999    1

B
sp|P41563|IDH3_BOVIN    ISOCITRATE DEHYDROGENASE (NAD),...   +3   131   3.1e-17   2
gp|U07681|HSU07681_1    NAD(H)-specific isocitrate dehy...   +3   131   3.5e-17   2
sp|P29696|LEU3_SOLTU    3-ISOPROPYLMALATE DEHYDROGENASE...   +3   110   1.2e-12   2
sp|P40495|YIJ4_YEAST    HYPOTHETICAL 40.1 KD PROTEIN IN...   +3    84   4.4e-08   2
pir|S20606|S20606       leuB protein - Salmonella typhi...   +3    98   1.4e-06   1
sp|P30125|LEU3_ECOLI    3-ISOPROPYLMALATE DEHYDROGENASE...   +3    98   1.4e-06   1
sp|P37412|LEU3_SALTY    3-ISOPROPYLMALATE DEHYDROGENASE...   +3    98   1.4e-06   1
pir|A43934|A43934       isocitrate dehydrogenase (NADP+...   +3    71   1.5e-06   2
sp|P33197|IDH_THETH     ISOCITRATE DEHYDROGENASE (NADP)...   +3    71   1.5e-06   2
gp|U07940|LPU07940_3    DlpA [Legionella pneumophila]       +3    72   2.0e-06   2
sp|P41564|IDHG_MACFA    ISOCITRATE DEHYDROGENASE (NAD),...   +3    96   2.7e-06   1
sp|P41565|IDHG_RAT      ISOCITRATE DEHYDROGENASE (NAD),...   +3    96   2.8e-06   1
pir|S39064|S39064       isocitrate dehydrogenase (NAD+)...   +3    96   2.8e-06   1
sp|Q02143|LEU3_LACLA    3-ISOPROPYLMALATE DEHYDROGENASE...   +3    78   1.1e-05   2
gp|M90761|LACLEUILV_2   leuB gene product [Lactococcus...   +3    78   1.1e-05   2
sp|P28241|IDH2_YEAST    ISOCITRATE DEHYDROGENASE (NAD),...   +3    61   5.8e-05   2
gp|Z46242|CEF35G12_2    F35G12.2, similar to isocitrate...   +3    84   0.00013   1
sp|P39126|IDH_BACSU     ISOCITRATE DEHYDROGENASE (NADP)...   +3    55   0.00045   2
sp|P29102|LEU3_BRANA    3-ISOPROPYLMALATE DEHYDROGENASE...   +3    79   0.00066   1
pdb|4ICD|               Phosphorylated Isocitrate Dehyd...   +3    53   0.00082   2
pdb|3ICD|               Isocitrate Dehydrogenase (E.C.1...   +3    53   0.00082   2
pdb|6ICD|               Isocitrate Dehydrogenase (E.C.1...   +3    53   0.00082   2
pdb|7ICD|               Isocitrate Dehydrogenase (E.C.1...   +3    53   0.00082   2
sp|P31958|LEU3_CLOPA    3-ISOPROPYLMALATE DEHYDROGENASE...   +3    77   0.0012    1
sp|P12010|LEU3_BACCO    3-ISOPROPYLMALATE DEHYDROGENASE...   +3    76   0.0017    1
gp|M33099|BACIPMD_1     B.coagulans 3-isopropylmalate d...   +3    76   0.0017    1
sp|P24404|LEU3_AGRTU    3-ISOPROPYLMALATE DEHYDROGENASE...   +3    76   0.0017    1
pir|A55591|A55591       isocitrate dehydrogenase (NADP+...   +3    53   0.0018    2
sp|P05644|LEU3_BACCA    3-ISOPROPYLMALATE DEHYDROGENASE...   +3    75   0.0023    1
sp|P24098|LEU3_THEAQ    3-ISOPROPYLMALATE DEHYDROGENASE...   +3    74   0.0032    1
pir|JX0286|JX0286       3-isopropylmalate dehydrogenase...   +3    74   0.0032    1
pdb|1IPD|               3-Isopropylmalate Dehydrogenase...   +3    73   0.0044    1
pir|S41223|S41223       3-isopropylmalate dehydrogenase...   +3    73   0.0044    1
gp|U25634|AVU25634_3    ttuC gene product [Agrobacteriu...   +3    52   0.0044    2
sp|P05645|LEU3_BACSU    3-ISOPROPYLMALATE DEHYDROGENASE...   +3    73   0.0044    1
sp|P41560|IDH1_VIBA1    ISOCITRATE DEHYDROGENASE (NADP)...   +3    48   0.0071    2
pir|B49341|B49341       isocitrate dehydrogenase (NADP+...   +3    48   0.0071    2
pir|S43888|S43888       leuB protein - Neisseria lactam...   +3    71   0.0082    1
sp|Q00412|LEU3_SPIPL    3-ISOPROPYLMALATE DEHYDROGENASE...   +3    71   0.0084    1
pir|A44851|A44851       3-isopropylmalate dehydrogenase...   +3    71   0.0084    1
sp|P24015|LEU3_LEPIN    3-ISOPROPYLMALATE DEHYDROGENASE...   +3    71   0.0084    1
sp|P41019|LEU3_BACME    3-ISOPROPYLMALATE DEHYDROGENASE...   +3    71   0.0084    1
gp|U00022|U00022_23     pbpC [Mycobacterium leprae]         -2    32   0.029     3
sp|P28834|IDH1_YEAST    ISOCITRATE DEHYDROGENASE (NAD),...   +3    67   0.030     1
sp|P00351|LEU3_THETH    3-ISOPROPYLMALATE DEHYDROGENASE...   +3    64   0.076     1
```

Fig. 4. Comparison of the output from **blast** searches of nucleotide vs protein databases. (**A**) Example of the output from a **blast** search using the nucleotide sequence in Figure 1 against a nucleotide database. (**B**) Same sequence used to search a protein database with **blast**.

3.5.2. Retrieve E-mail Server

Complete sequence entries can be retrieved directly from the NCBI by text terms or accession numbers. An example of how to obtain the complete GenBank sequence entry for entry U07681 follows:

> **mail retrieve@ncbi.nlm.nih.gov**
> **DATALIB** genbank
> **BEGIN**
> **U07681**

Send an e-mail message containing the word **help** to **retrieve@ncbi.nlm.nih.gov** to obtain the complete help file.

3.5.3. Entrez

Entrez is a program created at the NCBI that allows query and retrieval of an integrated set of databases consisting of nucleotide, protein, structure, and reference databases. Entries can be retrieved based on a number of criteria, including accession number, text terms, author, taxonomy, and similarity to other entries in the databases. Entries retrieved by entrez are all cross-referenced, such that starting with the accession number of a sequence of interest, one can quickly obtain the corresponding nucleotide and protein sequence entries, as well as those of related entries, literature references, abstracts, and links to structural information when available. The entrez program is available by anonymous ftp from **ncbi.nlm.nih.gov** and is available for most major computing platforms. Use of the network version of this program requires preregistration with the NCBI. A WWW interface to network entrez is accessible to anyone at **http://ncbi.nlm.nih.gov**.

3.6. Multiple Alignment of Homologous Sequences

Once a sequence has been identified as encoding a member of a potential gene family, important information concerning gene structure and function can be extracted from a multiple alignment of related sequences. There are several approaches to multiple-sequence alignment. Only two are presented here: the automated method of GCG's **pileup** and the interactive method of **macaw**.

3.6.1. Pileup

A multiple-sequence alignment is produced by **pileup** using a progressive alignment algorithm in which the sequences are clustered by pairwise comparison. This alignment method works best if the sequences are very similar and because this method requires that all the sequences be aligned; it fails when one or more of the sequences are not true homologs. The alignment produced

```
            1                                                       50
IDH_BACSU   ..........  .......MAQ  GEKITVSNGV  LNVPNNPIIP  FIEGDGTGPD
IDH_ECOLI   ..........  MESKVVVPAQ  GKKITLQNGK  LNVPENPIIP  YIEGDGIGVD
IDH1_YEAST  ....MLNRTI  AKRTLATAAQ  AE....RTLP  KKYGGRFTVT  LIPGDGVGKE
IDH2_YEAST  MLRNTFFRNT  SRRFLATVKQ  PSIGRYTGKP  NPSTGKYTVS  FIEGDGIGPE
IDH_THETH   ..........  .MPLITTETG  ......KKMH  VLEDGRKLIT  VIPGDGIGPE

            51                                                      100
IDH_BACSU   IWNAASKVLE  AAVEKAYKGE  KKITWKEVYA  GEKAYNKTGE  ..WLPAETLD
IDH_ECOLI   VTPAMLKVVD  AAVEKAYKGE  RKISWMEIYT  GEKSTQVYGQ  DVWLPAETLD
IDH1_YEAST  ITDSVRTIFE  AE.......N  IPIDW.E..T  INIKQ...TD  HKEGVYEAVE
IDH2_YEAST  ISKSVKKIFS  AA.......N  VPIEW.E..S  CDVSPIFVNG  LTTIPDPAVQ
IDH_THETH   CVEATLKVLE  AA.......K  APLAY.EVRE  AGASVFRRGI  ASGVPQETIE

            101                                                     150
IDH_BACSU   VIREYFIAIK  GPLTTPVGG.  GIRSLNVALR  QELDLFVCLR  PVRYFTGVPS
IDH_ECOLI   LIREYRVAIK  GPLTTPVGG.  GIRSLNVALR  QELDLYICLR  PVRYYQGTPS
IDH1_YEAST  SLKRNKIGLK  GLWHTPADQT  GHGSLNVALR  KQLDIYANVA  LFKSLKGVKT
IDH2_YEAST  SITKNLVALK  GPLATPIGK.  GHRSLNLTLR  KTFGLFANVR  PAKSIEGFKT
IDH_THETH   SIRKTRVVLK  GPLETPVG.Y  GEKSANVTLR  KLFETYANVR  PVREFPNVPT

            151                                                     200
IDH_BACSU   PVKRPEDTDM  VIFRENTEDI  YAGIEYAKGS  EEVQKLISFL  QNELNVNKIR
IDH_ECOLI   PVKHPELTDM  VIFRENSEDI  YAGIEWKADS  ADAEKVIKFL  REEMGVKKIR
IDH1_YEAST  RIP..DI.DL  IVIRENTEGE  FSGLEHESVP  GVVE......  ..........
IDH2_YEAST  TYE..NV.DL  VLIRENTEGE  YSGIEHIVCP  GVVQ......  ..........
IDH_THETH   PYAGRGI.DL  VVVRENVEDL  YAGIEHMQTP  SVAQ......  ..........

            201                                                     250
IDH_BACSU   FPETSGIGIK  PVSEEGTSRL  VRAAIDYAIE  HGRKSVTLVH  KGNIMKFTEG
IDH_ECOLI   FPEHCGIGIK  PCSEEGTKRL  VRAAIEYAIA  NDRDSVTLVH  KGNIMKFTEG
IDH1_YEAST  .......SLK  VMTRPKTERI  ARFAFDFAKK  YNRKSVTAVH  KANIMKLGDG
IDH2_YEAST  .......SIK  LITRDASERV  IRYAFEYARA  IGRPRVIVVH  KSTIQRLADG
IDH_THETH   .......TLK  LISWKGSEKI  VRFAFELARA  EGRKKVHCAT  KSNIMKLAEG

            251                                                     300
IDH_BACSU   AFKNWGYELA  EKEYGDKVFT  WAQYDRIAEE  QGKDAANKAQ  SEAEAAGKII
IDH_ECOLI   AFKDWGYQLA  REEFGGELID  GGPWLKV...  ..........  KNPNTGKEIV
IDH1_YEAST  LFRNIITEIG  QKEYPD....  ..........  ..........  .......ID
IDH2_YEAST  LFVNVAKELS  .KEYPD....  ..........  ..........  .......LT
IDH_THETH   ..PKRAFEQV  AQEYPD....  ..........  ..........  .......IE

            301                                                     350
IDH_BACSU   IKDSIADIFL  QQILTRPNEF  ..DVVATMNL  NGDYISDALA  AQVGG.IGIA
IDH_ECOLI   IKDVIADAFL  QQILLRPAEY  ..DVIACMNL  NGDYISDALA  AQVGG.IGIA
IDH1_YEAST  VSSIIVDNAS  MQAVAKPHQF  ..DVLVTPSM  YGTILGNIGA  ALIGGP.GLV
IDH2_YEAST  LETELIDNSV  LKVVTNPSAY  TDAVSVCPNL  YGDILSDLNS  GLSAGSLGLT
IDH_THETH   AVHIIVDNAA  HQLVKRPEQF  ..EVIVTTNM  NGDILSDLTS  GLIGG.LGFA

            351                                                     400
IDH_BACSU   PGANINYETG  HAIFEAT.HG  TAPKYAGLDK  VNPSSVILSG  VLLLEHLGWN
IDH_ECOLI   PGANIGDEC.  .ALFEAT.HG  TAPKYAGQDK  VNPGSIILSA  EMMLRHMGWT
IDH1_YEAST  AGANFGRDY.  .AVFEPGSRH  VGLDIKGQNV  ANPTAMILSS  TLMLNHLGLN
IDH2_YEAST  PSANIGHKI.  .SIFEAV.HG  SAPDIAGQDK  ANPTALLLSS  VMMLNHMGLT
IDH_THETH   PSANIGNEV.  .AIFEAV.HG  SAPKYAGKNV  INPTAVLLSA  VMMLRYLEEF
```

Fig. 5. Example of an alignment produced by **pileup**.

by **pileup** should be inspected and manually edited, if necessary, before proceeding to use this alignment in any further analysis. Figure 5 demonstrates a portion of an alignment produced by **pileup**.

3.6.2. Macaw

The **macaw** program offers a different approach to multiple sequence *(4)*. This program runs on both Macintosh and Windows operating systems, and allows interactive alignment of multiple sequences. Regions of high similarity can be aligned with high stringency and locked into place. Then intervening regions can be aligned with reduced stringency, until the entire sequence is aligned. Sequences that do not have similarity in multiple segments are quickly identified, allowing the alignment to be properly interpreted.

3.7. Identification of Functional Domains

3.7.1. Motifs

Conserved protein sequence motifs, such as nucleotide binding sites and protein modification sites, which have been identified in the PROSITE database, can be identified using the **motifs** program in the GCG package. The PROSITE database also contains detailed documentation about the proteins containing each motif and a list of known false-positive and false-negative sequences.

3.7.2. ProfileScan

Multiple alignments of related protein sequences have been used to calculate tables in which a score is given for every amino acid at each position based on its frequency of occurrence *(9)*. Sets of these profile tables can be used by **profilescan** to search a protein sequence for regions of high similarity with these protein families.

3.7.3. ProfileSearch

Profiles can also be created from multiple-sequence alignments produced by **pileup** or **macaw**, and used to search the protein sequence databases for more homologs. These searches have the potential to be much more sensitive than **blast** or **fasta** searches, but take much longer to run. Ideally, a profile should be created using only the most conserved regions of the alignment. This profile can then be used by **profilesearch** to search the database, and the results can be aligned with the query profile by the program **profilesegments**. As with any of these searching techniques, this can be an iterative process, with more sequences added to the alignment with each pass.

3.8. Submission of Completed Sequences to Databanks

New sequence data should be submitted to the genetic databanks when it is complete. There is rapid sharing of the data by the International Collaboration of Nucleotide Sequence Databases, and it is therefore only necessary to submit to one database: GenBank, EMBL, or DDBJ. Deposition of sequence data is preferably done either through WWW servers: **http://www.ncbi.nlm.nih.gov**, **http://www.ebi.ac.uk**, or e-mailing a completed submission form to **gb-sub@ncbi. nlm.nih.gov**, **datasubs@ebi.ac.uk**, or **ddbjsub@ddbj.nig.ac.jp**. Most journals require that sequence data be deposited and an accession number be obtained before publication. It is possible to request that the data not be released until the article is published in the scientific press.

4. Notes
4.1. GCG Introduction

1. Logging on and starting GCG: Users should have an account established by their system administrator on a machine that is running the GCG software. The exact instructions for starting GCG vary by system and should be obtained from the administrator.
2. GCG help: There are many more programs and program options than can be presented in this limited format. Details of these can be found in the printed GCG manual or found online by typing **genmanual**. Alternatively, entering **genhelp** will bring up a help file in which each program is listed in alphabetical order. A list of command line options for each program can also be obtained by entering the program name followed by **-check**.
3. Program prompting: Most GCG programs need, and ask for, at least one input file name. In addition, many programs write their output into a file and ask for an output file name. Usually the program suggests a default answer by presenting it within parentheses and asterisks, for example (* filename.out *). One can accept the default answer by pressing return or type another value then press return. Most programs are insensitive to unexpected responses and simply repeat the question if something unacceptable is entered.
4. Command line structure: There are three types of parameters that can be specified on the command line when executing a GCG command.
   a. Required parameters: Items such as input sequence name, beginning and ending positions, and output file name must be supplied or else the program will pause to allow input of any missing parameters. Use of **-default** disables this interactive input and forces the use of the default program values for any option that is not specified.
   b. Local data files: There is the option of changing the data file that the program will use by specifying a new file on the command line or by having a file of the same name as the default file in the local directory. These include translation tables, sequence comparison tables, restriction enzyme lists, and so forth.
   c. Optional parameters: Most programs have several options that can only be set on the command line. These include such options as the output line width, limits on the lengths of results, and so on. A list of optional parameters for each program can be found by using **genhelp** or by using the **-check** command line qualifier.

Table 1
General File Utility Commands in VMS and UNIX Operating Systems

Function	VMS command	UNIX command
List a file	**type** <filename>	**cat** <filename>
Delete a file	**delete** <filename>	**rm** <filename>
Copy a file	**copy** <filename1> <filename2>	**cp** <filename1> <filename2>
Rename a file	**rename** <filename1> <filename2>	**mv** <filename1> <filename2>

5. Graphic output: GCG programs support an array of graphic output drivers, including postscript, HPGL, tektronics, and XWindows. Users should see Appendix C of the Wisconsin Sequence Analysis Package for details.
6. Customizing GCG data files: GCG data files including restriction enzyme lists, and translation tables can be retrieved and edited. Type **fetch** <filename>, and a local copy of the file will be retrieved to the local directory. The local copies of these files will then be used as default.

4.2. UNIX vs VMS

7. GCG command differences: There are several major differences between the VMS and UNIX operating systems that are of concern to users of the GCG programs. The GCG command names are identical in both systems, but commands may be abbreviated and file names are case insensitive in VMS, whereas not in UNIX. A slash, "/" is used to identify a command-line option in VMS, whereas in UNIX, a dash, "–", is used. Commands may be extended over multiple lines by entering a "–" at the end of each line in VMS or a "\" in UNIX.
8. File utilities: File names in VMS follow the pattern DISK:[Dir]Filename.txt, whereas in UNIX, the pattern is /dir/filename.txt. *See* Table 1 for other general file utility commands.

4.3. Sequence Contamination

9. Vector sequence contamination: Unfortunately, a large number of sequences that are deposited into sequence databases either represent portions of the cloning vector or are contaminated with portions of the cloning vector sequence. A blast search performed at the earliest stages of the sequence will usually indicate whether the sequence is from a potentially interesting clone or from a cloning artifact. Precautions should be taken in the final stages of preparing the sequence to eliminate the deposition of any extraneous sequences into the databases.
10. Degenerate primer sequence contamination: The sequences that are complementary to degenerate primers used to amplify the DNA prior to cloning should be removed from the sequence prior to performing any sequence comparison search. These sequences, which often represent conserved regions and thus have a high likelihood of finding a match in the database, are not necessarily part of the origi-

nal coding sequence. These primer sequences should also be removed from the sequence before deposition into the sequence databases.

4.4. Sequence Searching Over the Internet

11. A number of resources are available for searching the sequence databases over the internet. Gateways to various search engines can be found at **http:// www.ncbi.nlm.nih.gov, http://www.ebi.ac.uk, http://www.blocks.fhcrc.org, http://dot.imgen.bcm.tmc.edu:9331.**

References

1. Genetics Computer Group (1994) *Program Manual for the Wisconsin Package, Version 8.* Madison, WI.
2. Griffin, H. G. and Griffin, A. M., eds. (1994) *Methods in Molecular Biology, vol. 24, Computer Analysis of Sequence Data, parts I and II*, Humana Press, Totowa, NJ.
3. Altschul, S. F., Boguski, M. S., Gish, W., and Wootton, J. C. (1994) Issues in searching molecular sequence databases. *Nat. Genet.* **6,** 119–129.
4. Schuler, G. D., Altschul, S. F., and Lipman, D. J. (1991) A workbench for multiple alignment construction and analysis. *Proteins: Struct. Funct., and Genetics* **9,** 180–190.
5. Gribskov, M., Devereux, J., and Burgess, R. R. (1984) The codon preference plot: graphic analysis of protein coding sequences and prediction of gene expression. *Nucleic Acids Res.* **12,** 539–549.
6. Fickett, J. W. (1982) Recognition of protein coding regions in DNA sequences. *Nucleic Acids Res.* **10,** 5303–5318.
7. Altschul, S. F., Gish, W., Miller, W., Myers, E. W., and Lipman, D. J. (1990) Basic local alignment search tool. *J. Mol. Biol.* **213,** 403–410.
8. Pearson, W. R. and Miller, W. (1992) Dynamic programming algorithms for biological sequence comparison. *Methods Enzymol.* **210,** 575–601.
9. Gribskov, M., Luethy, R., and Eisenberg, D. (1989) Profile analysis. *Methods Enzymol.* **183,** 146–159.

21

The TIGR Human cDNA Database

Ewen F. Kirkness and Anthony R. Kerlavage

1. Introduction

The Human cDNA Database (HCD) is a repository for human cDNA sequences and related data that is curated and maintained at The Institute for Genomic Research (TIGR). The foundation of the database is ~160,000 partial cDNA sequences that have been generated by TIGR and Human Genome Sciences (HGS). These expressed sequence tags (ESTs) were derived from 250 cDNA libraries that represent the expressed genes of 37 distinct human organs and tissues *(1)*. A combination of these ESTs and the human ESTs from dbEST has been assembled to yield ~56,000 consensus sequences and ~106,000 nonoverlapping ESTs. Of these ~162,000 distinct sequences, approx 18% display statistically significant similarity to previously known genes, whereas the remainder identify previously unknown cDNAs. The HCD data include nucleotide sequences, putative identifications of the sequences, and tissue-based expression information. New EST sequences that are acquired by TIGR are assembled and curated in an ongoing process that aims to provide a comprehensive database of human genes and their expression patterns. The development of HCD required the construction of a sister database, the Expressed Gene Anatomy Database (EGAD). This database contains a nonredundant dataset of human transcript sequences, together with information on expression patterns and cellular roles. Extensive links between HCD and EGAD permit browsing of sequence information, functional classifications, and tissue expression patterns for particular genes of interest. Owing to the random nature of large-scale EST sampling and the wide range of tissues that have been sampled, the data that can be extracted from HCD are of particular value for several common types of cDNA sequence analyses.

From: *Methods in Molecular Biology, Vol. 69: cDNA Library Protocols*
Edited by: I. G. Cowell and C. A. Austin Humana Press Inc., Totowa, NJ

1.1. Identification of Isologs

Gene isologs display nonexact, but significant, sequence homology with previously identified sequences. Since this homology is often only apparent at the level of protein sequence comparisons, HCD can reveal novel members of gene families that have not been detected previously by cross-hybridization of DNA sequences. Where possible, sequences in HCD have been assigned a putative identification by comparison to sequences in other nucleotide and peptide databases (e.g., GenBank and SwissProt). Investigators who wish to identify isologs of a specific gene family can query HCD by gene name. Alternatively, the sequence of a known gene or a minimal protein consensus sequence can be used to search six-frame translations of the sequences within HCD.

1.2. Tissue Expression Data

A random sampling of ESTs reflects the mRNA content of the source cells at the time of library construction. This expression information is a valuable component of the data that are assigned to each EST in HCD. Although the absence of a particular cDNA from a library cannot prove that a gene is not transcribed in the source material, the presence of an EST indicates that the tissue was expressing the corresponding transcript. The wide diversity of tissues and cell types that have been used for EST sampling at TIGR provides a good starting point to explore the expression patterns of specific cDNAs. This can be particularly informative when searching for genes within large stretches of sequenced genomic DNA (e.g., cosmid clones of disease loci). In addition, when multiple ESTs are derived from the same gene, the source of the EST can often reveal tissue-specific alternative splicing of the transcript.

1.3. Identification of Novel Genes

The majority of sequences in HCD display no homology with previously identified genes, and therefore, represent novel genes or new portions of known genes. Since many of these sequences contain large open reading frames, HCD provides a useful resource for investigators who wish to characterize cDNAs for which they have obtained only partial sequences (e.g., by exon trapping, differential display, or peptide sequencing). By searching HCD with a query sequence, it is possible to quickly obtain flanking or full-length cDNA sequences, together with information on where the relevant transcript is expressed.

2. HCD Data

2.1. Data Acquisition and Processing

With the aim of obtaining ESTs from the majority of human genes, TIGR and HGS have constructed approx 250 cDNA libraries from a variety of human tissues, tumor samples, cultured primary cells, and immortalized cell lines. The majority of these libraries were constructed directionally in the λ ZAP II vector (*2*; Stratagene, La Jolla, CA). Directional cloning simplified subsequent sequence analysis and permitted the majority of sequencing to be performed at the 5'-end of each clone. This provided the greatest likelihood of obtaining protein coding sequence and, hence, the best chance of assigning a putative identification to unknown cDNAs. The phage libraries were subsequently converted to a phagemid form, and individual clones were selected randomly for sequencing of their cDNA inserts. The sequencing reactions were analyzed using AB 373 DNA Sequencers.

The accuracy of EST sequencing at TIGR was controlled at two levels. First, the output from the sequencers was scanned visually to assess the general quality of the base calls. Subsequently, a series of software programs were used to assign "trash codes" to sequences that did not meet the minimum standards of accuracy (e.g., owing to low template concentration, mixed templates, or long poly-A sequences). Approximately 40% of the sequencing reactions were rejected because they did not meet TIGR's minimum standards of sequence quality. The remainder displayed >98% sequencing accuracy when compared to known sequences in GenBank.

Owing to the random nature of EST sequencing, abundant mRNAs in a particular library may be represented by many ESTs. These ESTs can therefore be treated as shotgun fragments for assembly of longer contiguous segments of sequence (contigs). At TIGR, such cDNA contigs have been assembled using stringent overlap criteria (*3*) and have been termed Tentative Human Consensus sequences (THCs). In addition to reducing the redundancy of the EST dataset, the process of assembling THCs has aided the assignment of putative identifications. Furthermore, the assembly process has helped to identify examples of mRNAs that undergo alternative splicing. THCs are derived from multiple independent clones and, therefore, have the potential to be chimeric. However, sufficient redundancy can ensure confidence in the accuracy of the assembled consensus sequence.

Having developed the appropriate software tools for manipulation of EST data, it has been possible for TIGR to incorporate all available data from dbEST into the analysis. This provides additional depth to the THC assemblies and more extensive expression data for each transcript. A semiautomated procedure has been developed to download daily updates from dbEST. Sequences are trimmed to remove ambiguous bases, and are checked for vector, polyA, or other sequence artifacts. After removal of ESTs that are derived from mitochondrial transcripts or ribosomal RNA, the sequences are finally subjected to the assembly process. Currently, it has been possible to assemble approx 40,000 THCs with the cluster size ranging from 2–2780 ESTs. The average length of the THCs is 492 bases, compared to 288 bases for individual ESTs.

In order to assign putative identifications to EST and THC sequences, a procedure has been developed that relies on both automatic processing of results and the evaluation of sequence comparisons by scientists at TIGR. Initially, sequences are prescreened by searching them against a nonredundant database of human transcript sequences (*see* Section 3.) using a modified FASTA *(4)* algorithm. This step uses a minimum match length of 50 bp and a minimum percent match of 97% over the aligned region. Each match is also evaluated automatically for evidence of alternative splicing or chimerism. For most cDNA libraries, 30–50% of the ESTs can be annotated automatically by this prescreen step. The ESTs and THCs that do not display matches after the automated annotation phase are searched against all nucleotide sequences in GenBank using a procedure that will produce optimal gapped alignment between two similar sequences *(1)*. Peptide searches are also performed, using all six possible translations of the THC or EST, against a composite database of sequences from GenPept, PIR, and SwissProt. The results of the nucleotide and peptide database searches are then inspected individually by scientists. If a match to a database sequence is considered to be significant, the match information and putative identification are saved. These putative identifications permit users of HCD to browse through the database for sequences of interest.

2.2. Accessing HCD

The information in HCD is available at two levels. The majority of the data (>87%), including all data that have been incorporated into THCs from dbEST, are in Level 1. Access to these data is available to all researchers at universities or other nonprofit research institutions on signature of a waiver of liability agreement. There are no intellectual property, confidentiality, or publication restrictions on the use of these data. The remainder of the data are available at Level 2, and can be accessed by researchers at nonprofit research institutions on signature of an agreement that grants HGS the option to negotiate a license to commercialize potential products that derive from the use of these data. The cDNA clones from which the TIGR and HGS sequences have been derived are available through the American Type Culture Collection (ATCC) as part of the TIGR/ATCC Human cDNA Special Collection. Level 1 clones can be obtained without intellectual property restrictions by all academic scientists. The Level 2 database agreement also permits access to Level 2 clones.

The data in HCD can be searched by browsing putative gene identifications or specific sequence identifiers (*see* Tables 9 and 10 of ref. *1*). Alternatively, users at both levels may submit a nucleotide or peptide sequence to search against the sequences in HCD. Searches are performed against six-frame translations of the HCD sequences using a modification of the Smith and Waterman algorithm *(5)* with frame-shift detection to permit more sensitive comparisons. Access to Level 1 data is by e-mail, or through the World Wide Web (WWW) at TIGR's WWW site (URL:http://www.tigr.org/tdb/hcd/ hcd.html). Access to Level 2 data is by e-mail only (information can be obtained from info@hcd.tigr.org). As an example of HCD output, a user could search HCD with short unidentified cDNA fragments that have been obtained from a differential display analysis *(6)*. The ESTs and THCs that are exact matches to the query sequences would provide additional sequence information that is required for identification of the fragment. These HCD sequences may represent novel cDNAs or longer versions of known cDNAs. The user may then obtain a report on each HCD sequence, as illustrated in Fig. 1. The report contains a putative identification, sequence data, and expression information that would be pertinent to the characterization of the cDNA in question.

3. EGAD

In order to permit consistent annotation of the cDNA sequences in HCD, a nonredundant set of human transcript sequences was constructed in EGAD. A combination of genomic, cDNA, and EST sequences, together with related information, has been extracted from GenBank and transformed into a format that can be accessed automatically for computational analyses.

In GenBank, a specific gene is often described by multiple entries (e.g., cDNAs of variable length, individual exons, or large stretches of genomic sequence). These data have been consolidated in EGAD by linking all GenBank accessions that describe the same gene. Transcripts that are alternatively spliced are represented individually and linked together with their gene of origin. In order to populate the database, candidate sequences were first searched against all human accessions in GenBank, using the BLAST algorithm *(7)*. GenBank sequences that displayed >98% similarity for >100 nucleotides were aligned, and the consensus sequence together with pointers to each component sequence were stored in EGAD. To date, over 31,000 GenBank accessions of human sequences have been linked to create approx 4400 EGAD transcripts. The dataset is updated with each new release of GenBank.

In addition to sequence data, EGAD provides information on the tissue expression patterns of the transcripts. This includes the relevant data that are recorded in GenBank accessions, together with information assigned by searching the transcripts against HCD sequences. Searches that result in an exact match indicate that the EGAD sequence is expressed in the tissue source of the cDNA library from which the matching EST was derived. As an additional aid to computational analyses of human genes, the transcript sequences in EGAD have also been assigned biological roles. A range of broad categories (e.g., cell signaling) each contain several subcategories (e.g., receptors) to which transcript sequences have been linked. At this time, the classification system is

```
>THC103362
TGGAGCTGTTGTTTTGTATGCTCAGCGAGGCCCGGAGAGACCCGGGAGAGAGCTAGGCCGAGTCCACCGCCCGAnTCTGC
TGCCCGAGCCCGCGTTACGCACAAAGCCGCCGATCCCCGGCCTGGGGTGAGCAGAGCAnCCACCGCCCGGGAGCAGCGCG
aCGAGACGCACGGTGCGCCCTATGCCCCCGCGGCCCCACCGCCCCCGCCGCGGCAnCGAAGCGCAGnGAGAACGCGCCAC
CGGGGGCCCGGGTGCAGCTAGCGACCCTCTCGCCACCTGCGCGCAGCCGAGGTGAGCAGTGAGCGGCGAGCGGAGGGCAG
CGAGGCGTTCGCGGGCCCCCTCCTGCTGCCCGGGCCCGGCCCGCTCATGGCGGCCATCCGCAAGAAGCTGGTGGTGGTGG
GCGACGGCGCGTGTGGCAAGACGTGCCTGCTGATCGTkTTCAGTAAGGACGAGTTCCCCGAGGTGTACGTGCCCACCGTC
TTCGAGAACTATGTGGCCGACATTGAGGTGGACGGCAAGCAGGTGgAGcTGGCGCTGTGGGACACGGCGGGCCAGGAGGA
CTACGACCGCCTGCgGCCGCTCTCCTACCCGGACACCGACGTCATTCTCATGTGCTTCTCGGTGGACAGCCCGGACTCGC
TGGAGAACATCCCCGAGAAGTGGGTCCCCGAGGTGAAGCACTTCTGTCCCAATGTGCCCATCATCCTGGTGGCCAACAAA
AAAGACCTGCGCAsGACGAGCATGTCCGCACAGAGCTGGCCCGCATGAAGCAGGAACCCGTGCGCACGGATGACGGCCGC
GCCATGGCCGTGCGCATCCAAGCCTACGACTACCTCGAGTGCTCTGCCAAGACCAAGGAAGGCGTGCGCGAGGTCTTCGA
GACGGCCACGCGCGcGyCGsgyTGCAGAAGCGCTACGGCTCCCAGAACGGCTkcATCAAcTGCTgCAAgGTGCTATGAGGGC
CGCGGCCGTTTGAGGCTnCCCCTGCCGGCACGGTTTCCCnTTCTTGGGACCAnTT
```

Putative ID: guanine nucleotide-binding protein rhoB

```
1==============================THC103362============================1015
-------1------>                <-------3-------- --------------14------------->
         -----------2------------->       --------------7------------->
                  ----------------------4---------------------->
                  <----5----- -------------8------------->
                  -----------6----------->
                           -----------------9----------------->
                           ----------10---------->
                           ----------11---------->
                           ----------12----------->
                           -----------13----------->
                           ----------15--------->
                           ---16-->
                           ---------17-------->
```

#	EST#		GB#	ATCC#	left	right	library
1	D		D31039		1	187	Lung
2	C	EST00363	M62289	37988	148	504	Brain
3	F		R21827		364	572	Placenta
4	E	HT1456			367	956	
5	F		R48827		383	522	Breast
6	A	EST55861			401	721	Adrenal gland
7	A	EST58502	T28876	104271	570	930	Brain
8	H		T15801		570	949	Brain
9	H		T03506	86139	572	1015	Brain
10	C	EST07832		85029	580	885	Brain
11	C	EST07982		85851	580	862	Brain
12	F		R50747		581	897	Breast
13	A	EST59362			586	916	Brain
14	F		R43032		596	964	Brain
15	A	EST56342			602	869	Brain
16	B	EST126303			632	733	Heart
17	F		H01643		654	897	Placenta

Sequence source codes:
D = GENBANK, C = NIH, F = WashU/Merck, E = EGAD, A = TIGR, H = U Colorado, B = HGS

Fig. 1. A THC Report from the HCD. The report consists of the THC sequence, its
putative identification, and information about the component ESTs. The ESTs are represented both pictorially and by their coordinates on the consensus sequence. The tissue
source of each EST together with the laboratory from which it was derived are also indicated. In the WWW representation of this report, each EST# is linked to a report in HCD
that is specific for each EST. Each human transcript (HT)# is linked to an HT report in
EGAD, whereas GenBank accession numbers are linked to a record in the Genome
Sequence DataBase (GSDB) at the National Center for Genome Resources. When available, ATCC accession numbers are linked to information in the ATCC Repository.

EGAD Report: HT3261

HT DATA

EGAD Gene ID:	HG3088
Sequence name:	splicing factor SC35, alt splice form 1
Genome:	nucleus
Coding sequence length:	666 nt
Transcript sequence length:	1879 nt
HCD Expression data:	THC87037

ACCESSION DATA
HT3261 is derived from accesssions(s):
GB:M90104 (Human splicing factor SC35 mRNA, complete cds.)

ALTERNATIVE SPLICE INFORMATION
Alternative splice forms for this gene:
HT3262 splicing factor SC35, alt splice form 2
HT3263 splicing factor SC35, alt splice form 3
HT3264 splicing factor SC35, alt splice form 4

MAPPING DATA
GDB accession(s) for this gene:
GDB ID: Symbol

G00-132-412   SFRS2

SEQUENCE
nucleotide:
cgctagcctgcggagcccgtccgtgctgttctgcggcaaggcctttcccagtgtccccac
gcggaaggcaactgcctgaaaggcgcggcgtcgcaccgcccagagctgaggaagccggcg
ccagttcgcggggctccgggccgccactcagagctatgagctacggccgcccccctcccg
atgtggagggtatgacctccctcaaggtggacaacctgacctaccgcacctcgcccgaca
cgctgaggcgcgtcttcgagaagtacaggcgcgtcggcgacgtgtacatcccgcgggatc
.....aaaaaaaaaaaaaaaaa

protein:
MSYGRPPPDVEGMTSLKVDNLTYRTSPDTLRRVFEKYRRVGDVYIPRDRYTKESRGFAFV
RFHDKRDAEDAMDAMDGAVLDGRELRVQMARYGRPPDSHHSRRGPPPRRYGGGGYGRRSR
SPRRRRRSRSRSRSRSRSRSRSRSRYSRSKSRSRTRSRSRSTSKSRSARRSKSKSSSVSRSR
SRSRSRSRSRSPPPVSKRESKSRSRSKSPPKSPEEEGAVSS

Fig. 2. A Human Transcript report from the EGAD. The report provides the name of the transcript, its length, the length of the coding region, and the nucleotide and translated protein sequences. Note that the nucleotide sequence has been shortened for this example. Links are available to expression data (in HCD), the GenBank accessions that were used to derive the transcript (in GSDB), alternatively spliced forms of the same gene (in EGAD), and mapping data (in Genome DataBase; GDB).

necessarily broad and flexible in order to permit the incorporation of new biological information.

An example of an EGAD report is illustrated in Fig. 2. A user, with an interest in RNA processing, could query the role categories of EGAD and obtain a report on splicing factor SC35 from the list of relevant genes. They may then link directly to a THC report for information on expression patterns, or to relevant splicing and mapping data. EGAD is accessible at TIGR's WWW site (URL:http://www.tigr.org/tdb/egad/egad.html), and can be searched by transcript accession number, gene name, or biological role.

References

1. Adams, M. D., Kerlavage, A. R., Fleischmann, R. D., Fuldner, R. A., Bult, C. J., Lee, N. H., et al. (1995) Initial assessment of human gene diversity and expression patterns based upon 83 million nucleotides of cDNA sequence. *Nature* **377(Suppl.),** 3–174.
2. Short, J. M., Fernandez, J. M., Sorge, J. A., and Huse, W. D. (1988) Lambda ZAP: a bacteriophage lambda expression vector with *in vivo* excision properties. *Nucleic Acids Res.* **16,** 7583–7600.
3. Fleischmann, R. D., Adams, M. D., White, O., Clayton, R. A., Kirkness, E. F., Kerlavage, A. R., et al. (1995) Whole-genome random sequencing and assembly of *Haemophilus influenzae* Rd. *Science* **269,** 496–512.
4. Pearson, W. and Lipman, D. (1988) Improved tools for biological sequence comparison. *Proc. Natl. Acad. Sci. USA* **85,** 2444–2448.
5. Smith, T. and Waterman, M. J. (1981) Identification of common molecular subsequences. *J. Mol. Biol.* **147,** 195–197.
6. Liang, P. and Pardee, A. B. (1992) Differential display of eukaryotic messenger RNA by means of the polymerase chain reaction. *Science* **257,** 967–971.
7. Altschul, S. F., Gish, W., Miller, W., Myers, E. W. and Lipman, D. J. (1990) Basic local alignment search tool. *J. Mol. Biol.* **215,** 403–410.

22

Searching the dbEST Database

Patricia Rodriguez-Tomé

1. Introduction

The accumulation and analysis of expressed sequence tags (ESTs) are now an important component of genome research *(1)*. An EST is a fragment of a sequence from a cDNA clone that corresponds to an mRNA. ESTs have applications in the discovery of new human genes, mapping of the human genome, and identification of coding regions in genomic sequences. Although the EST sequences are available in the EST section of both EMBL *(2)* and GenBank *(3)* nucleotide sequence databases, the National Center for Biotechnology Information (NCBI; National Institutes of Health, Bethesda, MD) has created a specialized database in which specific information related to these data can be found. This chapter describes this database and the different ways it can be accessed.

2. The Database

The dbEST database *(4)* is built on a relational database system. Data can be directly submitted to the database using a tagged "flat-file" format (*see* Note 1). The dbEST team at the NCBI also scans the daily updates of the nucleotide databases (EMBL/GenBank/DDBJ). Either way, the entries are assigned a specific dbEST accession number. Those received by direct submission to dbEST are given a GenBank accession number and included in the EST section of the nucleotide databases. These data are available immediately for public access, unless the submitters specify that they should be kept confidential until publication.

The public data are formatted in report files (*see* Note 2) that contain complete annotation for the sequences, including publication information (if any), contributor names and addresses, clone library information, organism and tis-

From: *Methods in Molecular Biology, Vol. 69: cDNA Library Protocols*
Edited by: I. G. Cowell and C. A. Austin Humana Press Inc., Totowa, NJ

sue description, mapping information, American Type Culture Collection (ATCC) identification numbers (if any), putative homologies assigned by the contributor as well as the result of homology searches against GenBank, and protein sequences, generated at the NCBI. A full report of the database is made available weekly and a FASTA format file (i.e., the sequences with descriptive header lines as described in ref. *5*) is placed in the NCBI Data repository. Files containing daily updates of the database are also placed in this repository.

3. Database Access

The dbEST output reports are mirrored from the NCBI at the European Molecular Biology Outstation in Hinxton, UK (the European Bioinformatics Institute [EBI]), which means that both sites contain the same files in their respective dbEST repository. You should access the site closest to you or the one with which your network connection is the most satisfactory.

Three methods of network access are provided by the NCBI and the EBI. These are the most popular ways of accessing information over computer networks. Therefore, there are a large number of different programs that can access those services. It is beyond the scope of this chapter to provide details of how to use all of these programs. Instead, this chapter provides the basic information that is necessary for navigating and accessing those services. The route that you choose will be dependent on the software that is available on your computer system. This chapter gives a general description of each of these tools, and then for each of the two sites, their specific use.

Examples of each of the tools include:

1. File transfer protocol (FTP): ftp terminal version, xftp for X-Window, winftp for windows, Fetch for Macintosh.
2. World Wide Web (WWW): Mosaic, Netscape, and Lynx run on most computers.
3. E-mail: Pine, Elm.

Most of the tools provide documentation for their use, windowing versions normally have a help button, and the terminal version usually provides a help command. Direct Internet access is required to use the FTP and WWW protocols.

3.1. E-Mail

The EST sequences can be searched using the BLAST *(6)* electronic mail server. For more information on how to use this server, you should send an e-mail message to **blast@ncbi.nlm.nih.gov** with the word "help" in the body of the message. This will return the most up-to-date help file.

The full reports on ESTs can be retrieved from the NCBI mail server at the address **retrieve@ncbi.nlm.nih.gov**. The first line in the body of the message must specify the database name. In this case it should read `datalib dbest`. The following lines will specify which keywords and parameters are being used to retrieve the data. The line containing the query parameters must begin with the keyword `RPT`. You can retrieve multiple entries by including their ID numbers on the same line, separating each one by a blank character.

Here are some examples of retrieving data:

```
%> mail retrieve@ncbi.nlm.nih.gov
Subject:
datalib dbest
RPT 556 22239 5476
```

This message will retrieve three different ESTs from dbEST. If you wish to use their GenBank accession number instead, you would write:

```
%> mail retrieve@ncbi.nlm.nih.gov
Subject:
datalib dbest
datatype GenBank
RPT M78440 T03211 Z14454
```

We have introduced in this example a new keyword, `datatype`. With this keyword you can specify the fields for querying the database. The values for these fields will be in a line beginning with `RPT`. The following fields can be accessed, in a line beginning with `datatype`:

NCBI	the identifier assigned by the NCBI
Source	the name given by the contributor
ATCC_H	the ATCC inhost number
ATCC _D	the ATCC pure DNA number
GenBank	the GenBank accession number
Clone	the clone id
Chrom	the chromosome
Mapstr	the mapping information

Another keyword, **org** (for organism), should appear in a line by itself, followed by the organism name. It is mandatory to use this keyword if either the datatype Chrom or Mapstr is used.

3.2. FTP

This is the main route for retrieving the full (or daily) reports file from the FTP archives. An FTP archive is simply a directory hierarchy that is made accessible to any anonymous user. Using FTP, you can navigate through directories and copy the files of interest to your computer.

3.2.1. A Basic ftp Session

Connect to the FTP server (for NCBI **ncbi.nlm.nih.gov** and for EBI **ftp.ebi. ac.uk**). Type anonymous at the login prompt, and enter your e-mail address as the password (e.g., john.doe@life.univ). This will take you to the top directory of the file archive. Since most of the large files are compressed tar files, it is a good idea to set the ftp file transfer to binary at this stage. The archive of databases is kept in different directories on both sites: **repository/dbEST** at the NCBI and **pub/databases/dbEST** at the EBI. Be warned: some of the files are very large. Change into the directory that has the dbEST database. If you are transferring large files using the terminal version of ftp, it is useful to set the hash mark (#) on, since this will give you a visual indication regarding how the transfer is progressing (each hash mark represents 1 kb of data being transferred). You should first transfer the README file, which contains additional information about the files in the current directory. Transfer the reports files you require. An example session is given in Fig. 1, using the EBI anonymous FTP server. Since the files are tape archives files (.tar) and compressed (.Z), you will need to uncompress them before dearchiving. The name of the programs to do this will depend on your local system.

3.3. WWW

This part focuses on access to the database using the WWW protocol. The system relies on software that is becoming commonplace in most laboratories (WWW browsers) and should be readily available on your system. The most common browsers (browsers are computer programs for navigating information that is available over computer networks) for both PCs and Macintoshs can be downloaded from the EBI ftp server **ftp://ftp.ebi.ac.uk/pub/software/ tools** (*see* Note 2). Start your WWW browser, and connect to the site you have chosen.

3.3.1. The NCBI WWW Server

The address of the NCBI WWW server that you must provide to your browser is **http://www.ncbi.nlm.nih.gov/**. Your browser is now connected to the "home page" of this site. Follow the appropriate links from this page to the access and query page for dbEST. In the list entitled "Other NCBI Resources" follow the link (i.e., click on the underlined words) dbEST: Database of Expressed Sequence Tags. You should now be at the dbEST page, with up-to-date information on the current dbEST release. Select the Search dbEST link from the list of options on this page. At the NCBI, the reports are indexed using the IRX software, a retrieval system that attempts to find the set of documents that best matches a user's question. It allows a question to be entered in plain

```
% ftp ftp.ebi.ac.uk
Connected to mercury.ebi.ac.uk.
220 mercury FTP server (Version wu-2.4(6) Thu Dec 15 15:22:30 GMT
1994) ready.
Name (ftp.ebi.ac.uk:doe): anonymous
331 Guest login ok, send your complete e-mail address as password.
Password: john.doe@life.univ
< Lines of login information deleted >
Remote system type is UNIX.
Using binary mode to transfer files.
ftp> binary
200 Type set to I.
ftp> cd pub/databases/dbEST
250-Please read the file README
250- it was last modified on Fri Nov 25 01:27:42 1994 - 243 days ago
250 CWD command successful.
ftp> hash
Hash mark printing on (1024 byes/hash mark).
ftp> get README
200 PORT command successful.
150 Opening BINARY mode data connection for README (11261
bytes)
#########.
226 Transfer complete.
< Lines of transfer information deleted >
ftp> quit
%
```

Fig. 1. Example of an ftp session to retrieve the README file.

English, but best results are obtained by querying the specific fields that have been indexed, and combining them using boolean operators. More information on IRX is available (follow the link here; *see* Fig. 2), but the following information will allow you to form simple queries.

3.3.1.1. THE FIELDS

Any term in a query can be restricted to a field using the form:

```
term [field list]
```

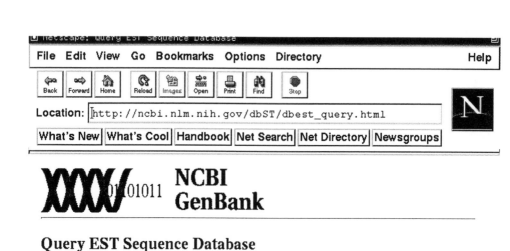

Query EST Sequence Database

This form allows users to enter ad-hoc *IRX* queries against the *dbEST* database. For information on how to formulate *IRX* queries for use against *dbEST*, please look here.

Fill out the form below and press "Submit Query" button to access *dbEST*.

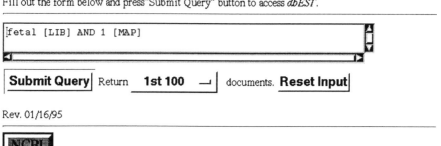

Rev. 01/16/95

Document: Done.

Fig. 2. The NCBI IRX query form.

The fields indexed in dbEST are:

NC	EST id number
ID	EST name, GenBank name, clone name
LIB	Library name and organism
DES	Library description
END	Putative protein identification, GDB D_segment number, citation info
SRC	Submitter and contact information
MAP	Mapping information
HOM	Homology information

3.3.1.2. BOOLEAN QUERIES

The Boolean operators supported are AND, OR, and NOT. They must be used in capital letters.

3.3.1.3. EXAMPLE

Figure 2 shows an example of a query using the fields LIB and MAP. The result of this query will give all the entries in which the number 1 appears anywhere in the mapping information (chromosome, physical map, and genetic map). You will then have to find the relevant entries in the list presented in Fig. 3.

3.3.2. The EBI WWW Server

If you connect your WWW browser to the home page of the EBI (**http://www.ebi.ac.uk**) you will be able to access the dbEST reports indexed using the Sequence Retrieval System (SRS) *(7)*, which allows entries to be retrieved based on a number of query forms. In this case follow the link **Database query/retrieval** and then **dbEST and dbSTS**. You will now have a page similar to the NCBI dbEST page. Follow the link (**search dbEST and dbSTS**) to the page listing all the databases available at the EBI for SRS searches. The dbEST button is one of the last in this page, and it leads you to the specific page for searching dbEST. You can search any combination of the four databases highlighted at the top of the form (*see* Fig. 4). Each database name is a link to the explanation on how the database is indexed at the EBI. All the fields in the dbEST report file are indexed and may be searched. This form allows the user to formulate simple queries, including wild cards and boolean operators (AND, OR, and BUTNOT). You can also choose which fields to include in the result (by default, all the fields are included). The result of each query is returned as a list of hits linked to the specific entries (*see* Fig. 5, p. 278). In each entry (*see* Fig. 6, p. 279), links to other entries in the various databases are highlighted as hypertext links and may also be traversed. To obtain related help on any part of the forms, just select the associated highlighted **?!**.

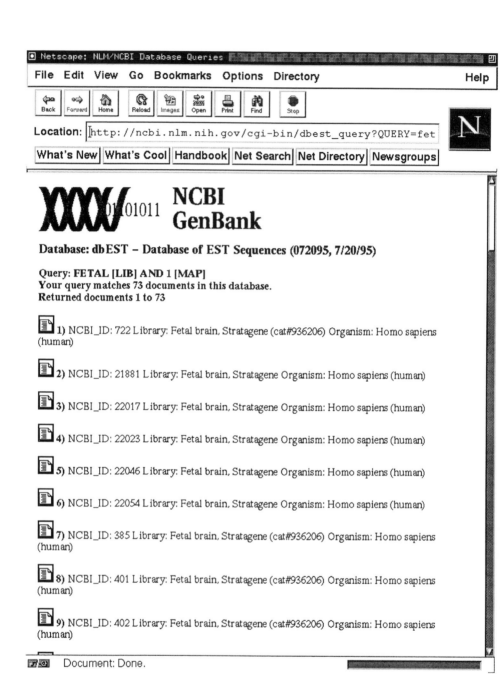

Fig. 3. The list of entries returned by the IRX query.

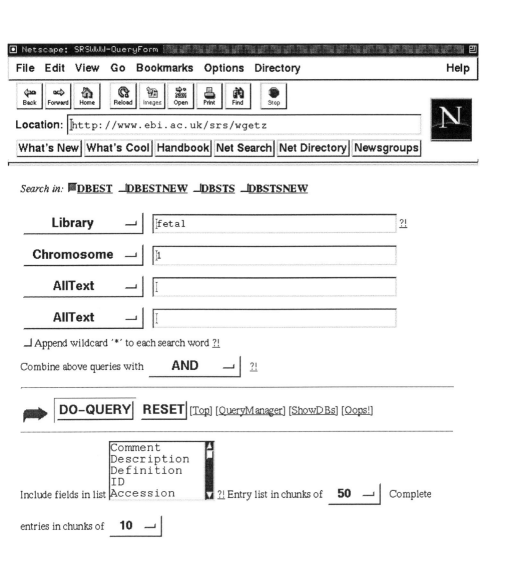

Fig. 4. The SRS query form at the EBI.

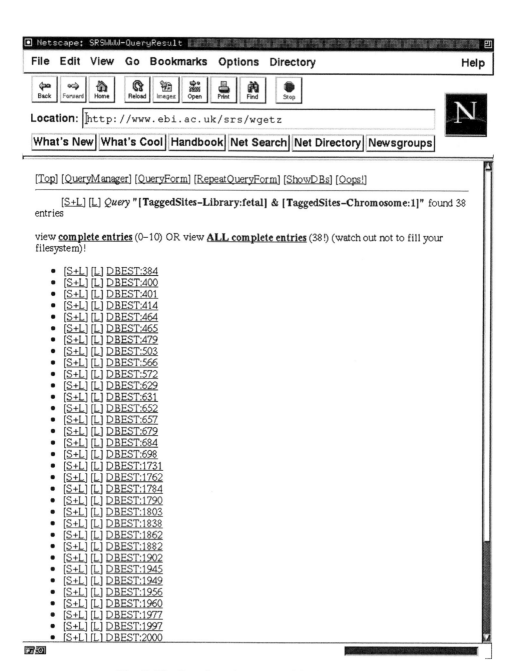

Fig. 5. The list of entries returned by the SRS query.

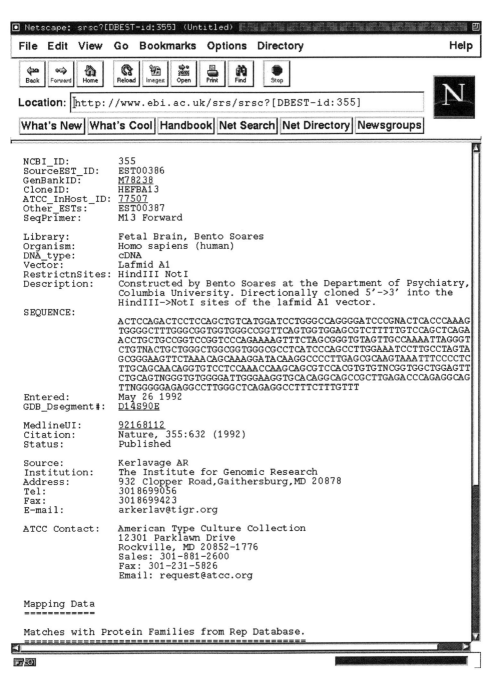

Fig. 6. Example of an entry with hypertext links.

279

If you need a more complex query combining the boolean operators, choose the option **expression** (the last of the boolean options in Fig. 4) and run your query. You will obtain the same form on the next page (*see* Fig. 7), but slightly expanded. Each input field is now associated with a query name (e.g., "Q1," "Q2"). You can use these names to write an expression into the newly added expression input field using the boolean expression tags: **&** for AND, | for OR, and **!** for BUTNOT.

Following the link to the **QueryManager** you obtain the list of the your different queries during this session (*see* Fig. 8). You can now rerun any of these queries or combine them. You may also combine them with your own SRS query, using the format [dbest-field:value].

An interesting feature of the EBI server is that the files created by your queries are kept for 2 d on the system. You can then save the URL (i.e., the address) in your HotList, quit your browser, and come back to the same queries and results later in the day.

3 3.2.1. Advanced Retrieval of Entries Using the EBI FTP Server

Using the WWW server is an interactive way of querying the database. It is possible to do more complicated queries through the EBI FTP server, using the get command:

```
get srs: [querystring] filename
```

As an example, here is the query from Fig. 4:

```
get srs:[dbest-Library:fetal] &
[dbest-chromosome:1] fetal_1.txt
```

The result of this query (all the entries from dbEST with the clone library containing the word "fetal" and on chromosome 1) will be written in the file "fetal_1.txt" in the user current directory.

No spaces or SRS command line parameters are allowed in the query. Consult the SRS documentation for more details on how to specify the query string. It is possible to run an FTP session in batch, which will allow you to submit queries at a time when your system, the EBI server, or the network are less used. The batch method will depend on the system you are using.

4. Notes

1. How to submit data: Since EST data are usually submitted as sets of many sequences, a tagged file format has been developed by the NCBI for data submission. This format is available by following the link **submit data** in the dbEST page of both NCBI and EBI WWW server. The format is evolving, so it is recommended that you read the documentation before submitting your sequences.

Fig. 7. How to expand a query.

List of Successful Queries

Here you can reinspect previous queries or combine them to new queries by entering an **expression**

Q1 *(retrieval)* := "[TaggedSites–Chromosome:1*] | [TaggedSites–FysicalMap:1*] |
[TaggedSites–GeneticMap:1*]" found 195 entries
    195 entries from DBEST
Q2 *(retrieval)* := "[TaggedSites–Chromosome:1*] & [TaggedSites–Organism:homo*]" found
186 entries
    186 entries from DBEST
Q3 *(retrieval)* := "[TaggedSites–Organism:homo*]" found 5443 entries
    5443 entries from DBEST
Q4 *(retrieval)* := "[TaggedSites–FysicalMap:1*] | [TaggedSites–GeneticMap:1*]" found 102
entries
    102 entries from DBEST
Q5 *(retrieval)* := "[TaggedSites–FysicalMap:1*] ! [TaggedSites–FysicalMap:10*]" found 43
entries
    43 entries from DBEST
Q6 *(retrieval)* := "[TaggedSites–Chromosome:1]" found 50 entries
    50 entries from DBEST

---

enter an expression (eg, "q1 | (q2 ! q3)" ...or just, eg, "q1" to repeat a query):

```
q3 & (q6 | q5 | [dbest-GeneticMap:1*])
```

⏚ Set output options before display. **DO-QUERY**

Fig. 8. The Query Manager.

2. The report format: You can obtain information on the report format by sending an electronic mail message to the following addresses, either at the NCBI **retrieve@ncbi.nlm.nih.gov** or at the EBI **retrieve@ebi.ac.uk**, containing the following lines in the body of the message:

```
datalib dbest
help
```

This will allow you to obtain the most up-to-date version of the report format.
3. WWW browsers (clients)
    a. Terminal Based (VT100)
        i. Lynx: **ftp://ftp2.cc.ukans.edu/pub/lynx/**
        ii. Emacs: **ftp://ftp.cs.indiana.edu/pub/elisp/w3/**
    b. MS Windows
        i. Mosaic: **ftp://ftp.ncsa.uiuc.edu/Mosaic/Windows/**
        ii. WinWeb: **ftp://ftp.einet.net/einet/pc/winweb/**
    c. Macintosh
        i. Mosaic: **ftp://ftp.ncsa.uiuc.edu/Mosaic/Mac/**
        ii. MacWeb: **ftp://ftp.einet.net/einet/mac/macweb/**
    d. X-Window
        i. Mosaic: **ftp://ftp.ncsa.uiuc.edu/Mosaic/Unix/**
    e. VMS
        i. Mosaic: **ftp://vms.huji.ac.il/www/www_client/**
4. SRS information: SRS information and documentation can be found at: **http://www.ebi.ac.uk/srs/srsman.html**.

References

1. Adams, M. D., Kelley, J. M, Gocayne, J. D., Dubnick, M., Polymeropoulos, M. H., Xiao, H., Merril, C. R., Wu, A., Olde, B., Moreno, R. F., Kervelage, A. R., McCombie, W. R., and Venter, J. C. (1991) Complementary DNA sequencing: expressed sequence tags and human genome project. *Science* **252**, 1651–1656.
2. Emmert, D. B., Stoehr, P. J., Stoesser, G., and Cameron, G. N. (1994) The European Bioinformatics Institute (EBI) databases. *Nucleic Acids Res.* **22**, 3445–3449.
3. Benson, D. A., Boguski, M., Lipman, D. J., and Ostell, J. (1994) GenBank. *Nucleic Acids Res.* **22**, 3441–3444.
4. Boguski, M. S., Lowe, T. M. J., and Tolstoshev, C. M. (1993) dbEST—database for "expressed sequence tags." *Nature Genetics* **4**, 332,333.
5. Pearson, W. R. and Lipman, D. J. (1988) Improved tools for biological sequence comparison. *Proc. Natl. Acad. Sci. USA* **85**, 2444–2448.
6. Altschul, S. F., Boguski, M. S., Gish, W., and Wootton, J. C. (1994) Issues in searching molecular sequences databases. *Nature Genetics* **6**, 119–129.
7. Eztold, T. and Argos, P (1993) SRS an indexing and retrieval tool for flat file data libraries. *Comput. Appl. Biosci.* **9**, 49–57.

23

Reference cDNA Library Facilities Available from European Sources

Michael P. Starkey, Yagnesh Umrania, Christopher R. Mundy, and Martin J. Bishop

1. Introduction
1.1. Objectives of cDNA Studies

The way cDNA libraries are used for finding genes of interest depends on the overall strategy adopted. Therefore, before exploring the options for cDNA library access, we briefly delineate these strategies and the place of cDNA libraries within them.

The present large-scale human sequencing efforts have focused largely on so-called expressed sequence tags (ESTs). These are partial sequences of cDNA clones that can be mapped as sequence tagged sites (STSs), or extended into "full-length" cDNA sequences. The value of cDNAs within the global genome mapping and sequencing effort stems from the fact that the vast regions of inter- and intragenic "junk" DNA can be ignored until a later date, whereas the expressed sequences which are responsible for biological effects can be identified.

"Bottom-up" approaches to gene finding make use of sequence data and coding sequence similarities between genes with similar functions, including between different organisms. Educated guesses about structure–function associations coupled with mutagenesis or knock-out studies can enable identification of gene function. cDNA libraries play a primary role in these studies because it is only in the form of cDNA clones that the coding sequence of most higher eukaryotic genes can be efficiently elucidated.

The "top-down" approach to gene finding relies on association of the biological effect of interest with a genetic map position, followed by physical

From: *Methods in Molecular Biology, Vol. 69: cDNA Library Protocols*
Edited by: I. G. Cowell and C. A. Austin Humana Press Inc., Totowa, NJ

mapping and candidate gene identification. Physical mapping in the positional cloning context means identifying a large genomic clone or clone contig, which contains the region of interest. Until recently, candidate gene identification within yeast artificial chromosome (YAC) clones has relied almost entirely on exon trapping *(1)*, direct selection *(2,3)*, and/or hybridization screening of cDNA clones, all of which are unreliable. The large-scale cDNA sequencing and mapping programs together with the continuing efforts to generate contiguous sets of YAC and bacterial (BAC *[4]*, PAC *[5]*, fosmid *[6]*, and cosmid) clones will yield a transcript map with most human genes located on large clones.

Most mRNAs in most tissues are in the low-abundance range, which is reflected by the abundance of cDNA clones within libraries, unless subtractive *(7,8)* or "normalization" procedures *(9,10)* are used. This limitation of cDNA libraries will be largely circumvented by the availability of repicked sets of unique clones, identified on the basis of their partial sequence. One of the current large-scale EST efforts has passed the 500,000 clone mark, so we can expect to see very low-abundance sequences represented (or those with access to the database [db] can).

Linking function to candidate gene remains, and will remain, the great challenge. In the absence of positional information, such techniques as differential display *(11)* are for the present being used to this end. Again, the repicked unique sets will have a major impact on this effort.

1.2. Library Format and Utilization

Screening methods for cDNA libraries are described in detail in Chapters 11–16. The key feature of cDNA, as with genomic libraries, is representation. Optimum representation of any library is attained without competitive growth of clones, so libraries are made available whenever possible as either ligation mixes or picked as individual clone collections in microtiter plates.

Very large libraries are required for representation of low-abundance sequences, but the sequence complexity of a higher vertebrate cDNA library is at least an order of magnitude less than that of a genomic library from the same organism. Consequently, the generation of pools for polymerase chain reaction (PCR) screening would be not only very arduous, but also unnecessary, because gene-specific probes can be generated from libraries by PCR relatively easily. Thus, gridded cDNA resources are usually screened by hybridization, following robot-assisted high-density filter spotting and colony processing.

Hybridization screening of high-density filter grids is improved very significantly if insert PCR-generated DNA is spotted rather than colonized, because the background caused by cell debris is eliminated. This can mean the difference between success and failure in such applications as YAC hybridization to a cDNA library. The extra effort required to produce insert-PCR products from very large numbers of clones is seldom expended, however, although one such resource is referred to in this chapter.

The technology for high-density gridding ($4 \times 4 \times 96 = 1536$, to as much as $10 \times 10 \times 96 = 9600$ clones/96 well microtiter plate area) is becoming more accessible, although it remains expensive and thus concentrated in a few centers. Labeling and imaging technology is also becoming more reliable, although much more slowly as concerning colony hybridization than many had hoped.

2. cDNA Libraries Available from the UK HGMP-RC

The UK HGMP Resource Centre (RC) currently provides libraries of two types—nominally "full length" libraries, and "normalized" fragmentary libraries (Table 1). The libraries are issued in two formats—either as aliquots of cDNA:vector ligation mixture for transformation or as high-density gridded cDNA filters for hybridization screening.

2.1. "Full-Length" cDNA Libraries

The RC supplies 50 "full-length" human and mammalian cDNA libraries (Table 1), constructed by D. L. Simmons (Cell Adhesion Laboratory, ICRF, Institute of Molecular Medicine, John Radcliffe Hospital, Oxford, UK). Second-strand cDNA was synthesized by the replacement procedure of Gubler and Hoffman *(12),* from oligo (dT)-primed first-strand cDNA. Nonself-compatible *Bst*XI adapters were ligated to size-selected cDNA fragments (larger than 500 bp), and the cDNAs were cloned nondirectionally into the *Bst*XI site of the constitutive mammalian expression vector pCDM8 *(13).*

cDNAs encoding specific cell-surface molecules have been isolated from the libraries by screening for transient expression in transfected COS cells *(13).* Since only cDNAs encoding the entire reading frame of the protein will produce properly folded and processed glycoproteins, the libraries have thus been demonstrated to contain at least a proportion of full-length clones. Full-length or near-full-length cDNAs corresponding to mRNAs ranging in size between 0.5 and 4.3 kb have been reported.

The "full-length" cDNA libraries are supplied as 100-ng cDNA:plasmid vector ligation mixture aliquots in 15 µL of 10 mM Tris-HCl (pH 8.0), 1 mM EDTA. The recombinant vector is propagated by transformation of *Escherichia coli* MC1061/p3 (supplied). Selection (12.5 µg/mL ampicillin and 7.5 µg/mL tetracycline) is mediated by a suppressor tRNA mutation (supF) in pCDM8, which suppresses amber stop codons in ampicillin and tetracycline resistance genes carried on the stable single-copy episome, p3, in strain MC1061/p3.

The following primers have been employed successfully for the PCR amplification/sequencing of cDNAs cloned between the *Bst*XI sites (map start posi-

Table 1
cDNA Libraries in the RC

HGMP ID.	Description	Format	N⁰ of subsets (256 ≡ 1 x genome)	Cloning vector	Cloning site	Enzyme/s for excision of cDNAs*	PCR primers for amplification of cDNAs	Availability
NA	**Human fetal adrenal** oligo (dT)-primed, non-directional, "normalised" fragmentary	4320 gridded cDNA clones	3	pBluescript II KS+	HindIII/ BamHI	ClaI/ BamHI	M13 reverse, M13 forward	**Filters:** Primary laboratories **Clones:** HGMP
H32	**Human DAUDI (B cells)** oligo (dT)-primed, non-directional, full length	cDNA:vector ligation mixture	NA	pCDM8	BstXI	BstXI	HGMP 5049, HGMP 5050	HGMP
H27	**Human LAD (leukocyte adhesion deficient patient EBV-B cells)** oligo (dT)-primed, non-directional, full length	cDNA:vector ligation mixture	NA	pCDM8	BstXI	BstXI	HGMP 5049, HGMP 5050	HGMP
NA	**Human fetal brain** oligo (dT)-primed, non-directional, "normalised" fragmentary	10704 gridded cDNA clones	27	pBluescript II KS+	HindIII/ BamHI	ClaI/ BamHI	M13 reverse, M13 forward	**Filters:** Primary laboratories **Clones:** HGMP
H16	**Human fetal brain, 15-16 weeks** oligo (dT)-primed, non-directional, full length	cDNA:vector ligation mixture	NA	pCDM8	BstXI	BstXI	HGMP 5049, HGMP 5050	HGMP
NA	**Human fetal brain, eye, intestine, lung, sternum, and tongue** oligo (dT)-primed, non-directional, "normalised" fragmentary	14,784 spotted PCR-amplified cDNAs	19	pCRII	between EcoRI	EcoRI	M13 reverse, M13 forward	**Filters & clones:** HGMP
H11	**Human bone marrow (aspirate, ALL +ve, 1st remission)** oligo (dT)-primed, non-directional, full length	cDNA:vector ligation mixture	NA	pCDM8	BstXI	BstXI	HGMP 5049, HGMP 5050	HGMP
H17	**Human normal colon** oligo (dT)-primed, non-directional, full length	cDNA:vector ligation mixture	NA	pCDM8	BstXI	BstXI	HGMP 5049, HGMP 5050	HGMP
H29	**Human HCT116 colon carcinoma** oligo (dT)-primed, non-directional, full length	cDNA:vector ligation mixture	NA	pCDM8	BstXI	BstXI	HGMP 5049, HGMP 5050	HGMP
H18	**Human HT29 colon carcinoma** oligo (dT)-primed, non-directional, full length	cDNA:vector ligation mixture	NA	pCDM8	BstXI	BstXI	HGMP 5049, HGMP 5050	HGMP
H34	**Human endometrium** oligo (dT)-primed, non-directional, full length	cDNA:vector ligation mixture	NA	pCDM8	BstXI	BstXI	HGMP 5049, HGMP 5050	HGMP
H23	**Human eosinophil** oligo (dT)-primed, non-directional, full length	cDNA:vector ligation mixture	NA	pCDM8	BstXI	BstXI	HGMP 5049, HGMP 5050	HGMP

HGMP ID.	Description	Format	N⁰ of subsets (256 ≡ 1 x genome)	Cloning vector	Cloning site	Enzyme/s for excision of cDNAs*	PCR primers for amplification of cDNAs	Availability
H5	**Human K562 (erythroleukaemia), full length** oligo (dT)-primed, non-directional, full length	cDNA:vector ligation mixture	NA	pCDM8	BstXI	BstXI	HGMP 5049, HGMP 5050	HGMP
H2	**Human HepG2 (hepato cellular carcinoma)** oligo (dT)-primed, non-directional, full length	cDNA:vector ligation mixture	NA	pCDM8	BstXI	BstXI	HGMP 5049, HGMP 5050	HGMP
H22	**Human SU-DH-L1 diffuse histiocytic lymphoma (non-Hodgkin's lymphoma)** oligo (dT)-primed, non-directional, full length	cDNA:vector ligation mixture	NA	pCDM8	BstXI	BstXI	HGMP 5049, HGMP 5050	HGMP
H15	**Human L920 Hodgkin's lymphoma line** oligo (dT)-primed, non-directional, full length	cDNA:vector ligation mixture	NA	pCDM8	BstXI	BstXI	HGMP 5049, HGMP 5050	HGMP
H6	**Human YT (HTLV-1 +ve adult leukaemia, T cell)** oligo (dT)-primed, non-directional, full length	cDNA:vector ligation mixture	NA	pCDM8	BstXI	BstXI	HGMP 5049, HGMP 5050	HGMP
H1	**Human HPBALL (peripheral blood, acute lymphocytic leukaemia)** oligo (dT)-primed, non-directional, full length	cDNA:vector ligation mixture	NA	pCDM8	BstXI	BstXI	HGMP 5049, HGMP 5050	HGMP
H19	**Human KG1 myeloblastic leukaemic line** oligo (dT)-primed, non-directional, full length	cDNA:vector ligation mixture	NA	pCDM8	BstXI	BstXI	HGMP 5049, HGMP 5050	HGMP
H20	**Human KG1A myeloblastic leukaemic line** oligo (dT)-primed, non-directional, full length	cDNA:vector ligation mixture	NA	pCDM8	BstXI	BstXI	HGMP 5049, HGMP 5050	HGMP
H21	**Human KG1B myeloblastic leukaemic line** oligo (dT)-primed, non-directional, full length	cDNA:vector ligation mixture	NA	pCDM8	BstXI	BstXI	HGMP 5049, HGMP 5050	HGMP
H3	**Human U937 (promonocytic leukaemia)** oligo (dT)-primed, non-directional, full length	cDNA:vector ligation mixture	NA	pCDM8	BstXI	BstXI	HGMP 5049, HGMP 5050	HGMP
H4	**Human U937 activated with phorbol 12-myristate 13-acetate** oligo (dT)-primed, non-directional, full length	cDNA:vector ligation mixture	NA	pCDM8	BstXI	BstXI	HGMP 5049, HGMP 5050	HGMP
H31	**Human APL promyelocytic leukaemia** oligo (dT)-primed, non-directional, full length	cDNA:vector ligation mixture	NA	pCDM8	BstXI	BstXI	HGMP 5049, HGMP 5050	HGMP
H7	**Human HL60 promyelocytic leukaemia** oligo (dT)-primed, non-directional, full length	cDNA:vector ligation mixture	NA	pCDM8	BstXI	BstXI	HGMP 5049, HGMP 5050	HGMP
H33	**Human small cell lung carcinoma** oligo (dT)-primed, non-directional, full length	cDNA:vector ligation mixture	NA	pCDM8	BstXI	BstXI	HGMP 5049, HGMP 5050	HGMP
H28	**Human DX3 melanoma** oligo (dT)-primed, non-directional, full length	cDNA:vector ligation mixture	NA	pCDM8	BstXI	BstXI	HGMP 5049, HGMP 5050	HGMP
H8	**Human G361 (amelanotic melanoma)** oligo (dT)-primed, non-directional, full length	cDNA:vector ligation mixture	NA	pCDM8	BstXI	BstXI	HGMP 5049, HGMP 5050	HGMP

(continued)

Table 1 (continued)

HGMP ID.	Description	Format	Nº of subsects (256 ≡ 1 x genome)	Cloning vector	Cloning site	Enzyme/s for excision of cDNAs*	PCR primers for amplification of cDNAs*	Availability
H30	Human THP1 (monocyte) oligo (dT)-primed, non-directional, full length	cDNA:vector ligation mixture	NA	pCDM8	BstXI	BstXI	HGMP 5049, HGMP 5050	HGMP
H24	Human fetal muscle oligo (dT)-primed, non-directional, full length	cDNA:vector ligation mixture	NA	pCDM8	BstXI	BstXI	HGMP 5049, HGMP 5050	HGMP
H25	Human natural killer cell oligo (dT)-primed, non-directional, full length	cDNA:vector ligation mixture	NA	pCDM8	BstXI	BstXI	HGMP 5049, HGMP 5050	HGMP
H9	Human placenta - full-term, normal pregnancy oligo (dT)-primed, non-directional, full length	cDNA:vector ligation mixture	NA	pCDM8	BstXI	BstXI	HGMP 5049, HGMP 5050	HGMP
H10	Human placental trophoblast (sorted 1st trimester) plus placental villi (1st trimester) oligo (dT)-primed, non-directional, full length	cDNA:vector ligation mixture	NA	pCDM8	BstXI	BstXI	HGMP 5049, HGMP 5050	HGMP
H26	Human tonsil oligo (dT)-primed, non-directional, full length	cDNA:vector ligation mixture	NA	pCDM8	BstXI	BstXI	HGMP 5049, HGMP 5050	HGMP
H12	Human HUVEC (umbilical vein endothelial cell line) oligo (dT)-primed, non-directional, full length	cDNA:vector ligation mixture	NA	pCDM8	BstXI	BstXI	HGMP 5049, HGMP 5050	HGMP
H13	Human HUVEC stimulated with IL1-B (4h) oligo (dT)-primed, non-directional, full length	cDNA:vector ligation mixture	NA	pCDM8	BstXI	BstXI	HGMP 5049, HGMP 5050	HGMP
H14	Human HUVEC stimulated with HT29 conditioned medium (48h) plus HUVEC stimulated with DX3 conditioned medium (48h) oligo (dT)-primed, non-directional, full length	cDNA:vector ligation mixture	NA	pCDM8	BstXI	BstXI	HGMP 5049, HGMP 5050	HGMP
G1	Gibbon MLA144 (T cells) oligo (dT)-primed, non-directional, full length	cDNA:vector ligation mixture	NA	pCDM8	BstXI	BstXI	HGMP 5049, HGMP 5050	HGMP
R5	Mouse bone marrow aspirate oligo (dT)-primed, non-directional, full length	cDNA:vector ligation mixture	NA	pCDM8	BstXI	BstXI	HGMP 5049, HGMP 5050	HGMP
R7	Mouse endothelioma cytokine-stimulated oligo (dT)-primed, non-directional, full length	cDNA:vector ligation mixture	NA	pCDM8	BstXI	BstXI	HGMP 5049, HGMP 5050	HGMP
R6	Mouse serum stimulated macrophages RAZ1 oligo (dT)-primed, non-directional, full length	cDNA:vector ligation mixture	NA	pCDM8	BstXI	BstXI	HGMP 5049, HGMP 5050	HGMP
R10	Mouse melanoma B16-F1 oligo (dT)-primed, non-directional, full length	cDNA:vector ligation mixture	NA	pCDM8	BstXI	BstXI	HGMP 5049, HGMP 5050	HGMP
R11	Mouse melanoma B16-black oligo (dT)-primed, non-directional, full length	cDNA:vector ligation mixture	NA	pCDM8	BstXI	BstXI	HGMP 5049, HGMP 5050	HGMP
R1	Mouse splenocyte B cells (LPS) oligo (dT)-primed, non-directional, full length	cDNA:vector ligation mixture	NA	pCDM8	BstXI	BstXI	HGMP 5049, HGMP 5050	HGMP

HGMP ID.	Description	Format	N° of subsets (256 ≡ 1 x genome)	Cloning vector	Cloning site	Enzyme/s for excision of cDNAs*	PCR primers for amplification of cDNAs*	Availability
R2	**Mouse splenocyte T cells (Con A)** oligo (dT)-primed, non-directional, full length	cDNA:vector ligation mixture	NA	pCDM8	BstXI	BstXI	HGMP 5049, HGMP 5050	HGMP
R12	**Mouse swiss 3T3 fibroblast (log phase)** oligo (dT)-primed, non-directional, full length	cDNA:vector ligation mixture	NA	pCDM8	BstXI	BstXI	HGMP 5049, HGMP 5050	HGMP
R13	**Mouse swiss 3T3 fibroblast (stationary phase)** oligo (dT)-primed, non-directional, full length	cDNA:vector ligation mixture	NA	pCDM8	BstXI	BstXI	HGMP 5049, HGMP 5050	HGMP
R3	**Mouse thymocytes** oligo (dT)-primed, non-directional, full length	cDNA:vector ligation mixture	NA	pCDM8	BstXI	BstXI	HGMP 5049, HGMP 5050	HGMP
R8	**Rat neonatal cerebellum** oligo (dT)-primed, non-directional, full length	cDNA:vector ligation mixture	NA	pCDM8	BstXI	BstXI	HGMP 5049, HGMP 5050	HGMP
R9	**Rat enterocyte** oligo (dT)-primed, non-directional, full length	cDNA:vector ligation mixture	NA	pCDM8	BstXI	BstXI	HGMP 5049, HGMP 5050	HGMP
R4	**Rat IC21 macrophage cell line, PMA stimulated** oligo (dT)-primed, non-directional, full length	cDNA:vector ligation mixture	NA	pCDM8	BstXI	BstXI	HGMP 5049, HGMP 5050	HGMP
R14	**Rat GH4 pituitary** oligo (dT)-primed, non-directional, full length	cDNA:vector ligation mixture	NA	pCDM8	BstXI	BstXI	HGMP 5049, HGMP 5050	HGMP
S1	**Sheep T cells** oligo (dT)-primed, non-directional, full length	cDNA:vector ligation mixture	NA	pCDM8	BstXI	BstXI	HGMP 5049, HGMP 5050	HGMP

NA - not applicable

*Where adaptored cDNAs are cloned, it is not evident whether flanking vector restriction enzyme sites are present in the cDNA inserts

291

tions: 2226 and 2582) within the pCDM8 (4356 bp) polylinker (map coordinates: 2198–2655):

> + strand (PCR)—HGMP primer ID. 5049.
>   5'-TCGTAACAACTCCGCCCCA-3'
> map position 2053   2071
>
> − strand (PCR)—HGMP primer ID. 5050.
>   5'-AGAGTCAGCAGTAGCCTCAT-3'
> map position 2912   2893
>
> + strand (sequencing)—HGMP primer ID. 5160.
>   5'-CGTGTACGGTGGGAGGTC-3'
> map position 2093   2110
>
> − strand (sequencing)—HGMP primer ID. 5161.
>   5'-TAAGGTTCCTTCACAAAG-3'
> map position 2677   2660

2.2. "Normalized" Fragmentary cDNA Libraries

The "normalized" cDNA libraries are fragmentary as a consequence of the method of "normalization" (Chapter 2). The "normalized" cDNA libraries are generated by a procedure in which a population of cDNA restriction fragments can be partitioned into 256 subpopulations. The 256 subsets combined represent the entire original cDNA population. Since an individual subpopulation is of relatively low complexity (as compared to the original population), the concentration of any given cDNA is higher than in the original population. The fragmentary libraries available consist of only a proportion of the possible 256 subpopulations and, consequently, do not purport to offer an entire genome coverage.

The RC has produced two oligo (dT)-primed fragmentary cDNA "libraries," both of which have been made available as high-density gridded filters for multiple hybridization screenings (Table 1). The first set of gridded filters were $10 \times 11.9 \times 7.8$ cm nylon filters of 4×4 arrayed *E. coli* colonies, comprising 10,704 fetal brain and 4320 fetal adrenal cloned cDNAs. cDNA fragments generated by the type IIs restriction enzyme *Fok*I were ligated (in a nondirectional manner) to specific adapters, which selected subsets of the fragment population. The adaptered cDNAs were amplified by PCR (utilizing adaptor primers), and cloned into pBluescript II KS+, containing appropriate adapters added between the *Hind*III and *Bam*HI sites of the vector polylinker. The *Hind*III site is destroyed during the adaptering procedure (it is therefore necessary to excise cDNAs using the enzymes *Cla*I and *Bam*HI, although one has no *a priori* knowledge of whether these sites are present in the cloned cDNAs). Although these cDNA filters are not currently available from the RC, a total of 46 sets have been issued to laboratories, on condition that the resource is made available to other investigators once no longer required. Advice on access will be given on request.

The second collection of fragmentary cDNAs has been derived from a mixture of fetal brain, eye, intestine, lung, sternum, and tongue tissues. Specific adapters were added (in a nondirectional manner) to *Fok*I cDNA restriction fragments, which were PCR-amplified (employing adapter primers), and cloned directly between the *Eco*RI sites within the polylinker of the "TA cloning vector," pCRII (Invitrogen, NV Leek, The Netherlands). A total of 14,784 cDNAs were amplified using flanking vector primers (M13 reverse and M13 forward), and spotted in duplicate in a 5 × 5 array on 13 × 11.9 × 7.8 cm nylon membranes.

3. cDNA Libraries Available from Other European Laboratories

The Max-Planck-Institute for Molecular Genetics (MPI, Ihnestrasse 73, 14195 Berlin (Dahlem), Germany; GENOME@ICRF.ICNET.UK) is a major European center for the distribution of cDNA libraries and cDNA clones. The institute manages the Reference Library System *(14)*, which was originally developed in the Genome Analysis Laboratory at the ICRF in London. cDNA libraries (in the form of high-density filter grids) are distributed free of charge to the scientific community. The Reference Library System is administered through the Reference Library Database (RLDB, accessible directly via both the EMBL ftp server at ftp.embl-heidelberg.de, and the World Wide Web (WWW) server: http://rzpd.rz-berlin.mpg.de), which also stores information pertaining to hybridization probes utilized, and clones requested by scientists participating in the scheme. The cDNA libraries collected by the MPI are detailed in Table 2. Since the MPI is also engaged in specific research projects with several collaborating laboratories, there is a paucity of information regarding many of the cDNA libraries listed in Table 2, which are not available through the Reference Library System. Table 2 also includes details of libraries studied by laboratories previously engaged in a European consortium identifying genes by partial sequencing of cDNAs.

4. The Integrated Molecular Analysis of Genomes Consortium

The IMAGE consortium *(15)* (info@image.llnl.gov) was founded in 1993 by four academic groups (C. Auffray, Genexpress/CNRS; G. Lennon, Lawrence Livermore National Laboratory; M. H. Polymeropoulos, National Center for Human Genome Research; M. B. Soares, Columbia University) with the intention of providing shared resources for gene sequencing, mapping, and expression studies. The underlying philosophy is that different laboratories

Table 2
cDNA Libraries in European Laboratories

Laboratory	Library ID. (Name/N°)	Description	Format	Cloning vector	Cloning site	Availability/Laboratory contact
Max-Planck Institute for Molecular Genetics	HFB cDNA 505	**Human fetal brain**		pBluescript KS+		Dr. Greg Lennon, University of California, Lawrence Livermore National Laboratory
Max-Planck Institute for Molecular Genetics	HFB cDNA 506	**Human fetal brain**		pBluescript KS+		Dr. Greg Lennon, University of California, Lawrence Livermore National Laboratory
Max-Planck Institute for Molecular Genetics	HFB cDNA 507	**Human fetal brain, 17 weeks** oligo (dT)-primed	gridded cDNA clones	pSPORT1		RLDB collection
Institute of Cell Biology, Rome		**Human brain (frontal lobe) meningioma** random-primed, non-directional, full length		pGEM1+*S. cerevisiae* pre-tRNA Leu3 gene containing *HpaI* site	*HpaI*	Professor G. Tocchini-Valentini
Max-Planck Institute for Molecular Genetics	EXONLIB 517	**Human COS cell line**		pAMP1		Dr. Marie-Laure Yaspo, Genome analysis Laboratory, ICRF
Genzentrum Laboratorium für Molekulare Biologie, Munich		**Human fibroblast W138 cell line (Stratagene)** oligo (dT)-primed, unidirectional		λZAPII		Dr. G.J. Arnold
Genzentrum Laboratorium für Molekulare Biologie, Munich		**Human gastric mucosa** oligo (dT)-primed, directional		Uni-ZAP XR	*EcoRI/XhoI*	Dr. G.J. Arnold
Genzentrum Laboratorium für Molekulare Biologie, Munich		**Human heart atrium (auriculi cordes)** oligo (dT)-primed, directional		λZAPII	*EcoRI/XhoI*	Professor H. Domdey

Laboratory	Library ID (Name/N°)	Description	Format	Cloning vector	Cloning site	Availability/Laboratory contact
Genzentrum Laboratorium für Molekulare Biologie, Munich		**Human heart ventricle (commercial)** oligo (dT) and random-primed, non-directional		λZAPII		Professor H. Domdey
Max-Planck Institute for Molecular Genetics	HFL cDNA 512	**Human fetal liver, 21 weeks** oligo (dT)-primed	gridded cDNA clones	pSPORT1		RLDB collection
Max-Planck Institute for Molecular Genetics	HPO cDNA 515	**Human fetal lung, 21 weeks** oligo (dT)-primed, directional	gridded cDNA clones	pSPORT1	*MluI*	RLDB collection
Max-Planck Institute for Molecular Genetics	XUD cDNA 513	**Human pancreas**		pSPORT1		Dr. Thomas Gress, Universität Ulm, Medizinische Universitätsklinik und Poliklinik
Max-Planck Institute for Molecular Genetics	503 518	**Human pancreas** oligo (dT)-primed, *NotI/SalI* adaptor linkers		pSPORT1	*NotI/SalI*	Dr. Thomas Gress, Universität Ulm, Medizinische Universitätsklinik und Poliklinik
Max-Planck Institute for Molecular Genetics	343 519	**Human pancreas** oligo (dT)-primed, *NotI/SalI* adaptor linkers		pSPORT1	*NotI/SalI*	Dr. Thomas Gress, Universität Ulm, Medizinische Universitätsklinik und Poliklinik
Max-Planck Institute for Molecular Genetics	XS2 cDNA 511	**Human pancreatic cancer cell line PATU 8988** oligo (dT)-primed, blunt-end ligation of *NotI/SalI* adaptor linkers		pSPORT1	*NotI/SalI*	Dr. Thomas Gress, Universität Ulm, Medizinische Universitätsklinik und Poliklinik
Max-Planck Institute for Molecular Genetics	PATU cDNA 514	**Human pancreatic cancer cell line PATU 8902** oligo (dT)-primed, *EcoRI/SalI* adaptor linkers		pBluescript KS+	*EcoRI/SalI*	Dr. Thomas Gress, Universität Ulm, Medizinische Universitätsklinik und Poliklinik
Max-Planck Institute for Molecular Genetics	HTE cDNA 508	**Human fetal thymus, 21 weeks** oligo (dT)-primed	gridded cDNA clones	pSPORT1		RLDB collection
Max-Planck Institute for Molecular Genetics	Xq28 cDNA 509	**Human fetal brain, fetal liver, and adult skeletal muscle** adaptored cDNAs hybridised in liquid against a pool of ~300 cosmids and 7 YACs specific for xq28		pAMP10		Dr. Bernhard Korn, DKFZ, Angewandte Tumor Virologie (ATV)

(continued)

Table 2 (continued)

Laboratory	Library ID. (Name/N°)	Description	Format	Cloning vector	Cloning site	Availability/Laboratory contact
Max-Planck Institute for Molecular Genetics	c21. cDNA 516	**Human chromosome 21-specific cDNAs from LLNL and ICRF libraries**		pSPORT1		Dr. Marie-Laure Yaspo, Genome analysis Laboratory, ICRF
Max-Planck Institute for Molecular Genetics	BFL 26hrs cDNA 531	*Branchiostoma floridae* (Amphioxus), 26h embryo oligo (dT)-primed, directional		pSPORT1		Mr. Matthew Clark, Max-Planck Institute for MolecularGenetics
Max-Planck Institute for Molecular Genetics	D_cDNA1 520	*Drosophila melanogaster* (Canton-S), 0-4h embryos oligo (dT)-primed, directional		pNB40		Dr. Joerg Hoheisel, DFKZ (ATV), Moleculargenetische Genomanalyse
Max-Planck Institute for Molecular Genetics	D_cDNA2 521	*Drosophila melanogaster* (Canton-S), 4-8h embryos oligo (dT)-primed, directional		pNB40		Dr. Joerg Hoheisel, DFKZ (ATV), Moleculargenetische Genomanalyse
Max-Planck Institute for Molecular Genetics	MBR cDNA 510	*Mus musculus* adult brain oligo (dT)-primed, directional	gridded cDNA clones	pSPORT1	*Notl/Sall*	RLDB collection
Max-Planck Institute for Molecular Genetics	RBMM7.5dECTO 529	*Mus musculus* (C57B16 x DBA), Ectoderm, 7.5 day embryo		pSPORT		Dr. Rosa Beddington, National Institute for Medical Research
Max-Planck Institute for Molecular Genetics	RBMM7.5dENDO 528	*Mus musculus* (C57B16 x DBA), Endoderm, 7.5 day embryo		pSPORT		Dr. Rosa Beddington, National Institute for Medical Research
Max-Planck Institute for Molecular Genetics	RBMM7.5dMESO 526	*Mus musculus* (C57B16 x DBA), Mesoderm, 7.5 day embryo		pSPORT		Dr. Rosa Beddington, National Institute for Medical Research
Max-Planck Institute for Molecular Genetics	RBMM7.5dPRST 527	*Mus musculus* (C57B16 x DBA), Primitive Streak, 7.5 day embryo		pSPORT		Dr. Rosa Beddington, National Institute for Medical Research
Max-Planck Institute for Molecular Genetics	RBMM7.5dES 525	*Mus musculus* (C57B16 x DBA), Embryonic stem cells, 7.5 day embryo		pSPORT		Dr. Rosa Beddington, National Institute for Medical Research

Laboratory	Library ID. (Name/N°)	Description	Format	Cloning vector	Cloning site	Availability/Laboratory contact
Max-Planck Institute for Molecular Genetics	RBMM7.5dTOT 530	*Mus musculus* (C57B16 x DBA), 7.5 day embryo (all tissues)		pSPORT		Dr. Rosa Beddington, National Institute for Medical Research
Max-Planck Institute for Molecular Genetics	9D MEMBR cDNA 522	*Mus musculus* , 9 day embryo		pSPORT1		Dr. Bernard Herrman, MPI für Immunobiologie
Max-Planck Institute for Molecular Genetics	12D MEMBR CDNA 523	*Mus musculus* , 12 day embryo		pSPORT		Dr. Bernard Herrman, MPI für Immunobiologie
Max-Planck Institute for Molecular Genetics	Zebrafish cDNA 524	*Danio rerio*, late somatogenesis stage directional		pSPORT1		Mr. Matthew Clark, Max-Planck Institute for MolecularGenetics
Max-Planck Institute for Molecular Genetics	ZFLIV 532	*Danio rerio*, late somatogenesis stage, liver directional		pSPORT1		Mr. Matthew Clark, Max-Planck Institute for MolecularGenetics

working on a common set of high-quality arrayed cDNA libraries, using different experimental approaches, can reduce unnecessary duplication of effort, and maximize the amount of information that one set of resources can provide. The data pertaining to clones in these high-density arrays are placed into the public domain as rapidly as possible. The ultimate aim is to utilize the information acquired to reorder the unique clones in a canonical array containing a cDNA clone representative of each gene of each genome studied.

The oligo (dT)-primed, directionally cloned plasmid cDNA libraries (Table 3) are arrayed at the Lawrence Livermore National Laboratory (LLNL), each clone receiving a Genome Database (GDB) accession number, in addition to an IMAGE number. Any researcher using these libraries and submitting information (including the IMAGE clone identifier) to a publicly accessible database (db) is considered a member of the consortium. The IMAGE cDNA clones are available without restriction from G. Lennon at LLNL. In addition, clones and high-density filters are available for a nominal charge from Research Genetics and the American Type Culture Collection (ATCC). The RC has been a distributor of IMAGE consortium clones since December 1995.

The majority of the IMAGE clones will be partially sequenced as part of the Washington University EST project, sponsored by Merck Pharmaceutical. The initiative, which involves groups at Washington University, the LLNL, and the University of Pennsylvania, aimed to produce 5' and 3' terminal sequences from 200,000 cDNA clones between October 1994 and March 1996. By July 1996, over 300,000 sequences had been submitted to dbEST and over 4000 had been assigned to chromosomes (as registered with GDB).

The Washington University–Merck EST project is largely duplicating the work of a partnership between Human Genome Sciences (HGS) and a non-profit organization, The Institute for Genomic Research (TIGR). TIGR was founded in 1992 by J. C. Venter, who instigated the large-scale EST approach to gene identification *(16)*.

5. Assessment of Library Quality

Unfortunately, most libraries studied have been employed for specific gene hunting exercises and have not been systematically characterized. Consequently, there is a paucity of information regarding their "quality." In spite of the efforts of the HGMP-RC, the MPI, and more recently the IMAGE consortium, the notion of widely utilized characterized "reference libraries" is, to date, true in concept alone. However, in this section, we have endeavored to compile the data that are available (Table 4, pp. 302–303).

The criteria that are broadly utilized to assess library quality are primarily the proportion of full length cDNAs, and the complexity of the library (number

Table 3
IMAGE Consortium cDNA Libraries

Library ID.	Originator	Description	Cloning vector	Cloning site	Details of library modifications	Mean insert size (kb)
INFLS	M.B. Soares & M. F. Bonaldo	**Human fetal liver and spleen, 20 weeks post conception** *Pac*I-oligo (dT)-primed, *Eco*RI adaptors added to ds cDNA, and cDNA directionally cloned	p17T3D-Pac (Pharmacia) with a modified polylinker	*Pac*I/*Eco*RI	single round of normalisation	
1NIB	M.B. Soares & M. F. Bonaldo	**Human infant brain, 73 days post natal (female)** *Not*I-oligo (dT)-primed, *Hind*III adaptors added to ds cDNA, and cDNA directionally cloned	Lafmid BA	*Not*I/*Hind*III	single round of normalisation (ss phagemid circles reassociated with short complementary strands, derived by controlled primer extension from the essentially unique non-coding 3'-ends of cDNA inserts cloned in the ss circles; range of abundance of mRNAs reduced from 4 to 1 order of magnitude)	
Nb2HP	M.B. Soares & M. F. Bonaldo	**Human placenta, full term (female)** *Not*I-oligo (dT)-primed, *Eco*RI adaptors added to ds cDNA, and cDNA directionally cloned	p17T3D-Pac (Pharmacia) with a modified polylinker	*Not*I/*Eco*RI	single round of normalisation	
2NbHBst	M.B. Soares & M. F. Bonaldo	**Human adult breast (female)** *Not*I-oligo (dT)-primed, *Eco*RI adaptors added to ds cDNA, and cDNA directionally cloned	p17T3D-Pac (Pharmacia) with a modified polylinker	*Not*I/*Eco*RI	single round of normalisation, to Cot230	
3NbHBst	M.B. Soares & M. F. Bonaldo	**Human adult breast (female)** *Not*I-oligo (dT)-primed, *Eco*RI adaptors added to ds cDNA, and cDNA directionally cloned	p17T3D-Pac (Pharmacia) with a modified polylinker	*Not*I/*Eco*RI	single round of normalisation, to Cot20	
N2b4HB55Y	M.B. Soares & M. F. Bonaldo	**Human adult brain** *Not*I-oligo (dT)-primed, *Eco*RI adaptors added to ds cDNA, and cDNA directionally cloned	p17T3D-Pac (Pharmacia) with a modified polylinker	*Not*I/*Eco*RI	single round of normalisation, to Cot50	1.5
N2b5HB55Y	M.B. Soares & M. F. Bonaldo	**Human adult brain** *Not*I-oligo (dT)-primed, *Eco*RI adaptors added to ds cDNA, and cDNA directionally cloned	p17T3D-Pac (Pharmacia) with a modified polylinker	*Not*I/*Eco*RI	single round of normalisation, to Cot50	0.5-1.5
937205	Stratagene	**Human fetal spleen** oligo (dT)-primed, cloned unidirectionally (5'- adaptor sequence: 5'-GAATTCGGCACGAG-3') (3'- adaptor sequence: 5'-CTCGAGTTTTTTTTTTTTTTTTT-3')	pBluescript SK-	*Eco*RI/*Xho*I		1.0
937224	Stratagene	**Human adult liver (49 year old male)** oligo (dT)-primed, cloned unidirectionally (5'- adaptor sequence: 5'-GAATTCGGCACGAG-3') (3'- adaptor sequence: 5'-CTCGAGTTTTTTTTTTTTTTTTT-3')	Uni-ZAP XR	*Eco*RI/*Xho*I		1.1
937210	Stratagene	**Human adult lung (72 year old male)** oligo (dT)-primed, cloned unidirectionally (5'- adaptor sequence: 5'-GAATTCGGCACGAG-3') (3'- adaptor sequence: 5'-CTCGAGTTTTTTTTTTTTTTTTT-3')	pBluescript SK-	*Eco*RI/*Xho*I		1.0
937217	Stratagene	**Human adult ovary (49 year old female)** oligo (dT)-primed, cloned unidirectionally (5'- adaptor sequence: 5'-GAATTCGGCACGAG-3') (3'- adaptor sequence: 5'-CTCGAGTTTTTTTTTTTTTTTTT-3')	pBluescript SK-	*Eco*RI/*Xho*I		0.8
937225	Stratagene	**Human placenta (male)** oligo (dT)-primed, cloned unidirectionally (5'- adaptor sequence: 5'-GAATTCGGCACGAG-3') (3'- adaptor sequence: 5'-CTCGAGTTTTTTTTTTTTTTTTT-3')	pBluescript SK-	*Eco*RI/*Xho*I		1.2

Table 3 (continued)

Library ID.	Originator	Description	Cloning vector	Cloning site	Details of library modifications	Mean insert size (kb)
2NbHM	M.B. Soares & M. F. Bonaldo	**Human melanocyte** *Not*I-oligo (dT)-primed, *Eco*RI adaptors added to ds cDNA, and cDNA directionally cloned	pT7T3D-Pac (Pharmacia) with a modified polylinker	*Not*I/*Eco*RI		
N2b4HR	M.B. Soares & M. F. Bonaldo	**Human adult retina** *Not*I-oligo (dT)-primed, *Eco*RI adaptors added to ds cDNA, and cDNA directionally cloned	pT7T3D-Pac (Pharmacia) with a modified polylinker	*Not*I/*Eco*RI	single round of normalisation, to Cot50	2.0
N2b5HR	M.B. Soares & M. F. Bonaldo	**Human adult retina** *Not*I-oligo (dT)-primed, *Eco*RI adaptors added to ds cDNA, and cDNA directionally cloned	pT7T3D-Pac (Pharmacia) with a modified polylinker	*Not*I/*Eco*RI	single round of normalisation, to Cot50	1.0
1NbHPG	M.B. Soares & M. F. Bonaldo	**Human adult pineal gland (male and female: ages 18, 20, 48)** *Not*I-oligo (dT)-primed, *Eco*RI adaptors added to ds cDNA, and cDNA directionally cloned	pT7T3D-Pac (Pharmacia) with a modified polylinker	*Not*I/*Eco*RI	single round of normalisation, to Cot40	
3NbHPG	M.B. Soares & M. F. Bonaldo	**Human adult pineal gland (male and female: ages 18, 20, 48)** *Not*I-oligo (dT)-primed, *Eco*RI adaptors added to ds cDNA, and cDNA directionally cloned	pT7T3D-Pac (Pharmacia) with a modified polylinker	*Not*I/*Eco*RI	single round of normalisation, to Cot38	
2NbHP8-9W	M.B. Soares & M. F. Bonaldo	**Human fetal placenta, 8 and 9 weeks post conception** *Not*I-oligo (dT)-primed, *Eco*RI adaptors added to ds cDNA, and cDNA directionally cloned	pT7T3D-Pac (Pharmacia) with a modified polylinker	*Not*I/*Eco*RI		
NbHOT	M.B. Soares & M. F. Bonaldo	**Human adult ovary tumour (36 year old female)** *Not*I-oligo (dT)-primed, *Eco*RI adaptors added to ds cDNA, and cDNA directionally cloned	pT7T3D-Pac (Pharmacia) with a modified polylinker	*Not*I/*Eco*RI	single round of normalisation, to Cot53	
2NbHMSP	M.B. Soares & M. F. Bonaldo	**Human adult multiple sclerosis lesions (46 year old male)** *Not*I-oligo (dT)-primed, *Eco*RI adaptors added to ds cDNA, and cDNA directionally cloned	pT7T3D-Pac (Pharmacia) with a modified polylinker	*Not*I/*Eco*RI	single round of normalisation, to Cot5	
NbHL19W	M.B. Soares & M. F. Bonaldo	**Human fetal lung, 19 weeks post conception** *Not*I-oligo (dT)-primed, *Eco*RI adaptors added to ds cDNA, and cDNA directionally cloned	pT7T3D-Pac (Pharmacia) with a modified polylinker	*Not*I/*Eco*RI	single round of normalisation	
NbHPA	M.B. Soares & M. F. Bonaldo	**Human adult parathyroid tumour** *Not*I-oligo (dT)-primed, *Eco*RI adaptors added to ds cDNA, and cDNA directionally cloned	pT7T3D-Pac (Pharmacia) with a modified polylinker	*Not*I/*Eco*RI	single round of normalisation	
NbHSF	M.B. Soares & M. F. Bonaldo	**Human senescent fibroblast** *Not*I-oligo (dT)-primed, *Eco*RI adaptors added to ds cDNA, and cDNA directionally cloned	pT7T3D-Pac (Pharmacia) with a modified polylinker	*Not*I/*Eco*RI	single round of normalisation	
NbHH19W	M.B. Soares & M. F. Bonaldo	**Human fetal heart, 19 weeks post conception** *Not*I-oligo (dT)-primed, *Eco*RI adaptors added to ds cDNA, and cDNA directionally cloned	pT7T3D-Pac (Pharmacia) with a modified polylinker	*Not*I/*Eco*RI	single round of normalisation	
10420-016	Life Tech.	**Human adult kidney (38 year old male)** *Not*I-oligo (dT)-primed, *Sal*I adaptors added to ds cDNA, and cDNA directionally cloned	pCMV-SPORT	*Not*I/*Sal*I		1.32
10421-014	Life Tech.	**Human adult leukocyte (male and female)** *Not*I-oligo (dT)-primed, *Sal*I adaptors added to ds cDNA, and cDNA directionally cloned	pCMV-SPORT	*Not*I/*Sal*I		1.3
10426-013	Life Tech.	**Human adult testis (60 year old male)** *Not*I-oligo (dT)-primed, *Sal*I adaptors added to ds cDNA, and cDNA directionally cloned	pCMV-SPORT	*Not*I/*Sal*I		1.56
Morton Fetal Cochlea	N. Robertson & C. Morton	**Human fetal cochlea, 16-22 week fetuses** oligo (dT)-primed, cloned unidirectionally (5'- adaptor sequence: 5'-GAATTCGGCACGAG-3' (3'- adaptor sequence: 5'-CTCGAGTTTTTTTTTTTTTTTTT-3')	pBluescript SK-	*Eco*RI/*Xho*I		37% <0.5kb, 56% 0.5-1.0kb, 7% > 1.0kb

Library ID.	Originator	Description	Cloning vector	Cloning site	Details of library modifications	Mean insert size (kb)
Takeda/Bell pancreatic islet	J. Takeda & G. Bell	**Human adult pancreatic islet** oligo (dT)-primed, cloned unidirectionally (5'- adaptor sequence: 5'-GAATTCGGCACGAG-3') (3'- adaptor sequence: 5'-CTCGAGTTTTTTTTTTTTTT-3')	pBluescript SK-	*Eco*RI/*Xho*I		
Weizmann Olfactory Epithelium	N. Walker & D. Lancet	**Human adult olfactory epithelium (35 year old female)** oligo (dT)-primed, cloned unidirectionally (5'- adaptor sequence: 5'-GAATTCGGCACGAG-3') (3'- adaptor sequence: 5'-CTCGAGTTTTTTTTTTTTTT-3')	pBluescript SK-	*Eco*RI/*Xho*I		0.8
NbME 13.5-14.5	M.B. Soares & M. F. Bonaldo	**Mus musculus C57BL/6J 13.5-14.5 days post conception** *Not*I-oligo (dT)-primed, *Eco*RI adaptors added to ds cDNA, and cDNA directionally cloned	pT7T3D-Pac (Pharmacia) with a modified polylinker	*Not*I/*Eco*RI	single round of normalisation	
p3NMF19.5	M.B. Soares & M. F. Bonaldo	**Mus musculus C57BL/6J 19.5 days post conception** *Not*I-oligo (dT)-primed, *Eco*RI adaptors added to ds cDNA, and cDNA directionally cloned	pT7T3D-Pac (Pharmacia) with a modified polylinker	*Not*I/*Eco*RI	single round of normalisation	
Life Tech. mouse adult brain	Life Tech.	**Mus musculus C57BL/6J adult brain** *Not*I-oligo (dT)-primed, *Sal*I adaptors added to ds cDNA, and cDNA directionally cloned	pCMV-SPORT2	*Not*I/*Sal*I		

Table 4
"Quality" of cDNA Libraries in European Laboratories

Laboratory	Library ID (Name/Nº)	Description	Nº of clones	% of full length cDNAs	Mean insert size (kb)	% of non-recombinant clones	% of "non-useful" clones	% of "unknown" cDNAs
Max-Planck Institute for Molecular Genetics	HFB cDNA 507	**Human fetal brain, 17 weeks** oligo (dT)-primed	120,000	60% ~1.2kb (as assessed by GAPDH clones)	1.4	<1	2 (ribosomal) 1-2 (repeat sequences)	
Institute of Cell Biology, Rome		**Human brain (frontal lobe) meningioma** random-primed, non-directional, full length			0.5	5	35	32
Genzentrum Laboratorium für Molekulare Biologie, Munich		**Human fibroblast W138 cell line** (Stratagene) oligo (dT)-primed, unidirectional					8 (mitochondrial) 7 (ribosomal)	55
Genzentrum Laboratorium für Molekulare Biologie, Munich		**Human gastric mucosa** oligo (dT)-primed, directional			0.828		8 (mitochondrial) 2 (ribosomal)	27
Genzentrum Laboratorium für Molekulare Biologie, Munich		**Human heart atrium (auriculi cordes)** oligo (dT)-primed, directional			0.8-1.2	16	16 (mitochondrial)	45
Genzentrum Laboratorium für Molekulare Biologie, Munich		**Human heart ventricle (commercial)** oligo (dT) and random-primed, non-directional			1.2-1.6	16	16 (mitochondrial) 12 (ribosomal)	48
Max-Planck Institute for Molecular Genetics	503 518	**Human pancreas** oligo (dT)-primed, *NotI/SalI* adaptor linkers	1000,000		1.8	1		
Max-Planck Institute for Molecular Genetics	343 519	**Human pancreas** oligo (dT)-primed, *NotI/SalI* adaptor linkers	800,000		1.8-2	1	10	40-50

Laboratory	Library ID. (Name/Nº)	Description	Nº of clones	% of full length cDNAs	Mean insert size (kb)	% of non-recombinant clones	% of "non-useful" clones	% of "unknown" cDNAs
Max-Planck Institute for Molecular Genetics	XS2 cDNA 511	**Human pancreatic cancer cell line PATU 8988** oligo (dT)-primed, blunt-end ligation of *NotI/SalI* adaptor linkers	500,000	50-60	1.9	1	10	40-50
Max-Planck Institute for Molecular Genetics	PATU cDNA 514	**Human pancreatic cancer cell line PATU 8902** oligo (dT)-primed, *EcoRI/SalI* adaptor linkers	1000,000	30-40	1.4	10	15	
Max-Planck Institute for Molecular Genetics	D_cDNA2 521	*Drosophila melanogaster* (Canton-S), **4-8h** embryos oligo (dT)-primed, directional			1.73			
Max-Planck Institute for Molecular Genetics	MBR cDNA 510	*Mus musculus* **adult brain** oligo (dT)-primed, directional			1.5			
Max-Planck Institute for Molecular Genetics	Zebrafish cDNA 524	*Danio rerio*, **late somatogenesis stage** directional	60,000		2			
Max-Planck Institute for Molecular Genetics	ZFLIV 532	*Danio rerio*, **late somatogenesis stage, liver** directional	90,000		2			

303

of different transcripts—a measure of the "representativeness" of the library). If more detailed information is available, then the following details are pertinent:

1. Average insert size;
2. Proportion of nonrecombinant clones;
3. Proportion of ribosomal, mitochondrial, and repetitive sequence clones; and
4. Proportion of both "known" ("significantly" similar to nucleic acid database sequences) and "unknown" cDNAs.

With the advent of the EST approach to the rapid characterization of expressed genes by partial DNA sequencing *(16)*, in the absence of non-anecdotal quantitative data on the complexity of a given library, a measure of the likely complexity can be acquired by measuring the redundancy of specific genes, i.e., the number of times specific genes are represented by ESTs in a given library.

6. Access to cDNA Resources from the HGMP-RC

The HGMP-RC offers cDNA libraries, and clones, to scientists registered with the organization. In addition, users are given access to the HGMP-RC cDNA db, which contains sequence and mapping data specific to individual cDNAs. Registration is free of charge to all UK and European community academic users. The resources can be obtained on completion of an appropriate request form.

cDNA resources are of three types:

1. "Full-length cDNA libraries";
2. High-density gridded cDNA libraries; and
3. cDNA clones (including IMAGE consortium clones).

A maximum of three "full-length" cDNA libraries can be requested at one time, whereas requests for a single given set of high-density gridded cDNA filters are permissible. cDNA clones are requested either on the basis of hybridization screening of a gridded filter set, or after screening the HGMP-RC cDNA, EMBL, or dbEST dbs with a sequence of interest using the *Blast Fasta* db searching programs. These two methods of analysis are reflected in the availability of cDNA clone request forms specific to "hybridization screening," or "db screening." A request for >20 cDNA clones, identified by hybridization screening, must be accompanied by a copy of the relevant autoradiograph.

Access to the HGMP-RC cDNA db, and to all cDNA resources is free to all academic UK and non-EC users. Commercial users are charged £100.00 for a "full-length" cDNA library, £1200.00 for a set of gridded cDNA filters, £110.00 for a sequenced cDNA clone, and £35.00 for a nonsequenced cDNA clone.

On registering with the HGMP-RC, the user enters into a commitment to return data pertaining to the cDNA resources requested. The acquired data are included in the HGMP-RC "cDNA Data Returned" db, which is accessible to all users. The information required comprises both that which forms the basis of a specific request and data relating to subsequent investigations of resources supplied. Information falling into the former category includes both clone-specific details, for instance, the probe(s) hybridizing to a given cDNA clone, or details of a db similarity search, and details of experimental procedures, such as a hybridization screening protocol. Data generated from materials issued, which is made public after a 6-mo period of confidentiality, include details of clone verification and facts, such as the mean size of cDNA inserts in clones from a "full-length" library.

Request forms and information concerning the cDNA resources available can be obtained directly from the RC (telephone: 01223-494500; facsimile: 01223-494512; e-mail: biohelp@hgmp.mrc.ac.uk) and via the WWW HGMP-RC data browser. Specific scientific inquiries pertaining to the resources available for gene isolation can be directed to gene.id@hgmp.mrc.ac.uk, whereas queries related to the HGMP-RC cDNA db can be communicated to support@hgmp.mrc.ac.uk.

7. HGMP-RC Computer System for Tracking Orders and Return of Data

The processing of requests and returned data at the HGMP-RC is performed under the auspices of the computing system illustrated in Fig. 1. The system enables requests to be initiated, or information to be returned, either electronically via a WWW interface, or through standard mail or fax channels. Data are stored in a Sybase relational db management system (named "LabApps"), with a USoft interface. The main WWW interface for the data browsing and resource request facilities available at the HGMP-RC is depicted in Fig. 2. This document not only directs the user toward the gamut of services available from the HGMP-RC, but offers the opportunity for registration as a user, and also provides access to a number of RC and public dbs.

The request forms pertaining to cDNA-related resources are of three types:

1. "Full-length" cDNA libraries;
2. cDNA clones identified by hybridization screening of high-density gridded *E. coli* cDNA clone colony filters; and
3. cDNA clones identified by similarity screening of either the HGMP-RC, or EMBL dbs.

A "full-length" cDNA library request form, available via the WWW interface, is shown in Fig. 3. A data return form, also accessible by the WWW interface, solicits information from the user regarding the manner in which each library issued has been utilized (e.g., method of screening) and the results of the library screening (including mean size of cDNAs isolated).

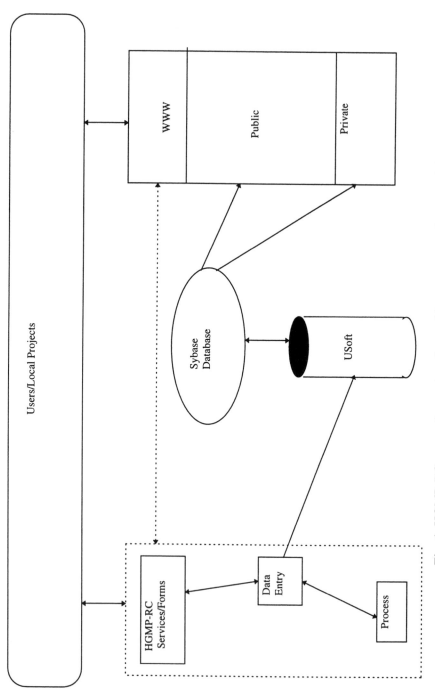

Fig. 1. HGMP-RC computing system for tracking requests and return of data.

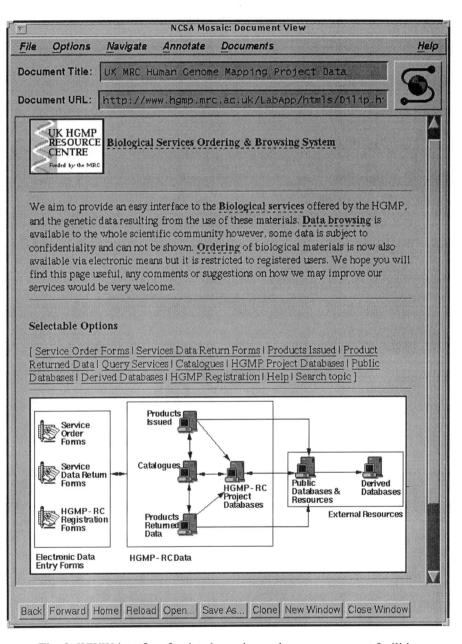

Fig. 2. WWW interface for data browsing and resource request facilities.

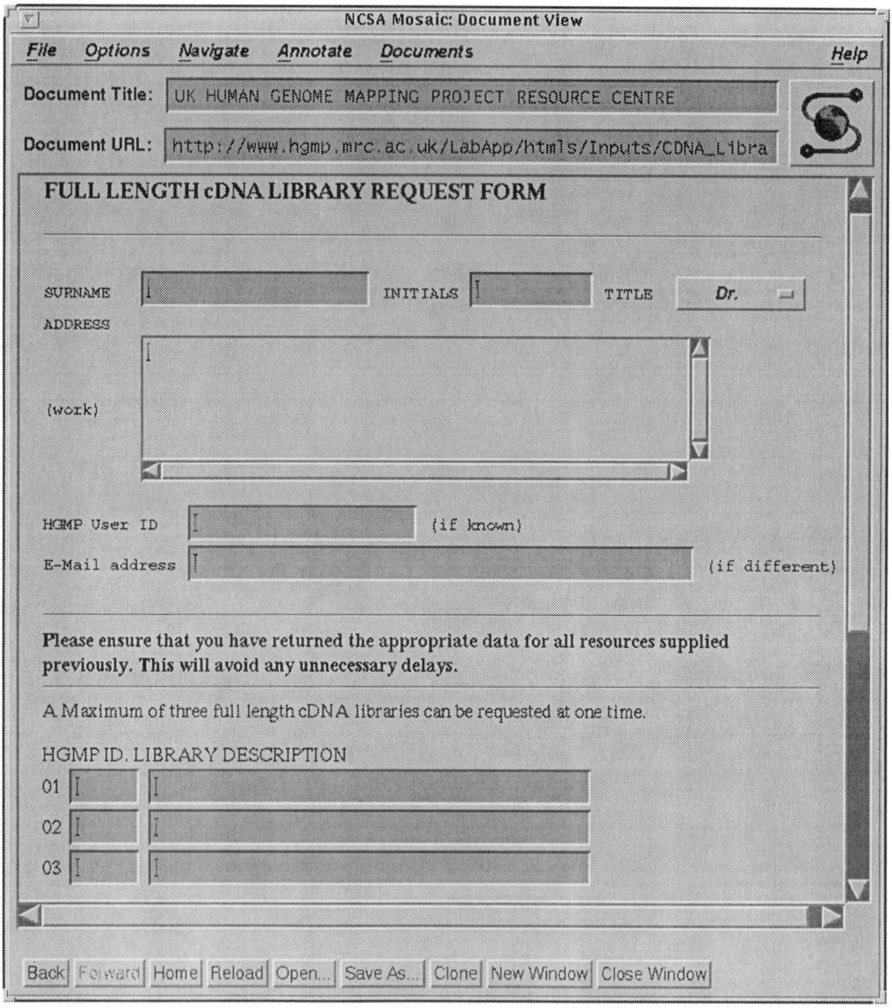

Fig. 3. "Full-length" cDNA library request form.

The computing structure enables the progress of requests to be monitored, and returned data to be scrutinized, either through the USoft interface to the Sybase db (Fig. 4), or via the WWW interface (Fig. 2). The Sybase db is linked to the various HGMP-RC project dbs (Fig. 2). Consequently, in the case of cDNA resources, data pertaining to a cDNA requested as a consequence of similarity screening of a sequence db are linked by the ID of the cDNA concerned to information (including sequence data) compiled about the same cDNA in the HGMP-RC cDNA db.

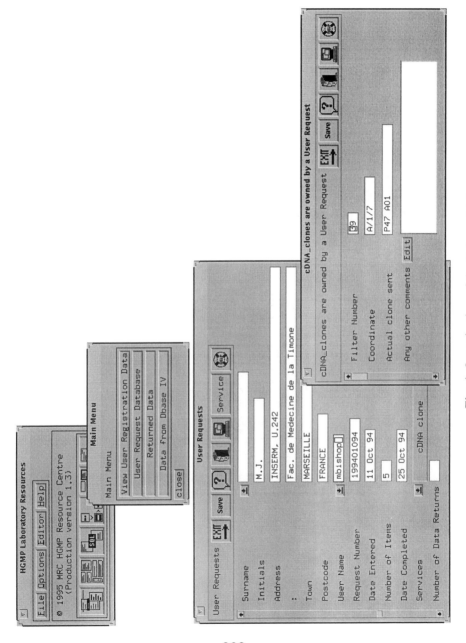

Fig. 4. Organization of the Sybase database.

309

8. HGMP-RC cDNA db

The cDNA db is a repository for information pertaining to cDNA clones and libraries studied at the HGMP-RC. The db also stores details of cDNAs sequenced in European laboratories, which have collaborated with the HGMP to identify genes by the partial sequencing of cDNAs.

The cDNA db is written in sybase and runs on the RC's unix network, interacting with several programs on this system, e.g., for the purpose of identifying similar sequences, the db can be screened with a sequence of interest using the *Blast Fasta* db searching programs via the HGMP molecular biology software main menu. In addition to sequence data, the db retains details about the origin of the sequences and the results obtained following a search of nucleic acid and protein dbs (Genbank, EMBL, Swissprot, and the HGMP cDNA db itself) for similarities to the sequences. A WWW interface has recently been developed in order to render the db more accessible to interrogation. The results of a db interrogation can be printed directly or saved to a file at any stage of an interrogation.

The main menu of the cDNA db WWW interface allows users to search for either cDNA sequences or for library details. Selection of either of these options calls up a form on which the user is invited to define parameters (from lists of valid options) to initiate a search.

Details of individual cDNAs can be obtained by selecting the "*Search for cDNA sequences*" from the WWW main menu page (Fig. 5). The "*Text in title*" option facilitates a search for cDNA sequences on the basis of the description of nucleic acid/protein db entries to which they are similar, e.g., entering "*oncogene*" will identify those cDNAs that are similar to a nucleic acid or protein sequence with the word "*oncogene*" in its description. The db can also be queried for cDNAs derived from a particular "*Species*" or "*Tissue.*"

Since the HGMP-RC archives cDNA clones in microtiter plates, the nomenclature ascribed to a cDNA sequence reflects not only the library from which the cDNA is derived, but also the specific location of the corresponding cDNA clone, i.e., a sequence is denoted by library code (one to two letters), plate number, and microtiter plate well location. In addition, a unique seven-letter code ("*Seq id*") is independently designated to each cDNA sequence on input into the db. Specific cDNA sequences can thus also be selected employing the library code and one or both of the other components of the sequence nomenclature, e.g., submitting a query for all the cDNAs in a given library generates a list of the sequences, in conjunction with details of the sequencing primer employed to generate each sequence, the accession number for sequences submitted to the EMBL db, and a description of the db entry to which each sequence is most significantly similar (Fig. 6).

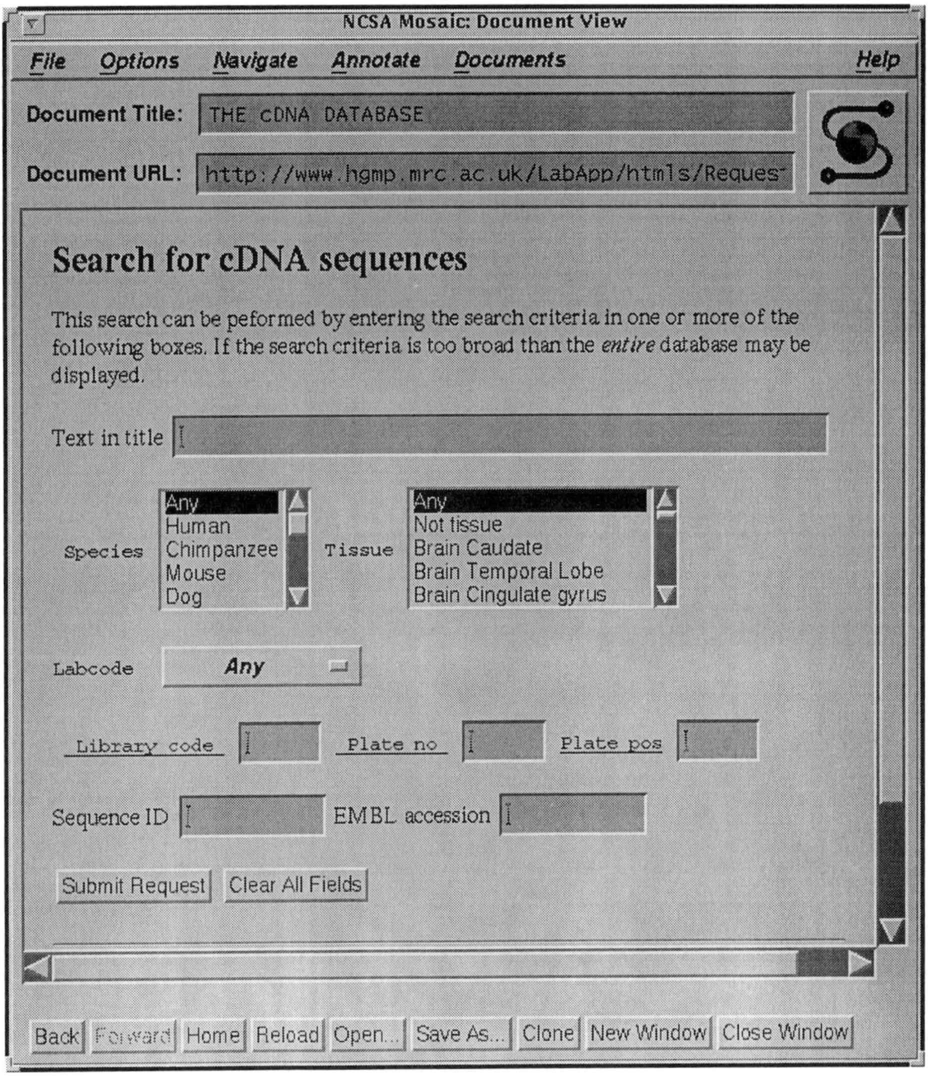

Fig. 5. WWW page facilitating cDNA sequence searching.

Once a cDNA of interest has been identified, additional information regarding the library from which the cDNA is derived can be acquired via the "*Search for libraries*" option of the WWW main menu page. The search is initiated by entry of the laboratory (originators of library) and library codes. The search retrieves details (Fig. 7) pertaining to the construction of the library, e.g., cloning vector, restriction enzyme(s), and so on. Additional information can be obtained from the "*Origin*" (details of the organ, tissue, or cell line from which

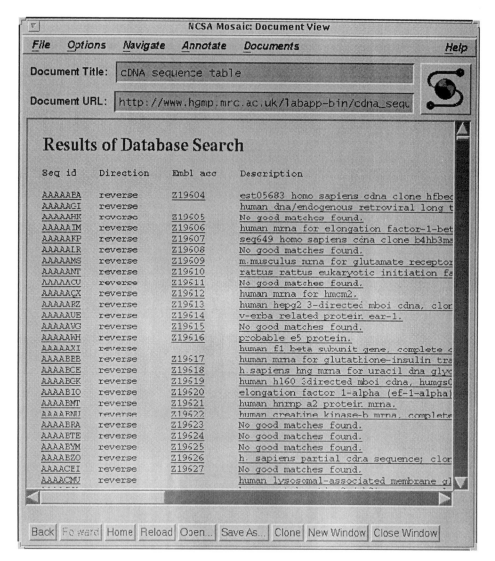

Fig. 6. Results of a search for cDNA sequences from a given library.

mRNA was extracted to generate cDNA), *"PCR details"* and *"Subcloning details"* options on the *"Library details"* menu. The option *"Contact"* reveals a named individual who can be contacted to ascertain cDNA clone availability.

9. ICAtools Software for Sequence Sorting

The Incremental Clustering Algorithm tools (*ICAtools*) *(17)* are a set of programs that may be used for adding information to otherwise amorphous sets of

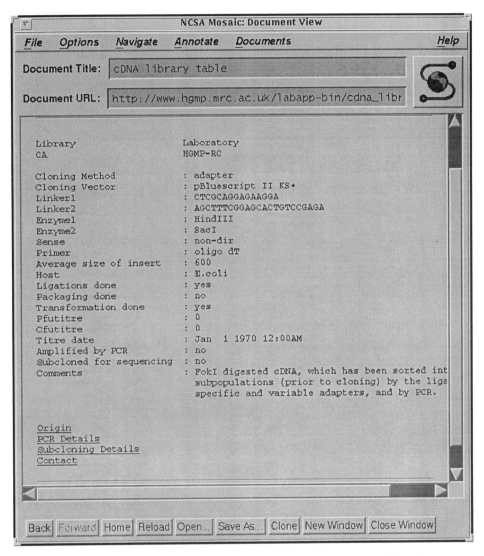

Fig. 7. Results of a search for information pertaining to a given cDNA library.

DNA sequences generated in large-scale sequencing projects. The programs process a file of DNA sequences to generate an index file in which sequences sharing similarities are grouped into clusters.

The programs *ICAtool* and *N2tool* are useful for identifying essentially dissimilar sequences containing regions of "local" similarity, e.g., identifying sequences from which cloning vector and/or linker sequences have not been removed. A separate cluster of matching sequences is produced for each unique "local" similarity (consequently, a given sequence can occur in more than a single cluster, according to the number of different subsequences it has in common with other sequences).

Conversely, *ICAass* forms clusters of sequences that share "global" similarities. The program accepts a file containing DNA sequences listed in order of decreasing size and group sequences that are approximately repeated within the length of another. The similarity score that defines the threshold at which one sequence is said to be an approximate "subsequence" of a "supersequence" can be varied (50–100%). It is this program that is appropriately applied to assess the composition of a given cDNA library. Each cluster ("family of sequences" or "unique sequence") can be characterized by similarity search of its member(s) against the nucleic acid dbs.

The aim of this type of analysis is to assess the complexity of a given library by generating a list of all the mRNA transcripts represented in the library. There is, however, a certain qualification to the utilization of *ICAass* to generate an index, in which each "cluster" represents a single transcript (all the cDNA sequences in a cluster represent a single mRNA species), and each mRNA species only appears once in the index. Essentially, difficulties arise as a consequence of the fact that, unless full-length (or near-full-length) cDNA sequences are compared, sequences derived from a given transcript do not necessarily originate from the same region of the transcript. This is of particular relevance for cDNAs cloned nondirectionally (in which case alternate strands of a given cDNA may be sequenced and analyzed for alignment) or directionally cloned cDNAs, derived by random-primed reverse transcription.

A more subtle problem, although with a similar outcome, is encountered when attempting to align sequences derived from the same region of a given strand of a particular transcript (e.g., when analyzing the 3'-ends of oligo-dT primed, directionally cloned cDNAs). Owing to the variable "quality" of each individual sequence (length of read, number of ambiguities removed by editing from the beginning and end of a sequence), two sequences of a given region of a transcript may begin and finish at a different position. As a consequence of this, there may be insufficient identity within the overlap between two such sequences for the *ICAass* program to index two such sequences in a single cluster. Therefore, from the output of the program, one would infer that the two homologous sequences represent different transcripts, when in fact they represent the same transcript.

An apparent way to circumvent this problem, by reducing the percentage of a "subsequence" that must be represented in a "supersequence" for the two sequences to be clustered together, increases the likelihood of a given sequence being indexed as a "subsequence" of several "supersequences," i.e., a single cDNA sequence may appear in more than one cluster, suggesting that the cDNA represents more than a single mRNA species.

Given the foibles of the *ICAass* program with regard to assessing cDNA library complexity, a complementary tool to employ in parallel is the *Fasta* program *(18),* which facilitates a search for a similarity between a sequence and one, or more, other sequences. *ICAass* and *Fasta* produce different estimates of the percentage identity between two sequences, since unlike *ICAass*, *Fasta* does not apply a penalty for insertions or deletions within the overlap between two sequences. Employing *Fasta*, it is possible to confirm by direct pairwise comparison that two sequences are of sufficient identity to represent the same transcript (and therefore appear in the same cluster). In addition, two sequences may be confirmed as being homologous if, on *Fasta* database similarity search, they display equivalent similarity to the same region of a specific nucleic acid database entry.

References

1. Duyk, G. M., Kim, S., Myers, R. M., and Cox, D. R. (1990) Exon trapping: a genetic screen to identify candidate transcribed sequences in cloned mammalian genomic DNA. *Proc. Natl. Acad. Sci. USA* **87,** 8995–8999.
2. Lovett, M., Kere, J., and Hinton, L. M. (1991) Direct selection: a method for the isolation of cDNAs encoded by large genomic regions. *Proc. Natl. Acad. Sci. USA* **88,** 9628–9632.
3. Parimoo, S., Patanjali, S. R., Shukla, H., Cahplin, D. D., and Weissman, S. M. (1991) cDNA selection: efficient PCR approach for the selection of cDNAs encoded in large chromosomal DNA fragments. *Proc. Natl. Acad. Sci. USA* **88,** 9623–9627.
4. Shizuya, H., Birren, B., Kim, U. J., Mancino, V., Slepak, T., Tachiiri, Y., and Simon, M. I. (1992) Cloning and stable maintenance of 300-kilobase-pair fragments of human DNA in *Escherichia coli* using an F-factor-based vector. *Proc. Natl. Acad. Sci. USA* **89,** 8794–8797.
5. Sternberg, N. (1990) A bacteriophage P1 cloning system for the isolation, amplification and recovery of DNA fragments as large as 100 kilobase pairs. *Proc. Natl. Acad. Sci. USA* **87,** 103–107.
6. Kim, U. J., Shizuya, H., Dejong, P. J., Birren, B., and Simon, M. I. (1992) Stable propagation of cosmid sized human DNA inserts in an F-factor based vector. *Nucleic Acids Res.* **20,** 1083–1085.
7. Sive, H. L. and St. John, T. (1988) A simple subtractive hybridisation technique employing photoreactivatable biotin and phenol extraction. *Nucleic Acids Res.* **16,** 10937.
8. Batra, S. K., Metzgar, R. S., and Hollingsworth, M. A. (1991) A simple, effective method for the construction of subtracted cDNA libraries. *Gene Anal. Tech.* **8,** 129–133.
9. Patanjali, S. R., Parimoo, S., and Weissman, S. M. (1991) Construction of a uniform-abundance (normalised) cDNA library. *Proc. Natl. Acad. Sci. USA* **88,** 1943–1947.

10. Soares, M. B., Bonaldo, de Fatima Bonaldo, M., Jelene, P., Su, L., Lawton, L., and Efstratiadis, A. (1994) Construction and characterization of a normalized cDNA library. *Proc. Natl. Acad. Sci. USA* **91**, 9228–9232.

11. Liang, P. and Pardee, A. B. (1992) Differential display of eukaryotic messenger RNA by means of the polymerase chain reaction. *Science* **257**, 967–971.

12. Gubler, U. and Hoffman, B. J. (1983) A simple and very efficient method for generating cDNA libraries. *Gene* **25**, 263–269.

13. Simmons, D. L. (1993) Cloning cell surface molecules by transient expression in mammalian cells, in *Cellular Interactions and Development* (Hartley, D., ed.), IRL Press, Oxford, UK, pp. 93–127.

14. Zehetner, G. and Lehrach, H. (1994) The Reference Library System: a service for sharing biological material and experimental data. *Nature* **367**, 489–491.

15. Lennon, G. G., Auffray, C., Polymeropoulos, M., and Soares, M. B. (1996) The I.M.A.G.E. Consortium: an integrated molecular analysis of genomes and their expression. *Genomics,* in press.

16. Adams, M. D., Kelley, J. M., Gocayne, J. D., Dubnick, M., Polymeropoulos, M. H., Xiao, H., Merril, C. R., Wu, A., Olde, B., Moreno, R. F., Kerlavage, A. R., McCombie, W. R., and Venter, J. C. (1991) Complementary DNA sequencing: expressed sequence tags and human genome project. *Science* **252**, 1651–1656.

17. Parsons, J. D., Brenner, S., and Bishop, M. J. (1992) Clustering of DNA sequences. *Comput. Applic. Biosci.* **8**, 461–466.

18. Pearson, W. R. and Lipman, D. J. (1988) Improved tools for biological sequence analysis. *Proc. Natl. Acad. Sci. USA* **85**, 2444–2448.

Index